Electronic Circuits for the Evil Genius™

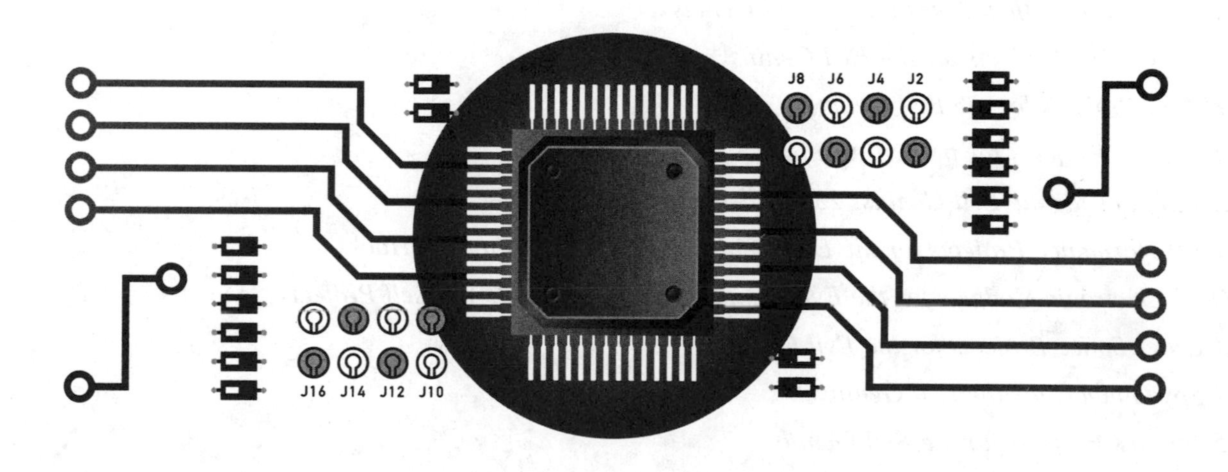

Evil Genius™ Series

Bike, Scooter, and Chopper Projects for the Evil Genius
Bionics for the Evil Genius: 25 Build-It-Yourself Projects
Electronic Circuits for the Evil Genius, Second Edition: 64 Lessons with Projects
Electronic Gadgets for the Evil Genius: 28 Build-It-Yourself Projects
Electronic Sensors for the Evil Genius: 54 Electrifying Projects
50 Awesome Auto Projects for the Evil Genius
50 Green Projects for the Evil Genius
50 Model Rocket Projects for the Evil Genius
51 High-Tech Practical Jokes for the Evil Genius
46 Science Fair Projects for the Evil Genius
Fuel Cell Projects for the Evil Genius
Holography Projects for the Evil Genius
Mechatronics for the Evil Genius: 25 Build-It-Yourself Projects
Mind Performance Projects for the Evil Genius: 19 Brain-Bending Bio Hacks
MORE Electronic Gadgets for the Evil Genius: 40 NEW Build-It-Yourself Projects
101 Outer Space Projects for the Evil Genius
101 Spy Gadgets for the Evil Genius
125 Physics Projects for the Evil Genius
123 PIC® Microcontroller Experiments for the Evil Genius
123 Robotics Experiments for the Evil Genius
PC Mods for the Evil Genius: 25 Custom Builds to Turbocharge Your Computer
PICAXE Microcontroller Projects for the Evil Genius
Programming Video Games for the Evil Genius
Recycling Projects for the Evil Genius
Solar Energy Projects for the Evil Genius
Telephone Projects for the Evil Genius
30 Arduino Projects for the Evil Genius
25 Home Automation Projects for the Evil Genius
22 Radio and Receiver Projects for the Evil Genius

Electronic Circuits for the Evil Genius™, Second Edition

64 Lessons with Projects

Dave Cutcher

New York Chicago San Francisco Lisbon London Madrid
Mexico City Milan New Delhi San Juan Seoul
Singapore Sydney Toronto

The McGraw-Hill Companies

Cataloging-in-Publication Data is on file with the Library of Congress

Electronic Circuits for the Evil Genius™, Second Edition: 64 Lessons with Projects

1 2 3 4 5 6 7 8 9 0 QDB QDB 1 0 9 8 7 6 5 4 3 2 1 0

ISBN 978-0-07-174412-6
MHID 0-07-174412-6

Sponsoring Editor
Roger Stewart

Editorial Supervisor
Jody McKenzie

Acquisitions Coordinator
Joya Anthony

Project Manager
Patricia Wallenburg

Copy Editor
Lisa McCoy

Proofreader
Paul Tyler

Indexer
Claire Splan

Production Supervisor
George Anderson

Composition
TypeWriting

Art Director, Cover
Jeff Weeks

Cover Designer
Todd Radom

About the Author

Dave Cutcher is a retired high school shop teacher. He always coaxed his students to believe in themselves and that success in life was not limited to school. He taught young people and electronics was just the topic. Currently living in British Columbia, he teaches night school courses and does volunteer work within the community. People comment that he always thinks outside of the box. His reply, "What box?" As a life long learner and adult with ADHD, he is interested in everything because everything is interesting.

Contents at a Glance

Contents

Acknowledgments

For a variety of reasons, there are many people I need to thank.

First are my current guinea pigs, who chose to be caged in a classroom with me for three years running. Andrew Fuller who put together the game "When Resistors Go Bad." He and André Walther, two very original Evil Geniuses. I hope they understand the molar concept in chemistry now and won't raise a stink about me mentioning them. Eric Raue and Eric Pospisal, both for being the gentler geniuses they are. And Brennen Williams, who was more patient with me at times than I was with him. It was a difficult year.

I've had only one formal class in electronics, taught by Gus Fraser. He let me teach myself. Bryan Onstad gave me a goal to work toward and a platform to work on. Don Nordheimer was the first adult who actually worked through my material outside of the classroom environment. At the same time, he proofed the material from the adult perspective. I owe heartfelt thanks for the encouragement from Pete Kosonan, the first administrator who enjoyed the creative flow of the students as much as I did. For Steve Bailey, the second administrator I found who wasn't threatened by kids who knew more than he did. For the many others like Paul Wytenbrok, Ian Mattie, Judy Doll, and Don Cann, who continually encouraged me over the five years it took to develop this material. For Brad Thode, who introduced me to the necessity of changing careers within teaching back in 1989. For Mrs. Schluter and Mrs. Gerard, who taught me to believe in myself and recognize that there was room for creativity, not just what they wanted to hear.

Then to Dave Mickie who understood that conditions like ADHD cannot be cured, only managed. I'll be forever grateful for the encouragement and support he provided as I moved forward with my work.

To my parents, who knew they couldn't change me, so they encouraged me.

Preface

We casually accept electronics in our everyday world. Those who don't understand how it works are casually obedient. Those who take the time to learn electronics are viewed as geniuses. Do you want to learn how to control the power of electronics?

This text provides a solid introduction to the field of electronics, both analog and digital. *Electronic Circuits for the Evil Genius* is based on practical projects that exercise the genius that exists in all of us. Components are introduced as you build working circuits. These circuits are modified and analyzed to help explain the function of the components. It's all hands-on. Analysis is done by observation, using a digital multimeter, and using your computer as an oscilloscope.

You will build two major projects in the first part:

- An automatic night light
- A professional-quality alarm

The remainder of the text focuses on three major projects, one per part:

- Building a digital toy using logic gates
- Designing and building an application using digital counting circuits
- Applying transistors and Op Amps as you build a two-way intercom system

The lessons and prototype circuits built in the book are focused on developing a solid foundation centered on each of these major projects. You work from ideas to prototypes, producing a final product.

Additional materials for this book, including lesson quizzes and answers keys, are available online at **www.mhprofessional.com/computingdownload**.

I hope you enjoy building the projects and reading the book as much as I enjoyed developing them.

Dave Cutcher

Common Components, Symbols, and Appearance

Name	Class	Purpose	Symbol	Photograph
Electrolytic Capacitor	Capacitor Micro Farad (μF)	1. RC timer 2. Isolate AC 3. Buffer/filter	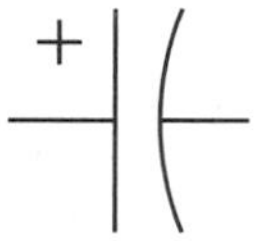 +	
Film Capacitor	Capacitor Nano Farad (nF)	1. RC timer 2. Isolate AC 3. Buffer/filter	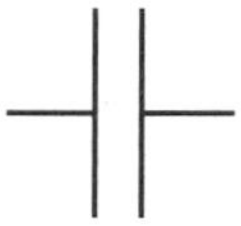	
Disk Capacitor	Capacitor Pico Farad (pF)	1. RC timer 2. Buffer/filter		
Power Diode	Diode	One way valve for high voltage	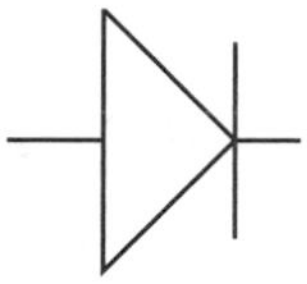	
Signal Diode	Diode	One way valve for low voltage	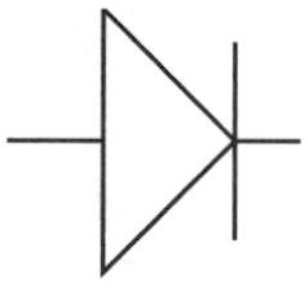	
Zener Diode	Diode	One way until voltage reaches preset breakdown	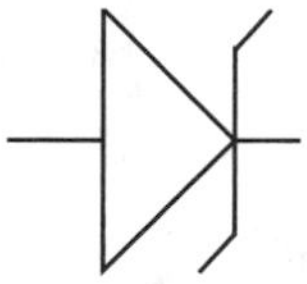	
Light Emitting Diode (LED)	Diode	1. Indicator 2. Light source 3. Signal transfer	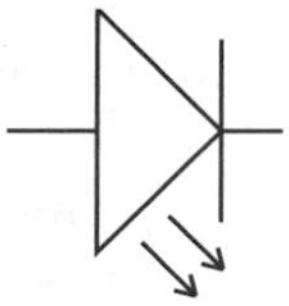	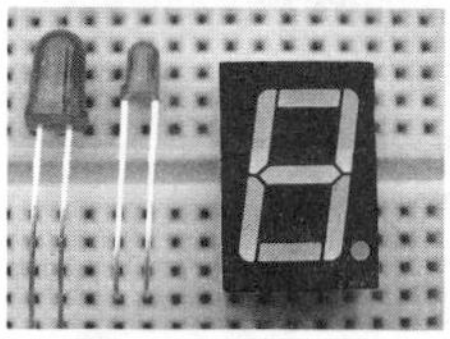

Name	Class	Purpose	Symbol	Photograph
Fixed Resistor	Resistor	Limits flow of current		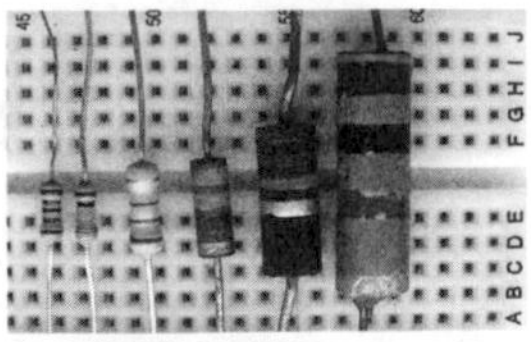
Potentiometer	Resistor	Adjustable resistor		
Light Dependent Resistor (LDR)	Resistor	General purpose light sensor		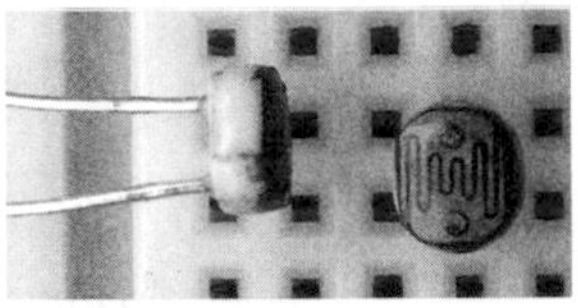
Push Button Normally Closed (PBNC)	Hardware	Momentary switch		
Push Button Normally Open (PBNO)	Hardware	Momentary switch	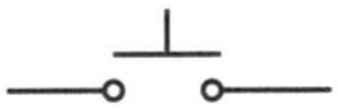	
Single Pole Single Throw (SPST)	Hardware	Simple open/ close switch		
Single Pole Double Throw (SPDT)	Hardware	Controls single connection one of two directions		
Double Pole Double Throw (DPDT)	Hardware	Matched control of two individual connections in two directions	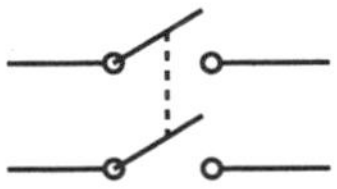	

Name	Class	Purpose	Symbol	Photograph
Relay	Switch	Secondary switch controlled by primary circuit	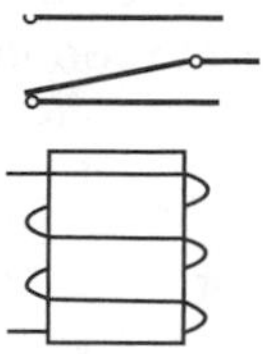	
Operational Amplifier	Amplifier	Very versitile, multipurpose component	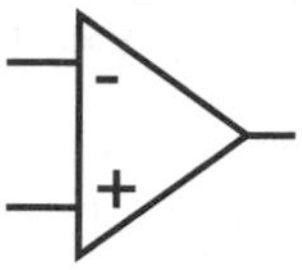 - +	Available DIP and SIP Various packages
NPN Transistor	Transistor	Simple analog electronic switch. Needs both current and voltage to operate. Acts like PBNO.	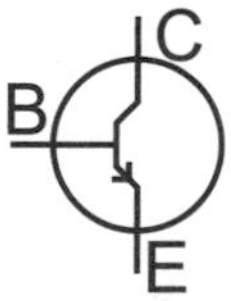 C B E	See Appendix A, Common Component Packaging
PNP Transistor	Transistor	Simple analog electronic switch. Needs both current and voltage to operate. Acts like PBNC.	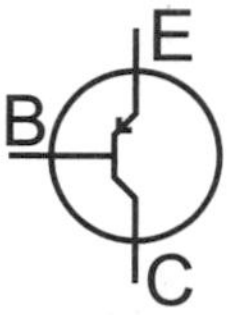 E B C	See Appendix A, Common Component Packaging
Photo Transistor	Transistor	Light sensitive analog and digital signal pickup.	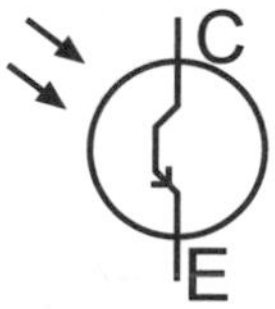 C E	Various packaging
Silicon Controlled Rectifier (SCR)	Transistor latching circuit	Single event	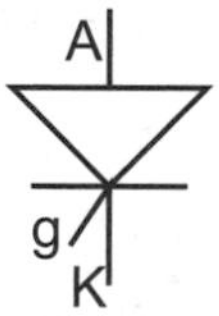 A g K	See Appendix A, Common Component Packaging
Power Regulator	Transistor	DC to DC power conversion.	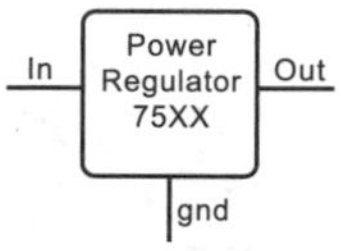 In Power Regulator 75XX Out gnd	See Appendix A, Common Component Packaging
Field Effect Transistor	Transistor	Available in PNP and NPN configuration. Needs only voltage to operate.	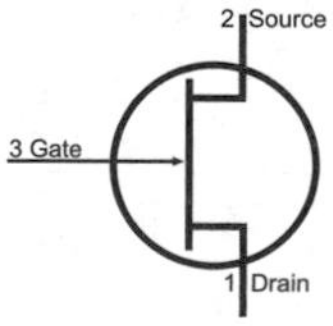 2 Source 3 Gate 1 Drain	See Appendix A, Common Component Packaging

Name	Class	Purpose	Symbol	Photograph
Electret Microphone	Microphone	Sound pickup		
Speaker	Speaker	Sound output		
Transformer	Transformer	Used to isolate or change AC voltage from a primary to secondary circuit.		See Appendix A, Common Component Packaging
AND Gate	Logic Gate	InA InB Out H H H H L L L H L L L L		4081
OR Gate	Logic Gate	InA InB Out H H H H L H L H H L L L		4071
NAND Gate	Logic Gate	InA InB Out H H L H L H L H H L L H		4011
NOR Gate	Logic Gate	InA InB Out H H L H L L L H L L L H		4001

PART ONE
Components

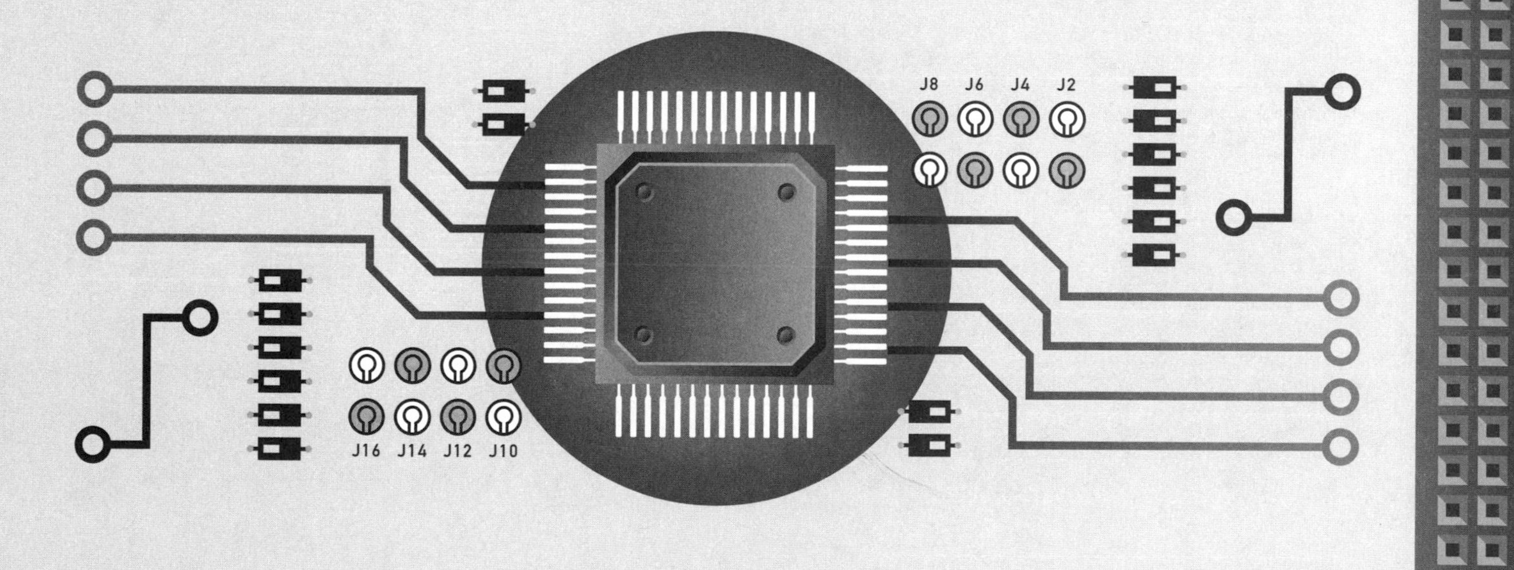

Building the Foundation

Imagine the solid foundation needed for the work being done on the construction shown here.

The Parts Bin on the following page has the complete parts list used in Part One. These are pictured in the front of the book in the section Common Components, Symbols, and Appearance.

Electronics is BIG. You need a solid foundation.

PARTS BIN FOR PART ONE

Description	Type	Quantity
Diode 1N4005	Semi(D)	3
LED	Semi(L)	3
2N-3906 PNP transistor	TO-92 case	1
2N-3904 NPN transistor	TO-92 case	1
Phototransistor	LTE 4206 E (darkened glass) 3mm diam: tuned to 940nm	1
Infra Red Diode	LTE 4206 (clear glass) 3mm diam: emits 940nm	1
SCR C106B	Various packages	1
7805 Power regulator	TO-220 case	1
Wall adapter 120vAC to 9vDC	Transformer	1
100 Ω	Resistor	1
470 Ω	Resistor	2
1000 Ω	Resistor	1
2,200 Ω	Resistor	1
10,000 Ω	Resistor	1
22,000 Ω	Resistor	1
47,000 Ω	Resistor	1
100,000 Ω	Resistor	1
220,000 Ω	Resistor	1
100,000 Ω 1/4 watt	Potentiometer	1
Light-dependent resistor	LDR	1
.1 μF film cap	Cap	1
10 μF Electrolytic	Cap	1
100 μF Electrolytic	Cap	1
1000 μF Electrolytic	Cap	1
470 μF Electrolytic	Cap	1
24 gauge wire	Hookup wire	Various colors
Battery clip	Hardware	2
Alligator clips (red and black)	Hardware	1 each
Buzzer 9v	Hardware	1
LED collars	Hardware	3
PCB for night light	Hardware	1
PCB for SCR alarm	Hardware	1
Solderless breadboard	Hardware	1
PBNC momentary	Switch	2
PBNC momentary	Switch	1

- Not all components will be consumed by project work.

SECTION 1

Components

In Lesson 1, you will be introduced to many common components that are always present in electronics and many of the bits and pieces you will use in the course. It starts out as a jumble. As you use the parts, the confused mass becomes an organized pile.

In Lesson 2, you will become acquainted with the two major tools that you will use throughout the course.

In Lesson 3, you will build your first circuit on the solderless breadboard, a platform that allows you to build circuits in a temporary format.

You use your digital multimeter and get voltage measurements when you set up and test your first circuits.

Lesson 1
Inventory of Parts Used in Part One

All components look the same if you don't know what they are. It's like when you first visit a different country. There's a pile of change, just like in Figure L1-1. You have to be introduced to the currency and practice using it, but you become comfortable with it quickly. Now you need to unjumble the pile and become familiar with your electronic components.

Figure L1-1

NOTE Do not remove the small integrated-circuit (IC) chips shown in Figure L1-2 from their antistatic packaging. They are packed in a special antistatic tube or special sponge material.

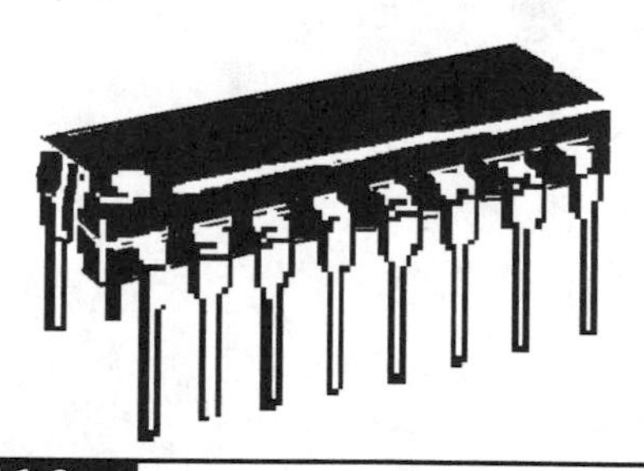

Figure L1-2

Semiconductors

These are the electronic components you will be using in Part One. As you identify them, set them aside into small groups.

Diodes

You will need three power diodes as shown in Figures L1-3 and L1-4.

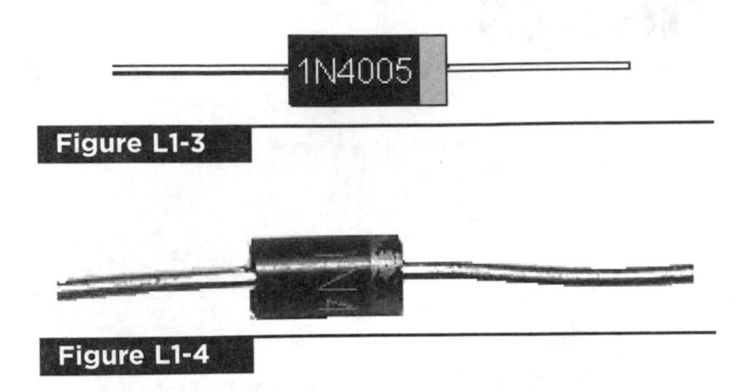

Figure L1-3

Figure L1-4

The number on the side reads 1N4005. If the last number is not 5, don't worry. Any diode of this series will do the job.

Light-Emitting Diodes

Light-emitting diodes are also known as LEDs. You will need three. An example is illustrated in Figure L1-5.

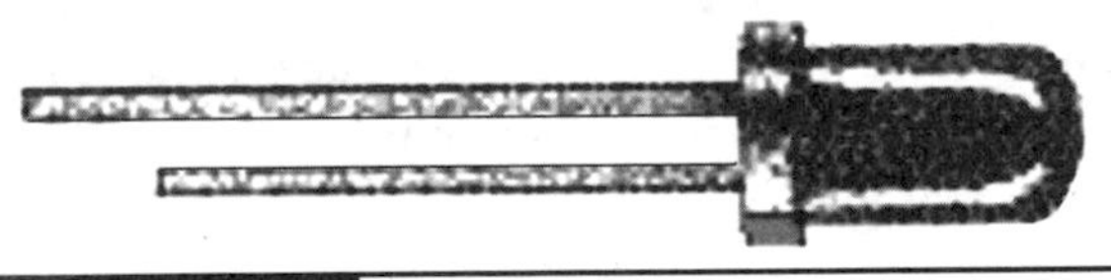

Figure L1-5

They can be any color. The most common colors are red, yellow, and green.

Resistors

There should be lots of colorful resistors, nearly all the same size. Notice that in Figure L1-6 each resistor has four color bands to identify it. If you know the colors of the rainbow, you know how to read resistors.

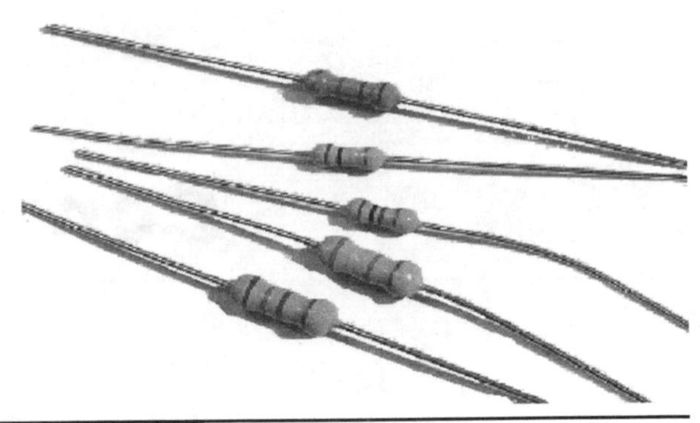

Figure L1-6

Find these resistors:

- One brown-black-brown-gold 100 Ω
- Two yellow-violet-brown-gold 470 Ω
- One brown-black-red-gold 1,000 Ω
- One brown-black-orange-gold 10,000 Ω
- One red-red-orange-gold 22,000 Ω
- One yellow-violet-orange-gold 47,000 Ω
- One brown-black-yellow-gold 100,000 Ω

Capacitors

As you see in Figure L1-7, the capacitor shown is black and white. The colors of capacitors are different, depending on the manufacturer. Then again, all pop cans look alike, but each brand has a different label. Locate four small capacitors, different in size. Written on each are different values and other mumbo jumbo. Look for the information that specifically lists 1 μF, 10 μF, 100 μF, and 1000 μF.

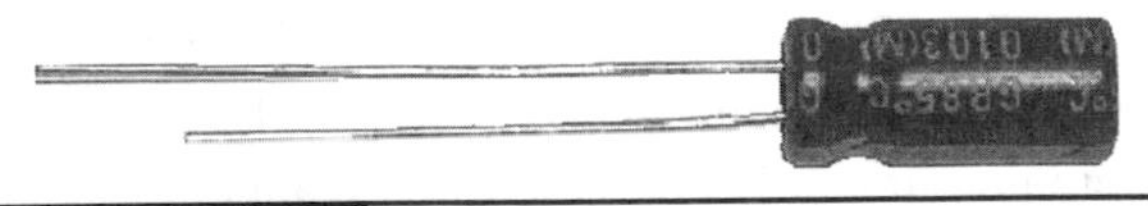

Figure L1-7

There is another capacitor of a different shape to locate. Figure L1-8 shows the other capacitor used in Part One. Again, it is presented in black and white, because the color will change as the manufacturer changes. It is a 0.1 μF capacitor. It may be marked as any of the following: 0.1, μ1, or 100 nF.

Figure L1-8

Silicon-Controlled Rectifier

The ID number 1067X for the *silicon-controlled rectifier* (SCR) is written on the face, as shown in Figure L1-9. This SCR comes in this particular package. Not everything with this shape is an SCR, just as not everything in the shape of a pop can is your favorite flavor.

Figure L1-9

Transistors

You need two transistors, like that illustrated in Figure L1-10. They are identical except for the number 3904 or 3906. All other writing and marks are the manufacturer telling us how great they are.

Figure L1-10

Hardware

The solderless breadboard is shown in Figure L1-11.

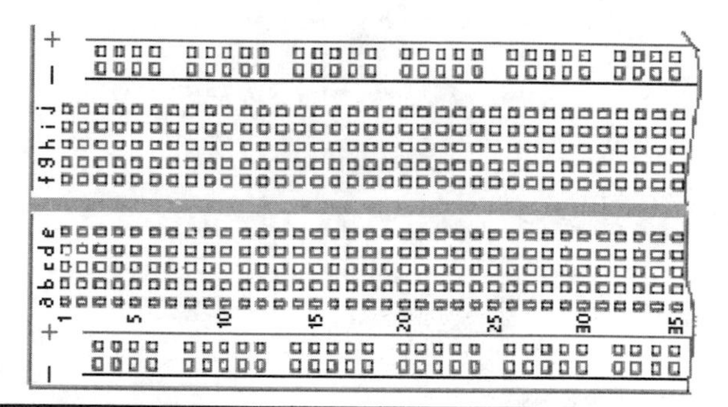

Figure L1-11

Figures L1-12 and L1-13 illustrate two push buttons—they are different, but you can't tell this by looking at them. Figure L1-12 is the normally open push button (push to close the contacts), and Figure L1-13 shows the normally closed push button (push to open the contacts).

Figure L1-12

Figure L1-13

You should have lots of 24-gauge solid wire with plastic insulation in many different lengths.

Two battery clips are shown in Figure L1-14.

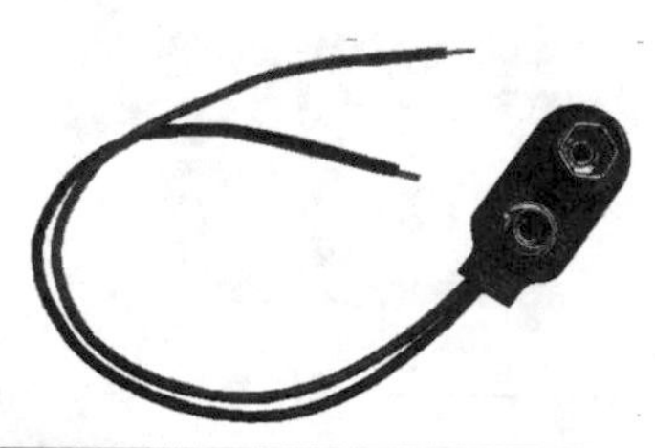

Figure L1-14

A 9-volt buzzer is shown in Figure L1-15.

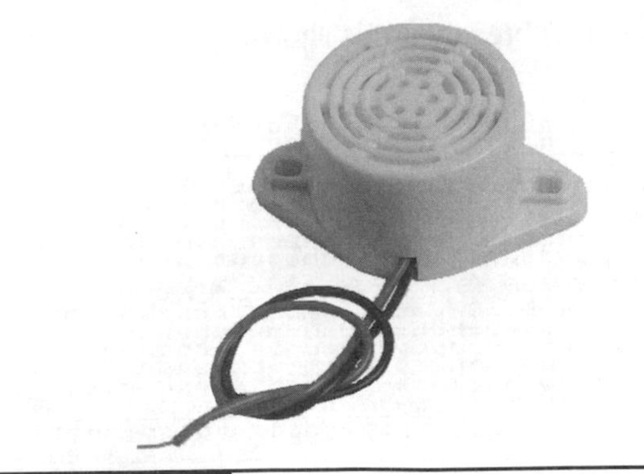

Figure L1-15

Two printed circuit boards are premade for your projects: Figure L1-16 shows the one that will be used for the night-light project; Figure L1-17 shows the one that will be used for your SCR alarm project.

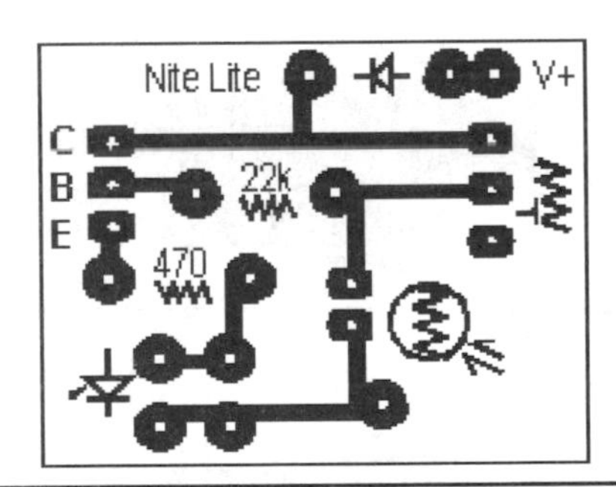

Figure L1-16

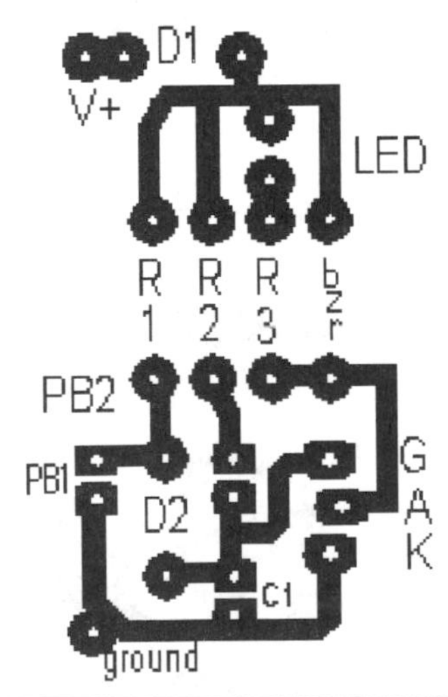

Figure L1-17

Two adjustable resistors are also supplied: The *light-dependent resistor* (LDR) is shown in Figure L1-18 and the potentiometer is shown in Figure L1-19.

Figure L1-18

Figure L1-19

Lesson 2
Major Equipment

The solderless breadboard and digital multimeter are two of the most common tools used in electronics. Let's introduce you to them now.

The Solderless Breadboard

When smart people come up with ideas, first they test those ideas. They build a prototype. The easiest way to build prototypes and play with ideas in electronics is on the solderless breadboard, shown here in Figure L2-01.

The main advantage of the solderless breadboard is the ability to exchange parts easily and quickly.

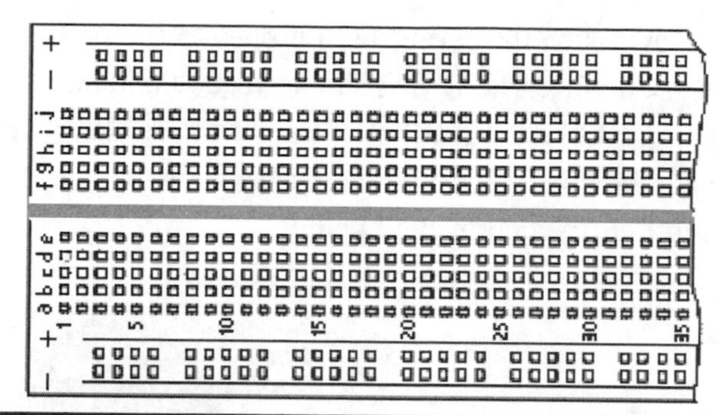
Figure L2-1

The top view in Figure L2-1 shows the many pairs of short five-hole rows and a pair of long rows down each side; each of these lines is marked with a strip of paint.

The Digital Multimeter

I recommend the Circuit Test DMR2900 displayed in Figure L2-2. The autoranging digital multimeter (DMM) offers beginners the advantage of being easier to learn. The second style of DMM is not autoranging. This style is easy to use after you become familiar with electronics, but it tends to be confusing for the beginner. A typical dial of a nonautoranging multimeter is confusing, as you can see in Figure L2-3.

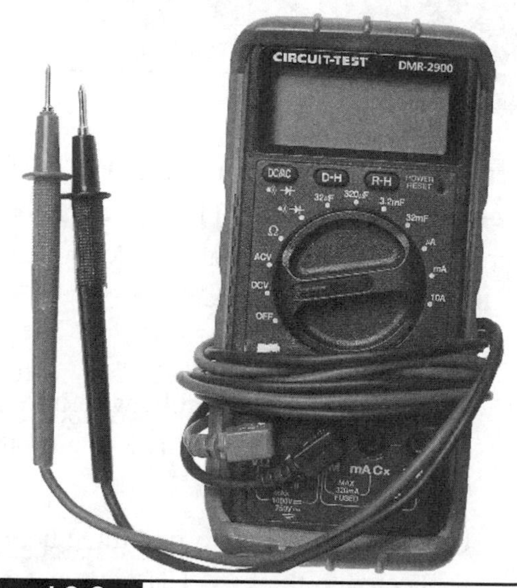

Figure L2-2

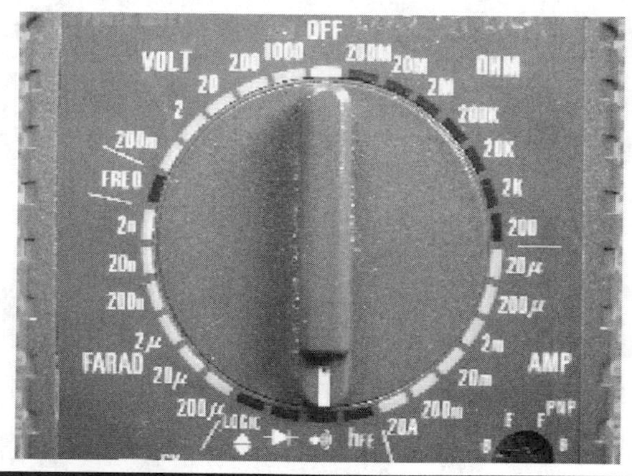

Figure L2-3

I discourage the use of outdated whisker-style multimeters for this course. Figure L2-4 shows an example of what to avoid.

Figure L2-4

Connection Wire

A box of wire provided in the kit is displayed in Figure L2-5.

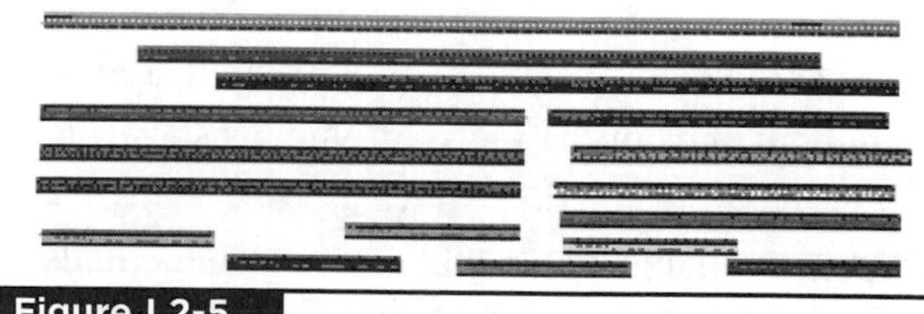
Figure L2-5

These are different lengths convenient for the solderless breadboard. However, if you need to cut the wire, wire clippers will work perfectly. Old scissors work as well.

Set the dial of the DMM to CONTINUITY. This setting is shown in Figure L2-6.

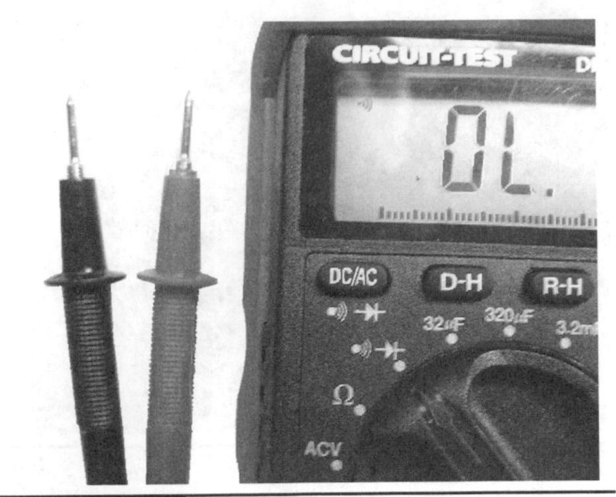

Figure L2-6

Touch the end of both red and black probes to the colored covering. The DMM should be silent and read OL, as in the readout illustrated in Figure L2-7, because the resistance of the insulation prevents any current from passing.

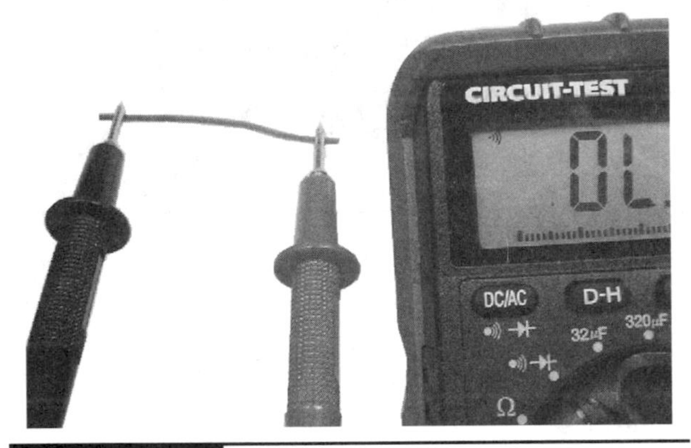

Figure L2-7

Be sure the strip of insulating plastic is removed from both ends of the piece of wire, as shown in Figure L2-8. If you don't have a proper wire stripper available, use a knife or your fingernails to cut the insulation. Be careful not to nick the wire inside the insulation.

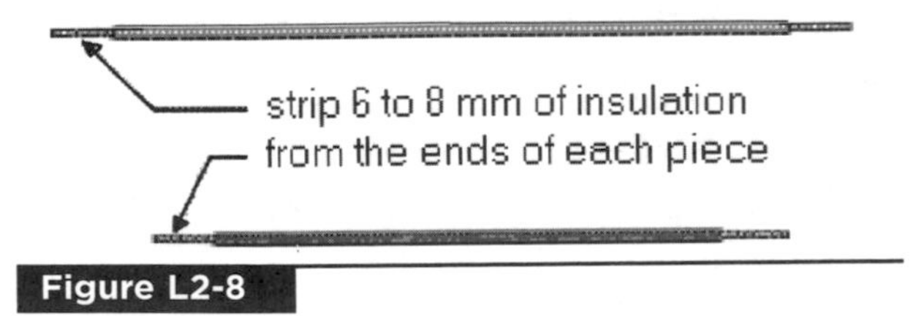

Figure L2-8

Now touch the end of both probes to the exposed wire. The DMM should read "00" and beep, just like the readout in Figure L2-9. The wire is a good conductor, and the DMM shows "continuity," a connected path.

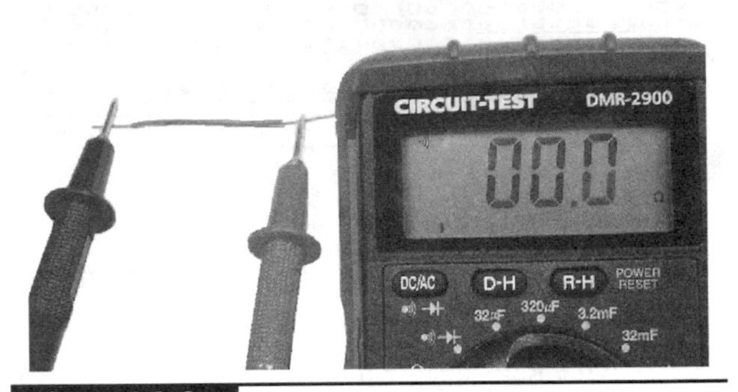

Figure L2-9

Exercise: Mapping the Solderless Breadboard

Strip the end of two pieces of wire far enough to wrap around the DMM probes on one end and enough to insert into the solderless breadboard (SBB) on the other end, as shown in Figure L2-10.

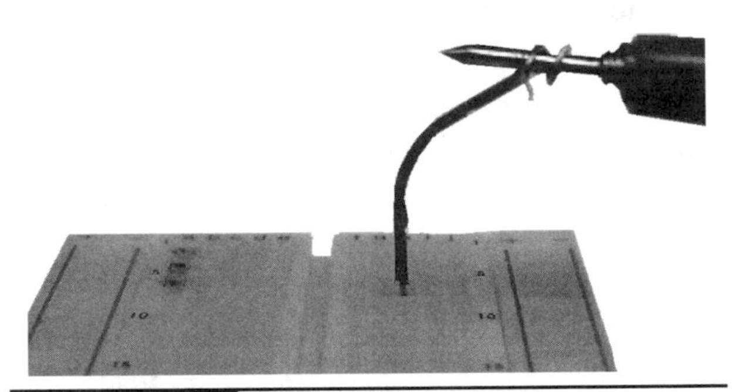

Figure L2-10

1. Set your digital multimeter to CONTINUITY. Now refer to Figure L2-11. Notice the letters across the top and the numbers down the side of the solderless breadboard.
2. Probe placement:
   a. Place the end of one probe wire into the SBB at point "h3" and mark that on the drawing.
   b. Use the other probe to find three holes connected to the first. The multimeter will indicate the connection.
   c. Draw these connections as solid lines.

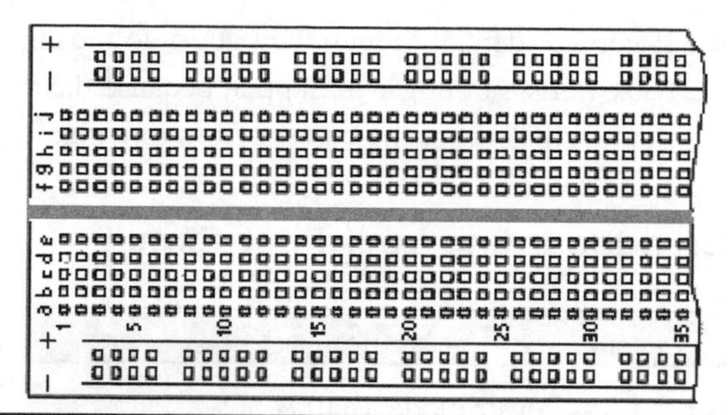

Figure L2-11

3. Base points:
    a. Create four more base points at e25, b16, f30, and c8.
    b. Use the other probe to find three holes connected to each of these points.
    c. Again draw these connections as solid lines.
4. Additional base points:
    a. Choose two more base points on the outside long, paired lines. These lines are not lettered or numbered but have a stripe of paint along the side. Mark them on the previous diagram.
    b. Find three holes connected to each of these points.
    c. Again draw these connections as solid lines.
5. Be sure that you can define the terms *prototype*, *insulator*, and *conductor*.
6. With your multimeter set on CONTINUITY, walk around and identify at least five common items that are insulators and five common materials that are conductors.

Lesson 3
Your First Circuit

You build an actual circuit on the breadboard, then measure and observe how the voltage is used while getting more experience with your multimeter.

The solderless breadboard has a definite layout, as shown in Figure L3-1. One strip of the spring metal in the breadboard connects the five holes. You can easily connect five pieces in one strip. The two long rows of holes allow power access along the entire length of the breadboard.

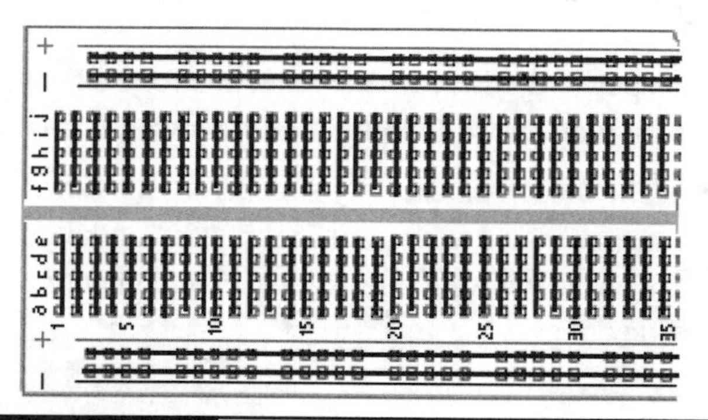

Figure L3-1

Setting Up the Solderless Breadboard

You will have a standard setup for every circuit. The battery clip is connected to one of the first rows of the breadboard, and the diode connects that row to the outer red line (see Figure L3-2).

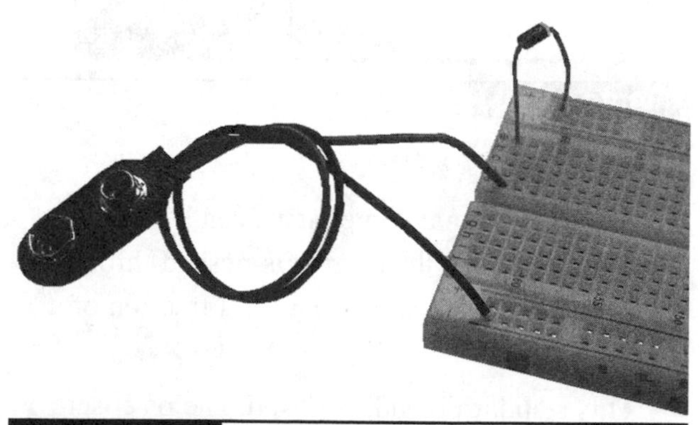

Figure L3-2

Notice the gray band highlighted in Figure L3-3 on the diode. It faces in the direction that the voltage is pushing.

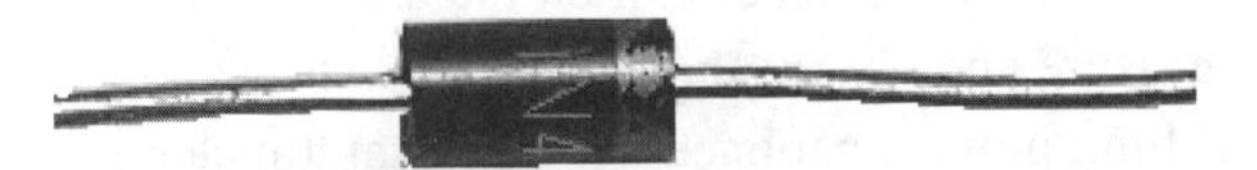

Figure L3-03

The voltage comes through the red wire, through the diode, and then to the power strip on the breadboard.

Why Bother?

This power diode provides protection for each circuit that you build in the following ways:

- The diode is a one-way street. You can view the animated version of Figure L3-4 at the website www.mhprofessional.com/computing download.

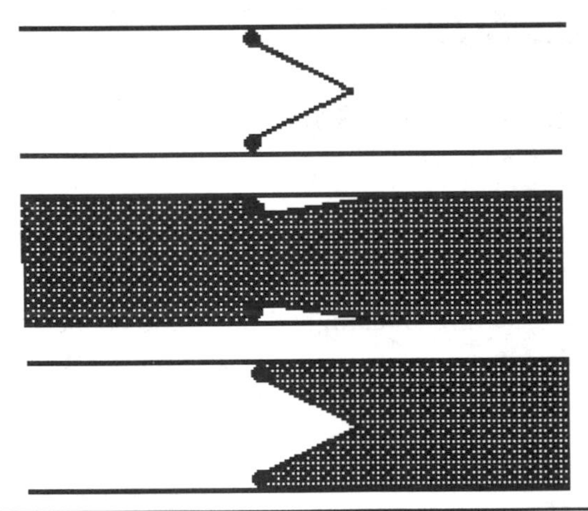

Figure L3-4

- Many electronic components can be damaged or destroyed if the current is pushed through them the wrong way, even for a fraction of a second.
- This standard breadboard setup helps ensure that your battery will always be connected properly.
- If you accidentally touch the battery to the clip backwards, nothing will happen because the diode will prevent the current from moving.

Breadboarding Your First Circuit

PARTS BIN
■ D1—Power diode 1n400x
■ LED1—LED any color
■ R1—470-ohm resistor

Your LED is a light-emitting diode. That's right, a diode that emits light. It has the same symbol as a diode, but it has a "ray" coming out, as shown here in Figure L3-5.

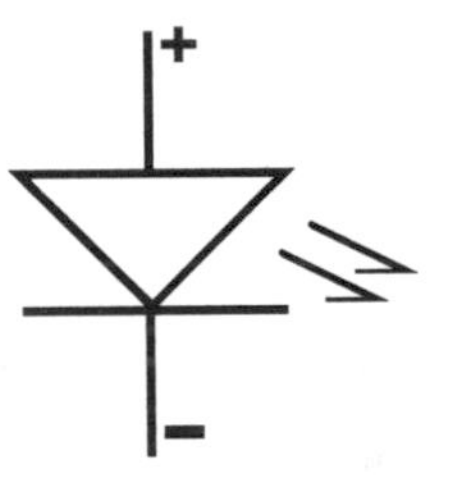

Figure L3-5

Figure L3-6 is a picture of an LED. Never touch your LED directly to your power supply. A burned-out LED looks just like a working LED. Note in the picture how to identify the negative side.

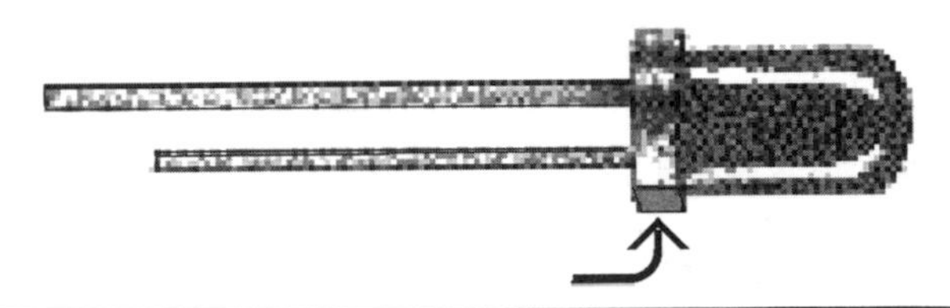

Figure L3-6

The shorter leg: This is always reliable with new LEDs, but not with ones that you have handled in and out of your breadboard. As you handle the components, the legs can get bent out of shape.

The flat side on the rim: This is always reliable with round LEDs, but you have to look for it.

Remember that the LED, as a diode, is a one-way street. It will not work if you put it in backward.

Figure L3-7 shows several resistors. The resistor symbol is illustrated in Figure L3-8. The resistor you need is the 470-ohm yellow-violet-brown-gold.

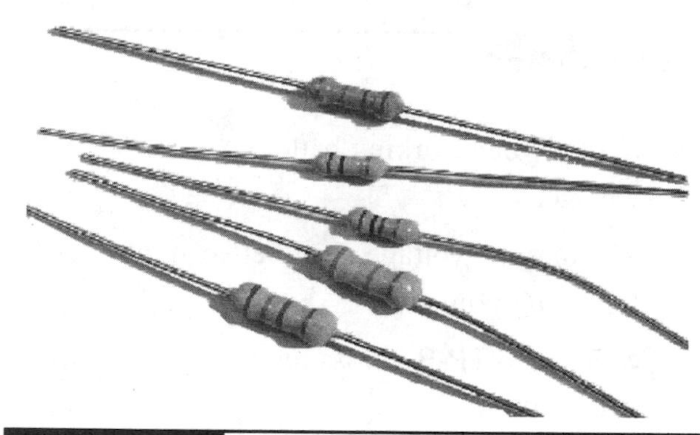

Figure L3-7

Figure L3-8

Resistance is measured in ohms. The symbol for ohms is the Greek capital letter omega: Ω.

The schematic is shown in Figure L3-9. Set up your breadboard as shown in Figure L3-10. Note that this picture shows the correct connections. The red wire of the battery clip is connected to the power diode that in turn provides voltage to the top of the breadboard. The black wire is connected to the blue line at the bottom of the breadboard.

NOTE

1. **Always complete your breadboard before you attach your power to the circuit.**
2. **Attach your battery only when you are ready to test the circuit.**

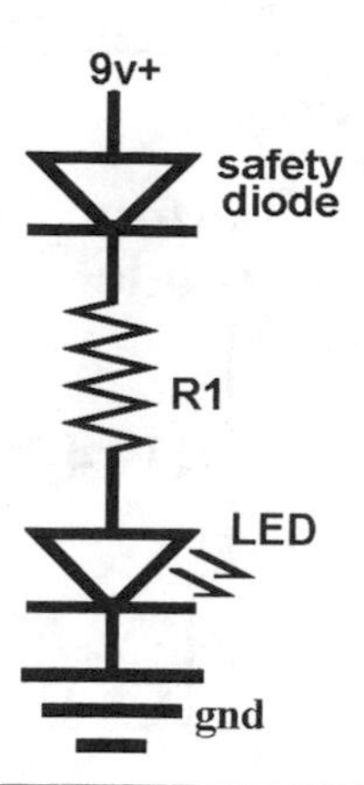

Figure L3-9

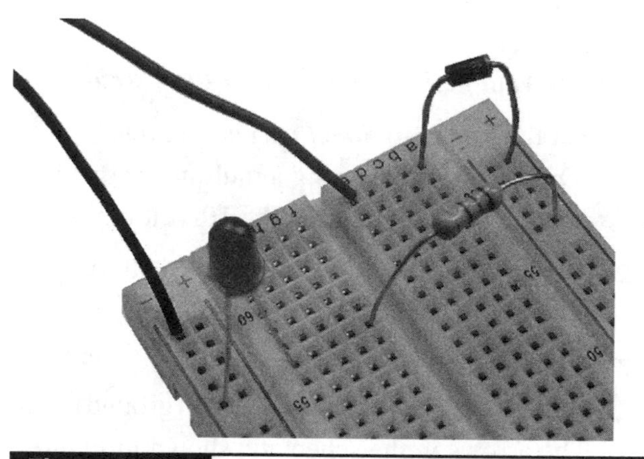

Figure L3-10

3. **When you have finished testing your circuit, take your battery off.**

Exercise: Measuring Voltage on Your First Circuit; Your First Circuit Should Be Working

Figure L3-11 shows what is happening. Like a waterfall, all of the voltage goes from the top to the bottom. The resistor and LED each use up part of the voltage. Together, they use all the voltage. The 470-ohm resistor uses enough voltage to make sure the LED has enough to work, but not so much that would burn it out.

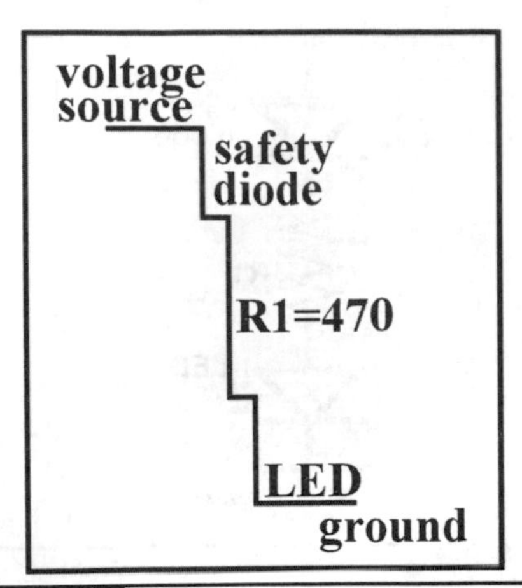

Figure L3-11

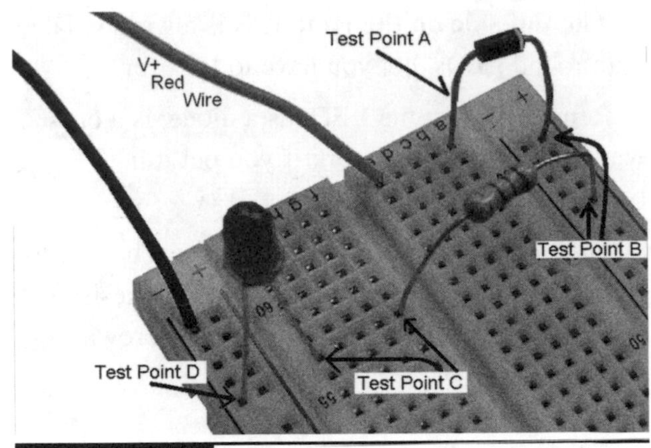

Figure L3-13

How the Voltage Is Being Used in the Circuit

1. Set the DMM to *direct current voltage* (DCV). If you are using a multimeter that is not autoranging, set it to the 10-volt range.
2. Measure the voltage of the 9-volt battery while it is connected to the circuit.
3. Place the red (+) probe at test point A (TP-A) and the black (–) probe at TP-D (ground). The arrows in the schematic shown in Figure L3-12 indicate where to attach the probes. Corresponding test points have been noted in Figure L3-13 as well.

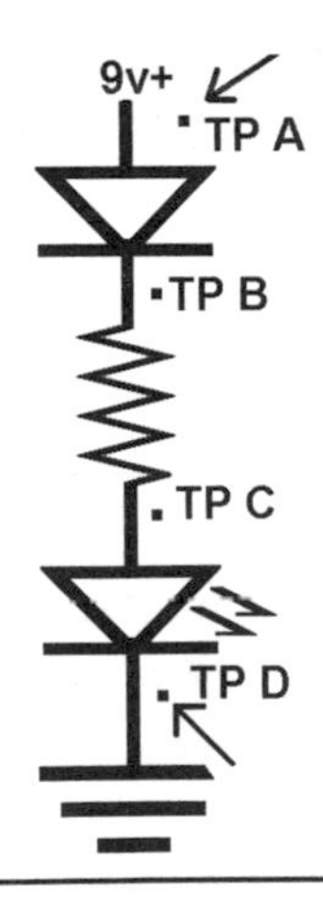

Figure L3-12

4. Record your working battery voltage. ____V
5. Measure the voltage used between the following points:
   - TP-A to TP-B across the safety diode ____V
   - TP-B to TP-C across the 470-ohm resistor ____V
   - TP-C to TP-D across the LED ____V
6. Now add the voltages from #5. ____V
7. List working battery voltage (recorded in item 2). ____V
8. Compare the voltage used by all of the parts to the voltage provided by the battery.

The voltages added together should be approximately the same as the voltage provided by the battery. There may be only a few hundredths of a volt difference.

SECTION 2

Resist If You Must

RESISTORS ARE ONE OF the fundamental components within electronics. They are funny little things and come in all different colors. And just like a rainbow, they come in all sizes too.

To master electronics, you must first master the secret color code, unlocking the mystery of how to tell one resistor from another.

But beware! Can you handle the knowledge and power that lies beyond this task?

Lesson 4
Reading Resistors

Fixed resistors are the most common electronic components. They are so common because they are so useful. Most often, these are identified using their color code (Table L4-1). If you think the secret code is hard to remember, just ask any six-year-old to name the colors in the rainbow.

The gold bands are always read last. They indicate that the resistor's value is accurate to within 5 percent.

When using the digital multimeter to measure resistance, set the dial to Ω. Notice the two points of detail shown in Figure L4-1.

The first point is that when the dial is set directly to the Ω symbol to measure resistance, it also appears on the readout. Second, notice the M next to the Ω symbol. That means the resistor being measured is 0.463 MΩ, which is 0.463

TABLE L4-1 Resistor Band Designations

Color Band	First Band: Value	Second Band: Value	Third Band: Number of Zeros	Units
Black	0	0	No zeros	Tens ##
Brown	1	1	One zero "0"	Hundreds ##0
Red	2	2	Two zeros "00"	Thousands (k) #, #00
Orange	3	3	Three zeros "000"	Ten thousands (k) ##,000
Yellow	4	4	Four zeros "0,000"	Hundred thousands (k) ##0,000
Green	5	5	Five zeros "00,000"	Millions (M) #,#00,000
Blue	6	6	Six zeros "000,000"	Ten millions (M) ##,000,000
Violet	7	7	Not available	
Gray	8	8	Not available	
White	9	9	Not available	

Figure L4-1

million ohms, or 463,000 ohms. When the M is there, *never* ignore it.

As you use resistors, you quickly become familiar with them. The third band is the most important marker. It tells you the range in a power of 10. In a pinch, you could substitute any resistor of nearly the same value. For example, a substitution of a red-red-orange could be made for a brown-black-orange resistor. But a substitution of a red-red-orange with a red-red-yellow would create more problems than it would solve. Using a completely wrong value of resistor can mess things up.

Exercise: Reading Resistors

If you have an autoranging multimeter, set the digital multimeter (DMM) to measure resistance. If you do not have an autoranging DMM, you have to work harder because the resistors come in different ranges. Set the range on your DMM to match the range of the resistor. That means that you should have an idea of how to read resistor values before you can measure them using a DMM that is not autoranging. Thus, as you can see, an autoranging DMM really does make it much easier.

Your skin will conduct electricity, and if you have contact with both sides of the resistor, the DMM will measure your resistance mixed with the resistor's. This will give an inaccurate value.

Proper Method to Measure Resistor's Value

Figure L4-2 shows how to measure a resistor. Place one end of the resistor into your solderless breadboard and hold the probe tightly against it, but not touching the metal. You can press the other probe against the top of the resistor with your other finger.

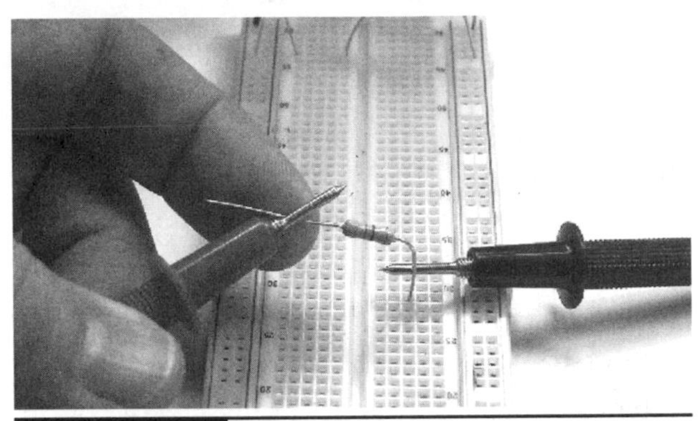
Figure L4-2

1. Table L4-2 lists some of the resistors that you will need to be able to identify, because you use them soon.
2. Don't be surprised if the resistor value is not exactly right. These resistors have a maximum error of 5 percent. That means that the 100-ohm resistor can be as much as 105 ohms or as little as 95 ohms. Plus or minus 5 ohms isn't too bad. What is 5 percent of 1,000,000?
   - What is the maximum you would expect to see on the 1,000-ohm resistor? _____ Ω
   - What is the minimum you would expect to see on the same 1-kilo-ohm resistor? ____ Ω
3. Measure your skin's resistance by holding a probe in each hand. It will bounce around, but try to take an average. _______ Ω
   - Did you know that this can be used as a crude lie detector? A person sweats when they get anxious. Have a friend hold the

TABLE L4-2 Resistors Needed

First Band: Value	Second Band: Value	Third Band: Number of Zeros	Resistor Value	DMM Value
Brown	Black	Brown	100 Ω	_____Ω
1	0	0		
	Violet	Brown	470 Ω	_____Ω
4	7	0		
Brown	Black	Red	1,000 Ω	_____Ω
____	____	00		
Brown	Black	Orange	10 kΩ	_____Ω
____	____	000	10,000Ω	
Red	Red	Orange	22 kΩ	_____Ω
____	____	____	22,000 Ω	
Brown	Black	Yellow	100 kΩ	_____Ω
____	____	____	100,000	

probes. Then ask them an embarrassing question. Watch the resistance go down for a moment.

4. Write each of these values as a number with no abbreviations.
   - 10 kΩ = ________ Ω
   - 1 kΩ = ________ Ω
   - 0.47 kΩ = ________ Ω
   - 47 kΩ = ________ Ω

Lesson 5
The Effect Resistors Have on a Circuit

Throughout electronics, resistors are used to control the voltage and flow of the current. Even though this lesson is not very long, it does take time. Do it properly and you will get proper results. You will observe, chart, and describe the effects of different strength resistors when they are all set up in identical circuits.

Let's go back to the breadboard and see how different resistors affect a simple circuit. Both the resistors and LEDs are loads. The resistor uses most of the voltage, leaving just enough for the LED to work. The LEDs need about two volts.

What would happen if you changed resistors on the circuit you just built, shown in Figure L5-1?

You measured the voltage used across the resistor from TP-B to TP-C and measured the voltage used across the LED from TP-C to TP-D.

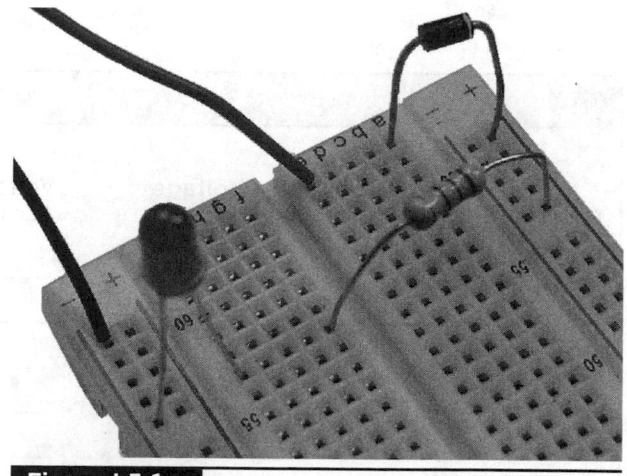

Figure L5-1

Figure L5-2 is the schematic of the circuit.

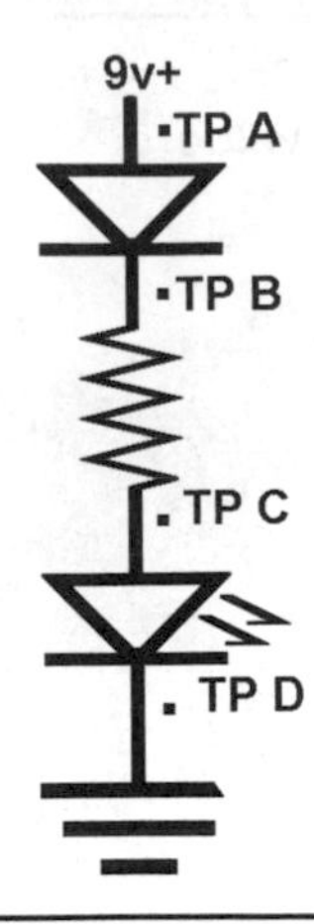

Figure L5-2

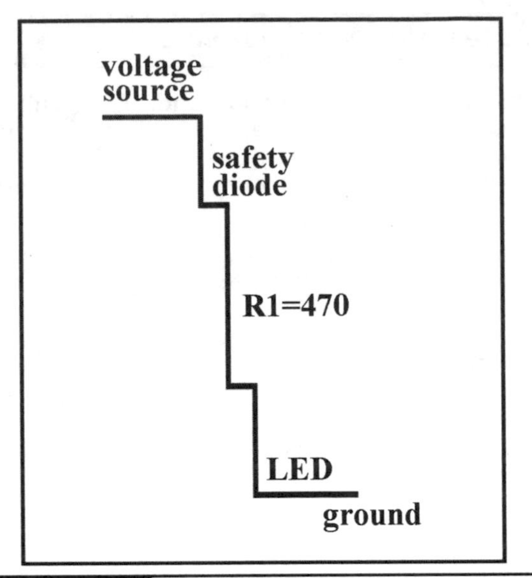

Figure L5-3

Figure L5-3 shows a waterfall. A waterfall analogy explains how voltage is used up in this circuit. The water falls over the edge. Some of the force is used up by the first load, the safety diode. More of the voltage is then used by the second load, the resistor. The remaining voltage is used by the LED.

This "waterfall" shows how the voltage is used by a 470-ohm resistor. If the resistor wasn't there, the LED would be hit with the electrical pressure of more than eight volts. It would burn out.

Remember, all the water over the top goes to the bottom, and all of the voltage is used between source and ground. Each ledge uses some of the force of the falling water. Each component uses part of the voltage.

What happens if there is more resistance? More of the voltage is used to push the current through that part of the circuit, leaving less to power the LED.

This is represented visually in Figure L5-4.

Exercise: The Effect Resistors Have on a Circuit

Your setup should look like Figure L5-5. Have your resistors arranged from lowest to highest value as presented in Table L5-1.

TABLE L5-1 Exercise Sheet

Resistor Value	Total Voltage Available	Voltage Drop Across Resistor	Voltage Drop Across the LED	LED Brightness (compared to 470 Ω)
100 Ω	_____V	_____V	_____V	__________
470 Ω	_____V	_____V	_____V	Normal
2,200 Ω	_____V	_____V	_____V	__________
10,000 Ω	_____V	_____V	_____V	__________
47,000 Ω	_____V	_____V	_____V	__________
220,000 Ω	_____V	_____V	_____V	__________

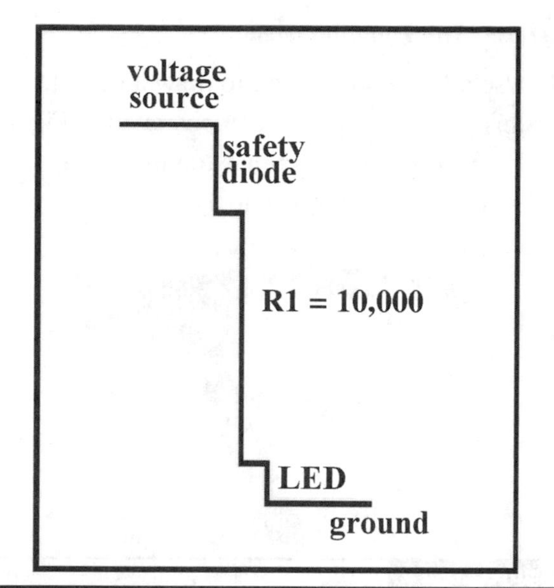

Figure L5-4

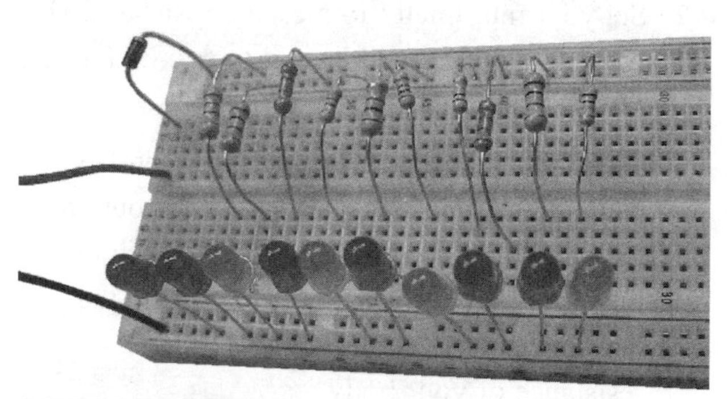

Figure L5-5

Lesson 6
The Potentiometer

Some resistors change resistance over a wide range. You use potentiometers daily as volume controls. Potentiometers are still widely used, though they are being replaced by digital push buttons.

Not all resistors are "fixed" like the small color-banded ones that you've already been introduced to. A common variable resistor is the potentiometer, pictured in Figure L6-1.

This useful device is often simply referred to as a *pot*. A smaller version is also shown. These are called *trim pots*. You have often used potentiometers as volume controls. The maximum resistance value is usually stamped onto the metal case.

Figure L6-1

Figure L6-2 shows a picture of a potentiometer taken apart. The potentiometer works because the sweep arm moves across the carbon ring and connects that to the center. The leg on the left is referred to as A, the center leg as C (center), and the right leg as B.

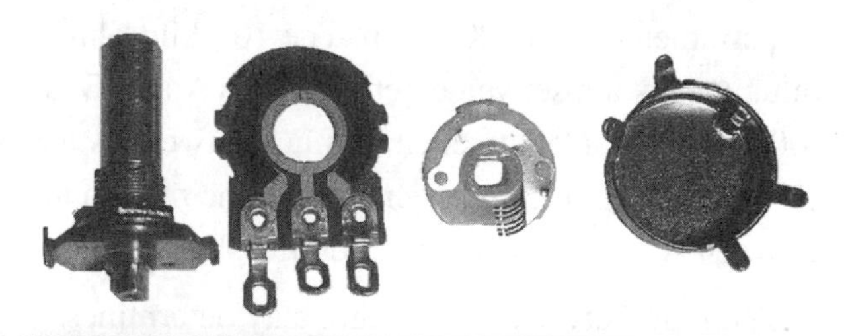

Figure L6-2

The carbon ring shown in Figure L6-3 is the heart of the potentiometer. It is made of carbon mixed with clay. Clay is an insulator. Carbon is the conductor.

The action of the potentiometer is the sweep arm (copper on white plastic) moving across the carbon ring (Figure L6-4). The sweep arm allows the current to move between A and C as its

Figure L6-3

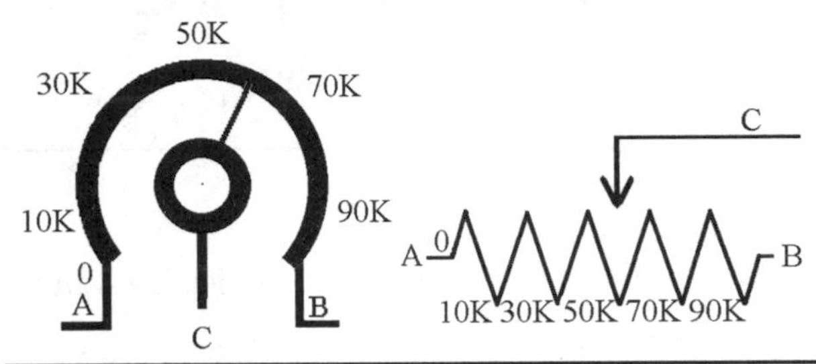

Figure L6-4

position changes. The resistance between A and C also changes with distance.

The distance between A and B is always the same, so the resistance between A and B is always the same. The value for this demonstration potentiometer is 100,000 ohm. The 100-kilo-ohm value means the set value between legs A and B is 100 kilo-ohm. Ideally, the minimum between A and C is 0 ohm (directly connected), and the maximum between A and C should be 100 kilo-ohm.

The ratio between carbon and clay determines how easily electrons pass through the resistor. More clay means less carbon. Less carbon means less conducting material. That creates higher resistance.

The carbon in the ring is similar to the carbon in a pencil. The pencil lead is also made of a mixture of carbon and clay. Soft pencils have less clay and more carbon. A mark by a soft pencil will have less resistance. Hard pencils have lead that contains more clay and less carbon. These provide higher resistance.

Exercise: The Potentiometer

1. Use a No. 2 soft pencil to draw a thick line on this piece of paper as demonstrated in Figure L6-5. A harder pencil has too much clay and will not give good results.

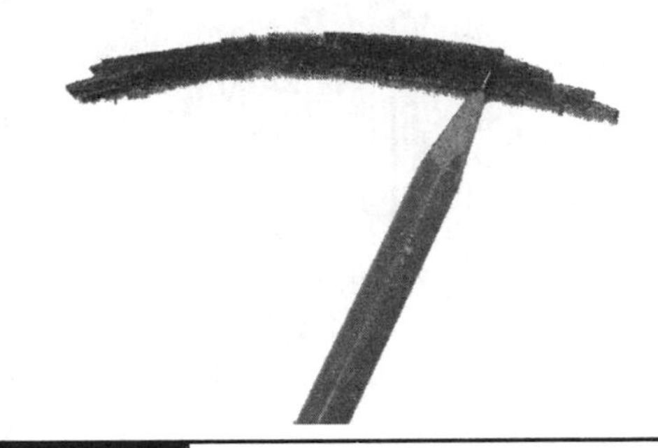

Figure L6-5

2. Set your multimeter to measure resistance Ω. If it is not autoranging, set it to maximum resistance.
3. As shown in Figure L6-6, press the probes down hard against the pencil trace about an inch apart. Be sure that you don't touch the tips of the probe. You want to measure the resistance of the pencil trace, not the resistance of your body.

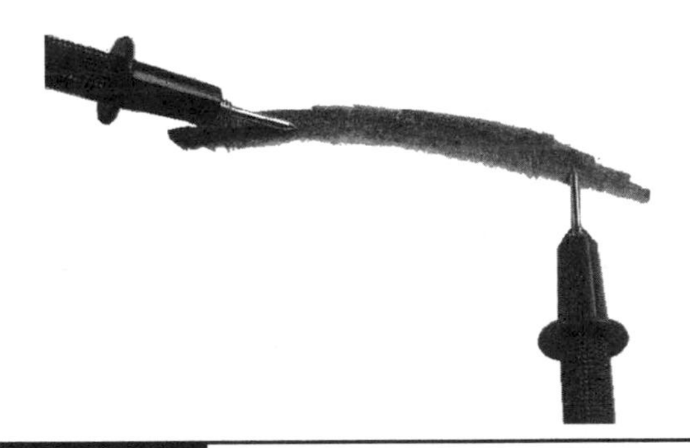

Figure L6-6

   a. Now record the resistance from the multimeter: ________ Ω. If the DMM says the resistance is out of range, move the probes together until you get a reading.
   b. Move the probes closer together and then farther apart. Write down what you observe.

4. Use the 100-kilo-ohm potentiometer. Record your results.
   a. Measure the resistance between the two outer legs A and B. ______ Ω
   b. Adjust the knob and check the resistance between A and B again. ______ Ω
   c. Adjust the knob about halfway. Measure the resistance between the left and middle legs—A and C. ______ Ω
   d. Turn the knob a bit and check again. Note any change. ______ Ω

   Explain what is happening, relating that to the carbon ring shown in Figure L6-3.

   ____

   ____

5. Make sure that you have the battery hooked up properly through the power diode as noted on the schematic.
6. As you turn the shaft of the potentiometer, the LED should brighten and dim. Explain what is happening.

   ____

   ____

7. Why is there a 470-ohm fixed resistor in this circuit?________________________

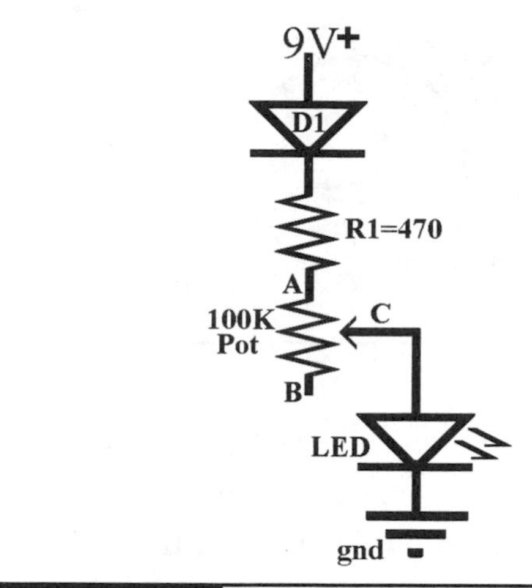

Figure L6-7

Figure L6-8

Breadboarding the Circuit

Note the similarities of the schematic shown in Figure L6-7 and the picture of the circuit displayed in Figure L6-8.

Lesson 7
Light-Dependent Resistors

Another variable resistor is the *light-dependent resistor* (LDR). The LDR changes its ability to conduct electrons with the change of light. It is commonly used to turn equipment on automatically as night falls. Some cars use it as the input to the switch that turns on headlights as conditions change, even as they drive through a tunnel. The symbol for the LDR is shown here in Figure L7-1.

There is no room to place a value on most LDRs. They are ordered and supplied in specific values. An easy way to measure the maximum resistance is to measure it in darkness.

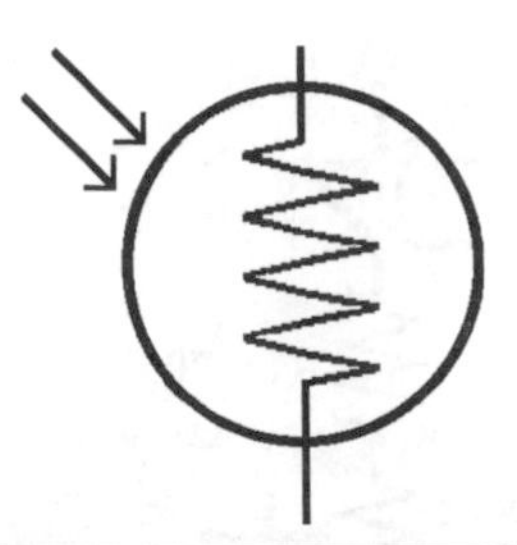

Figure L7-1

Insert the LDR onto the breadboard so the legs are not connected, as shown here in Figure L7-2. Measure the resistance using your DMM. The readout may be jumping around because LDRs are sensitive.

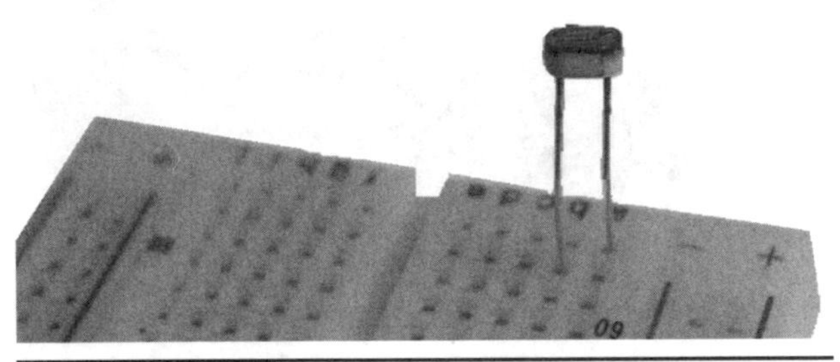

Figure L7-2

Look at Figure L7-3. Place the lid of a black pen over the LDR and measure the resistance again.

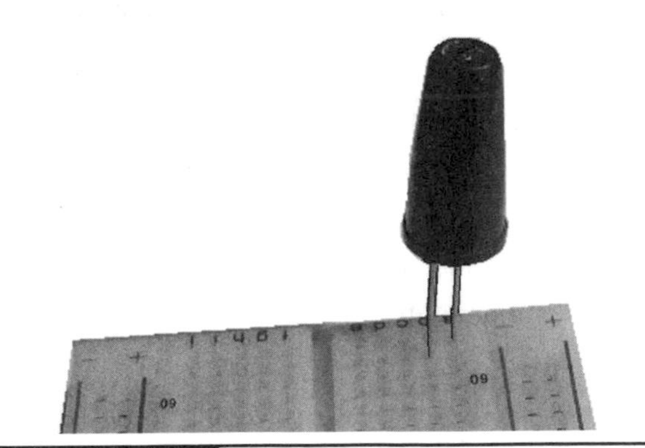

Figure L7-3

Breadboard the Circuit

Note the similarities of the schematic in Figure L7-4 and the breadboard layout in Figure L7-5.

PARTS BIN
■ D1—Power diode
■ LDR—1 MW dark
■ LED—5 mm round

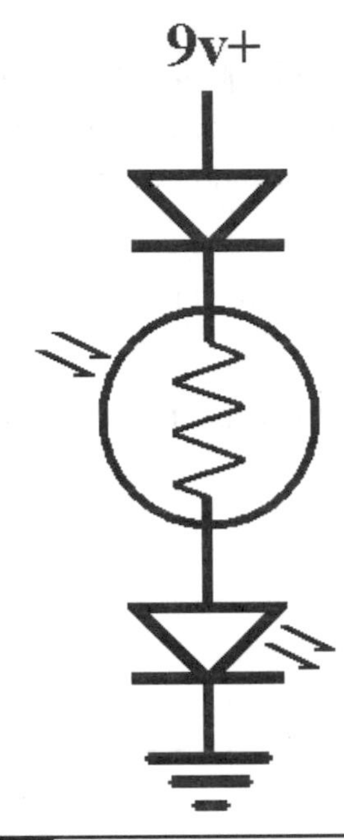

Figure L7-4

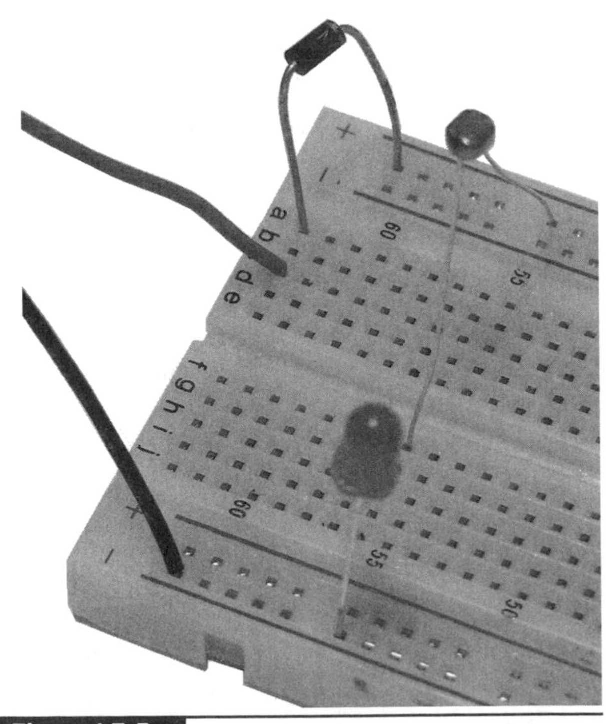

Figure L7-5

What to Expect

1. Attach the battery and note the brightness of the LED. It should be fairly bright.
2. Place the lid of the pen over the LDR again. The LED should dim to nearly nothing.
3. Consider this. What is the relationship between the amount of light on the LDR and the LDR's resistance?

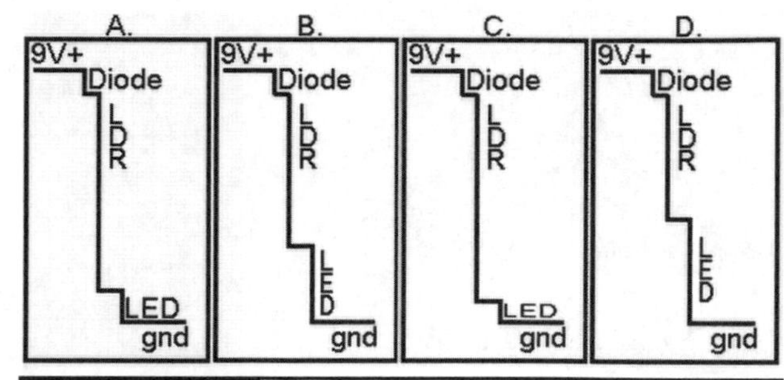

Figure L7-6

Exercise: Light-Dependent Resistors

1. Disconnect the power supply. Measure and record the resistance of the LDR in the light. It may be necessary to take a rough average because it will be jumping around wildly.
2. Place a dark black pen lid over the LED and measure the resistance again. Remember that your fingers can affect the readout.
3. Attach the power supply and note the brightness of the LED. Place the lid of the pen over the LDR again. State the relationship between the amount of light on the LDR and the resistance of the LDR.
4. Note the minimum resistance that occurs on the LDR in the light. Why is the 470-ohm resistor not used in this circuit?
5. Consider the "waterfall" diagrams presented in Figure L7-6. From brightest to darkest conditions, what would be the best order of these diagrams regarding the LDR's effect on the brightness of the LED?

SECTION 3

More Components and Semiconductors

SIMPLE PLUMBING SYSTEMS can be used to explain many components in electronics.

- Water flows like electrons flow.
- We can put pressure behind water.
- There are different size pipes.
- Valves control the flow of water.
- We can fill containers with water.
- We can drain the water from those containers.

Lesson 8
Capacitors and Push Buttons

Yes, there is more to electronics than resistors and LEDs. Capacitors are used to store small charges. Push buttons allow you to control connections to voltage. This lesson introduces both capacitors and push buttons. You then build a circuit that applies them together.

Capacitors

A capacitor has the capacity (ability) to store an electric charge. You can see in Figure L8-1 that the symbol of the capacitor represents two plates.

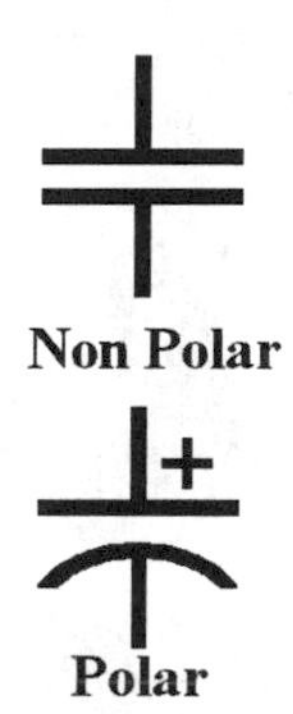

Figure L8-1

In Figure L8-2, the opened capacitor clearly shows that the capacitor is made of just two metal plates, with a bit of insulation between them. They come in three basic shapes and all sizes.

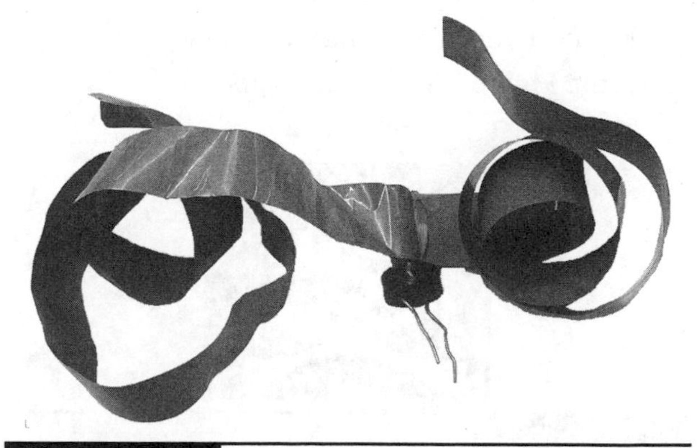

Figure L8-2

Capacitors in the upper range, 1 microfarad and higher, are electrolytic capacitors. They must be connected in the right direction. There are two indicators of the negative side. First, there is a colored stripe down the side that indicates polarity, and second, if both legs come out of the same side, one leg is shorter. That is the negative leg. It is "minus" some length. Only the electrolytic capacitors have a positive and negative side. The disk and film capacitors do not have a positive or negative side. A variety of capacitors are represented in Figure L8-3.

Figure L8-3

Remember that a backwards electrolytic is a dead electrolytic. These must be connected correctly. Figure L8-4 helps remind us.

Figure L8-4

When electricity was first being defined over 200 years ago, the measurements were done with crude instruments that were not sensitive. The people who defined the units missed the mark, but we still use them today. The farad is the basic unit of capacitance. One farad is so huge that today, the standard unit in electronics is one-millionth of a farad. The Greek letter μ (mu) represents micro for the unit. That is 0.000001 F or 1×10^{-6} farads, and is commonly written as 1 μF (1 μF = 1 microfarad = 0.000001 F = 1×10^{-6} F).

We'll go back to using the water analogy. If you think of the electric charge like water, the capacitors can be compared to containers able to hold that water. The amount of charge capacitors can hold depends on their purpose, just like varying size containers used to hold water. Such containers are pictured in Figures L8-5, L8-6, and L8-7.

Figure L8-5

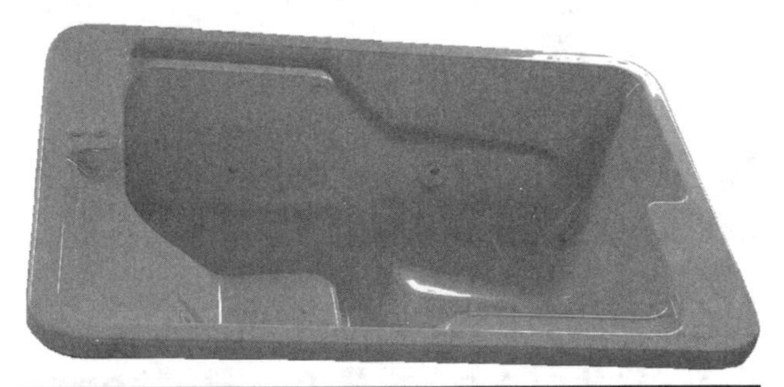

Figure L8-6

As mentioned before, capacitors come in three standard types. Disk capacitors hold the smallest amount. They have a common shape shown in Figure L8-8. They are so small that their capacitance is measured in trillionths of a farad,

Figure L8-7

called *picofarads*. Their general range is from 1 picofarad to 1,000 picofarads. To look at that another way, that is one-millionth of a microfarad to one-thousandth of a microfarad.

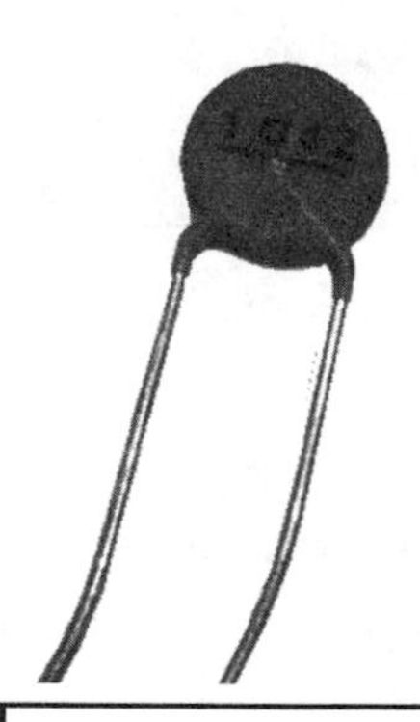

Figure L8-8

To visualize the size of charge they are able to hold, think of water containers ranging from a thimble (Figure L8-9) up to a mug (Figure L8-10).

Figure L8-9

Figure L8-10

Film capacitors are box-shaped, as shown in Figure L8-11. They are midrange. They hold between a thousandth of a microfarad and a full microfarad.

Figure L8-11

Their capacitance range is 1,000 times that of the disk capacitor. A good analogy for the relative size of charge a film capacitor holds is to think of a range from a sink, shown in Figure L8-12, up to the size of a large bathtub, shown in Figure L8-13.

Electrolytic capacitors are small and can-shaped. Find the electrolytic capacitors in your inventory. They should look similar to the electrolytic capacitors pictured in Figure L8-14. There might be various colors.

Figure L8-12

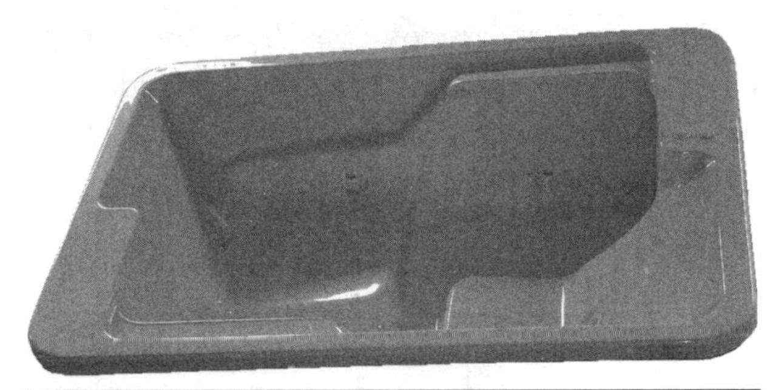

Figure L8-13

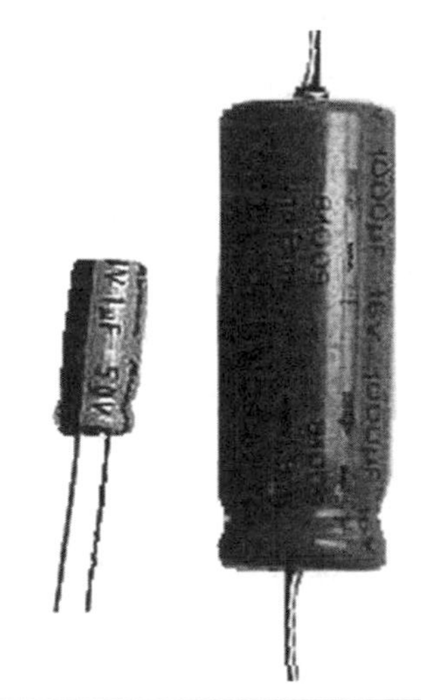

Figure L8-14

These hold the larger amounts of 1 microfarad and above. Their capacitance abilities can be thought of in larger dimensions, from swimming pools (Figure L8-15) to lakes (Figure L8-16).

Figure L8-15

Figure L8-16

Push Buttons

There are two main types of push buttons, and they can look identical to the picture in Figure L8-17.

Figure L8-17

Push Button Normally Open (PBNO)

Push the button; a piece of metal connects with two metal tabs inside, as you can see in Figure L8-18. It creates a temporary path, and the charge can flow. Set your digital multimeter (DMM) to CONTINUITY and put a probe to each contact for the push button. CONTINUITY should show only when you are pushing the plunger down.

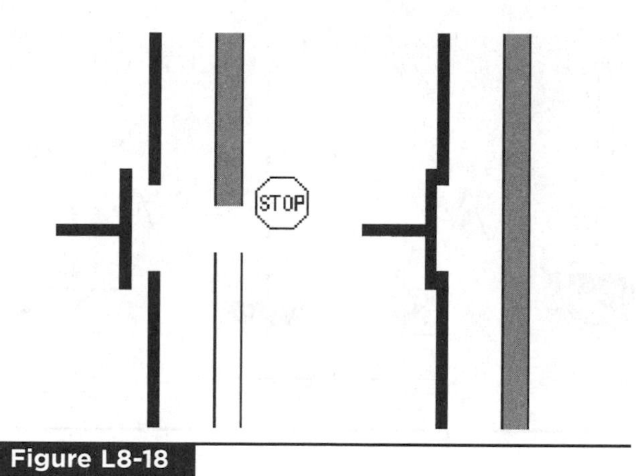

Figure L8-18

Push Button Normally Closed (PBNC)

Push the button; a piece of metal disconnects from the two metal tabs inside, as depicted in Figure L8-19. It creates a temporary break, and the charge cannot flow. Set your DMM to CONTINUITY and put a probe to each contact for the push button. CONTINUITY will show all the time, except when you are pushing the button down.

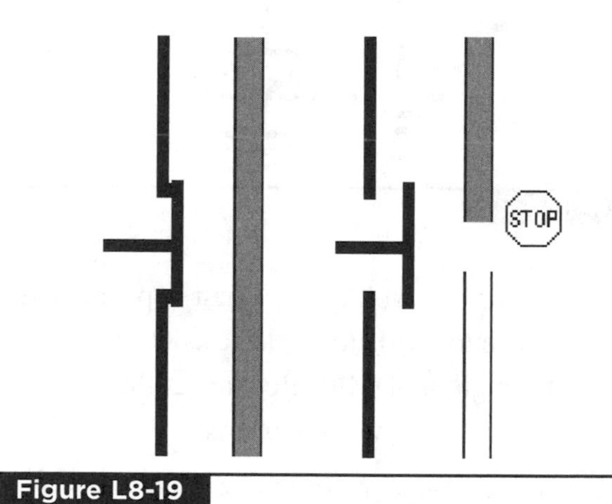

Figure L8-19

Build This Circuit

Build the circuit shown in Figure L8-20 (see also the Parts Bin). Note the similarity between the schematic in Figure L8-20 and the photograph in Figure L8-21.

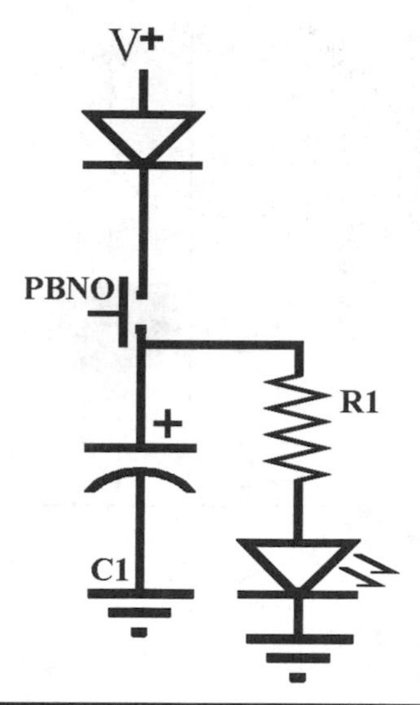

Figure L8-20

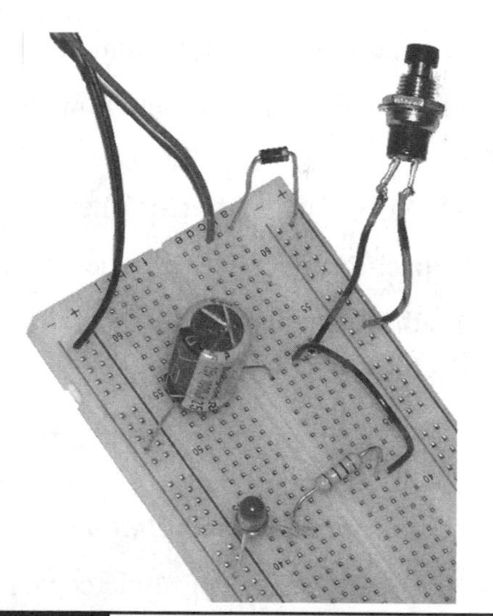

Figure L8-21

PARTS BIN

- PB1—Normally open solder connecting wire to the legs
- C1—1,000 μF electrolytic
- LED—5 mm round
- R1—470 Ω

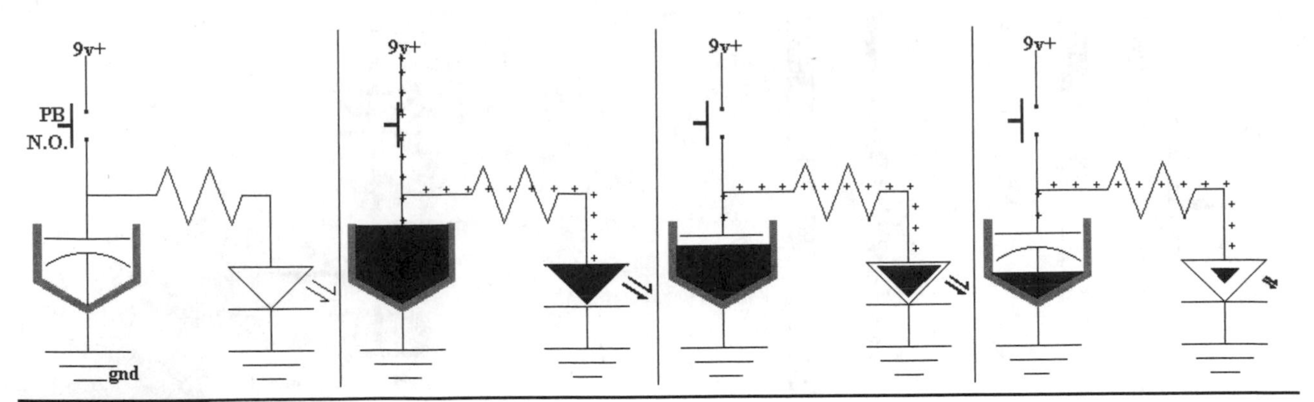

Figure L8-22

How It Works

Carefully note the sequence of actions in Figure L8-22.

1. The normally open push button closes.
2. Voltage fills the capacitor and powers the LED.
3. The PBNO opens, cutting off the voltage.
4. The capacitor drains through the LED.
   a. As the capacitor drains, the voltage decreases.
   b. As the voltage decreases, the LED dims.

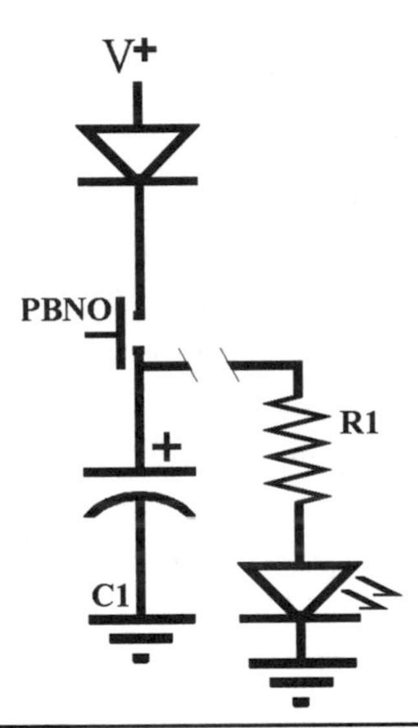

Figure L8-23

Exercise: Capacitors and Push Buttons

1. Look closely at the electrolytic capacitors. Be sure to note the stripe and the short leg that marks the polarity.
2. Describe what happens in your circuit as you push the button, then let go.
3. Disconnect the wire indicated in Figure L8-23 between the capacitor and R1.
   a. Push the button to charge the capacitor. Now wait for a minute or so.
   b. Set your DMM to the proper voltage range. Put the red probe to the positive side of the cap and the black probe to ground.
   c. Record the voltage that first appears. The capacitor will slowly leak its charge through the DMM. Reconnect the wire and describe what happens.
4. Use Table L8-1 to record your information as you play with your circuit.

TABLE L8-1 Information Record

Cap Value	Time
1,000 μF	
470 μF	
100 μF	
10 μF	
1 μF	

a. As you replace each capacitor and record the time, the LED stays on. Don't expect the time to be very exact.

b. Describe the pattern that you see here.

____

5. Briefly describe what capacitors do.

____

6. Place the 1,000-microfarad capacitor back into its original position. Now replace the normally open push button (PBNO) with the normally closed push button (PBNC). Describe the action of this circuit.

Lesson 9
Introducing Transistors

NOTE **Learning electronics is not hard. It is lots of new information, but it is not hard. Think about it, but not as hard as the guy in Figure L9-1.**

Figure L9-1

Considering that it has only been a bit more than 100 years since the first transatlantic radio message, electronics is a young technology. The invention of the transistor in 1947 was the first step towards the microsizing of all electronics we use today. The NPN (negative-positive-negative) is truly electronic. It acts like a normally open push button, but has no moving parts. The transistor is the basic electronic switch. It does have an interesting history that makes for good outside reading. Our entire electronic age is dependent on this device.

Transistors are commonly packaged in the TO-92 case shown in Figure L9-2. Notice how the legs correspond to the schematic symbol in Figure L9-3.

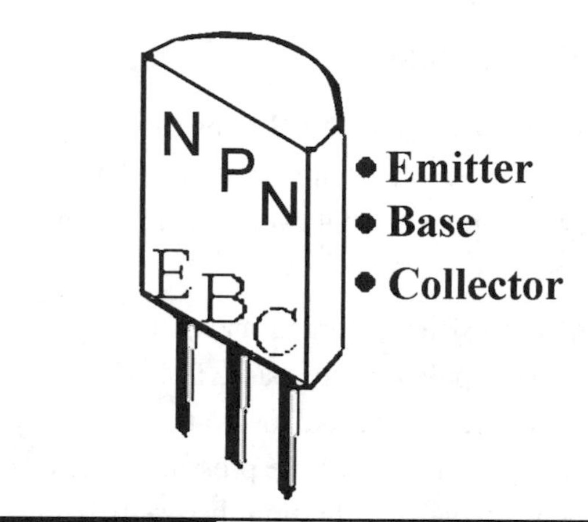

Figure L9-2

Figure L9-3

Note the arrow inside the schematic symbol. It indicates two things: First, it points in the direction of the current, towards ground. Second, it is always on the side of the emitter. It is important to identify the legs of the transistor. For this package, it is easy to remember. Hold the transistor in your fingers with the flat face toward you. Think of a high mountain in the rugged back country of British Columbia—a cliff face reaching skyward. Now, reading left to right, whisper "Enjoy British

Columbia." You have just identified the three legs. Cute, but it helps.

There are thousands of different types of transistors. The only way to identify them is to read the numbers printed on the face of the package itself. But even with thousands, there are only two basic types of transistors, the NPN transistor and the PNP transistor.

The NPN Transistor

This lesson introduces the NPN transistor, using the 3904 NPN. Lesson 10 introduces the 3906 PNP. They are opposites but evenly matched in their properties.

The NPN transistor is turned on when voltage and current are applied to the base. The NPN transistor acts very much like the water faucet pictured in Figure L9-4. A little pressure on the handle opens the valve, releasing the water under pressure.

Figure L9-4

As you can see in Figure L9-5, a little voltage and current on the base of the NPN transistor leads to a large increase in the flow of current through the NPN transistor from the collector to the emitter.

Another way of thinking about it—the force needed to open the gates on the Grand Coulee Dam, pictured in Figure L9-6, is small compared to the amount of force that moves through those gates.

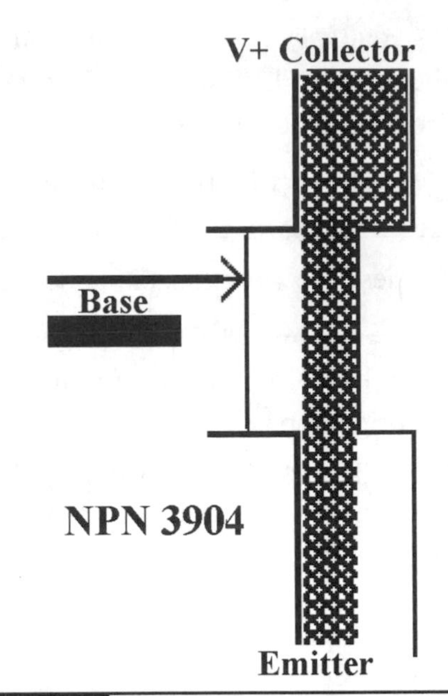

Figure L9-5

Figure L9-6

Build the NPN Transistor Demonstration Circuit

You have used the capacitor to store small amounts of electricity. It powered the LED directly, but could only do that for a brief moment. Here, we use the capacitor to power the transistor. Again, you need to note the similarity between the schematic in Figure L9-7 and the way the circuit is pictured on the solderless breadboard in Figure L9-8 (see also the Parts Bin).

PARTS BIN

- D1—Safety diode
- PB—PBNO
- C1—10 μF
- R1—22 kΩ
- R2—470 Ω
- Q1—NPN 3904
- LED—5 mm round

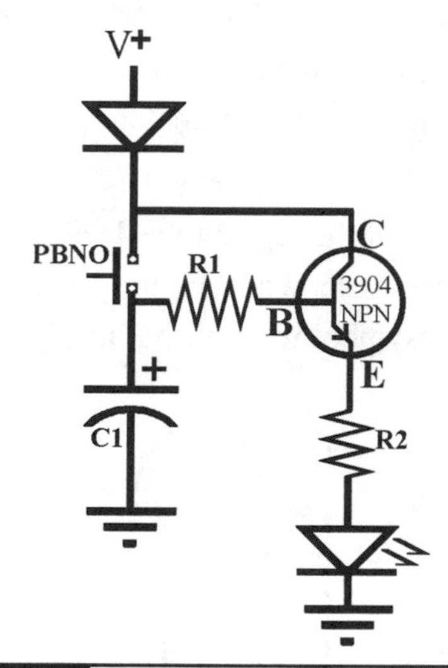

Figure L9-7

Figure L9-8

What to Expect

The LED stays off as you attach your battery. Push and release the push button. The LED will turn on immediately. It will dim and turn off. This action is faster with smaller capacitors.

How This Circuit Works

You are using the charge held in the capacitor to power the transistor. The transistor provides a path for the current to the LED.

Because the base of the transistor uses much less power than the LED, the voltage drains from the capacitor very slowly. The higher-value resistor of 22,000 ohms slows the drain from the capacitor significantly.

The LED staaaaaays on muuuuuuch loooooonger.

Exercise: Introducing Transistors

1. Briefly describe the purpose of the transistor.

   ____

2. What do you think? Anything that looks like a transistor is a transistor.
3. Describe how to tell which leg of the transistor is the emitter.

   ____

4. Which leg of the transistor is the base?

   ____

5. What two separate things does the arrow inside the transistor symbol indicate?

   a. ______________________________

   ____

   b. ______________________________

   ____

6. What is the only way to tell the type of transistor? ______________________________

   ____

7. Regarding the water faucet analogy, is the water pressure provided by the water system or the handle? The pressure is provided by the

   ____

8. Press and release the push button. After you release the push button, what part provides the power to the base of the transistor?

____

____

9. Describe the path of the current that provides the power to the LED. Here is something to consider regarding the answer. The capacitor is not powering the LED. It is only powering the transistor.

10. Record three time trials of how long the LED stays on with the 10-microfarad capacitor.

Time 1	Time 2	Time 3	Average
____ s	____ s	____ s	____ s

Replace C1 with the 100-microfarad capacitor. Time the LED here for three times as well, and find the average.

Time 1	Time 2	Time 3	Average
____ s	____ s	____ s	____ s

Roughly stated, how much more time did the 100-microfarad capacitor keep the LED on for than the 10-microfarad capacitor?

a. Three times longer

b. Five times longer

c. Eight times longer

d. Ten times longer

Write down your prediction of how much time the 1,000-microfarad capacitor would *keep* the LED working.

Okay, now put in your 1,000-microfarad capacitor. Try it out three times, and average the time.

Time 1	Time 2	Time 3	Average
____ s	____ s	____ s	____ s

How accurate was your prediction?

11. Describe in detail how this circuit works. Consulting Figure L8-22 of the capacitor powering the LED once the push button is released, the voltage pressure to the base is provided by the______________________.

Lesson 10

The PNP Transistor

We use only the NPN 3904 and PNP 3906. These represent the two general classifications of transistors. They are evenly matched but opposite in their action.

The identity of the legs on the TO92 package stays the same, as shown in Figure L10-1. But look closely at the symbol for the PNP in Figure L10-02.

Figure L10-1

Figure L10-2

Note that the schematic symbol of the transistor holds some important information. The arrow inside the schematic symbol still points in the direction of the common current, but is on the top side now. Because it is always on the side of the emitter, that means the PNP emitters and collectors have reversed positions relative to the NPN. The legs on the package are still the same, though. The emitter and collector have traded positions relative to the current flow.

Not only are the emitter and collector positions reversed, but the action is reversed as well. The PNP transistor's action is the opposite of the NPN's. As you increase the voltage to the base, the flow decreases; as the voltage to the base decreases, the PNP transistor is turned on more. The valve starts in an open position.

The PNP transistor still acts very much like the water faucet as shown in Figures L10-3 and L10-4.

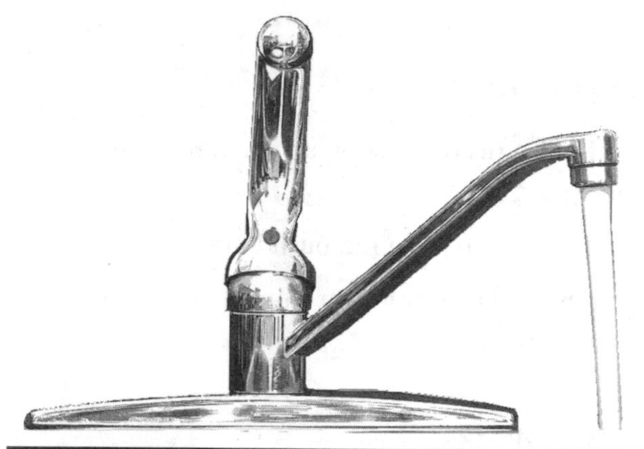

Figure L10-3

But now, a little pressure on the handle closes the valve, stopping the water. No pressure on the handle allows the water to push through the faucet. No pressure (voltage) on the base of the PNP

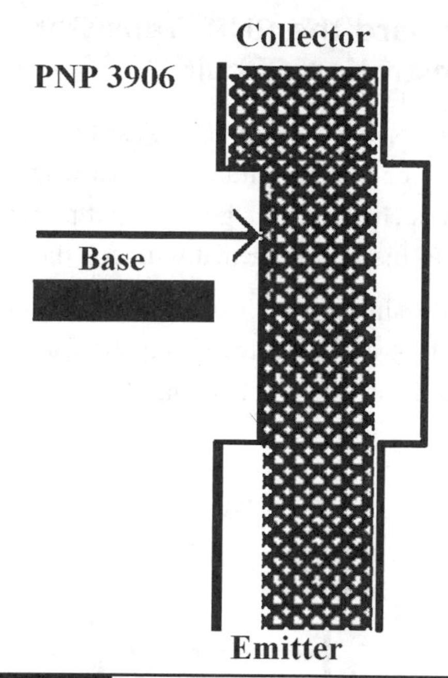

Figure L10-4

transistor allows the voltage and current to push through the transistor.

But just like turning the water faucet's handle will decrease the water flow, voltage to the base of the transistor will decrease the flow of current through the transistor. Enough pressure on the handle of the water faucet will shut it off. Enough voltage at the base will turn the PNP transistor off completely as well.

Surprisingly, the base has the same action for both the NPN and PNP transistors. It just has a different starting position, as shown in Figure L10-5.

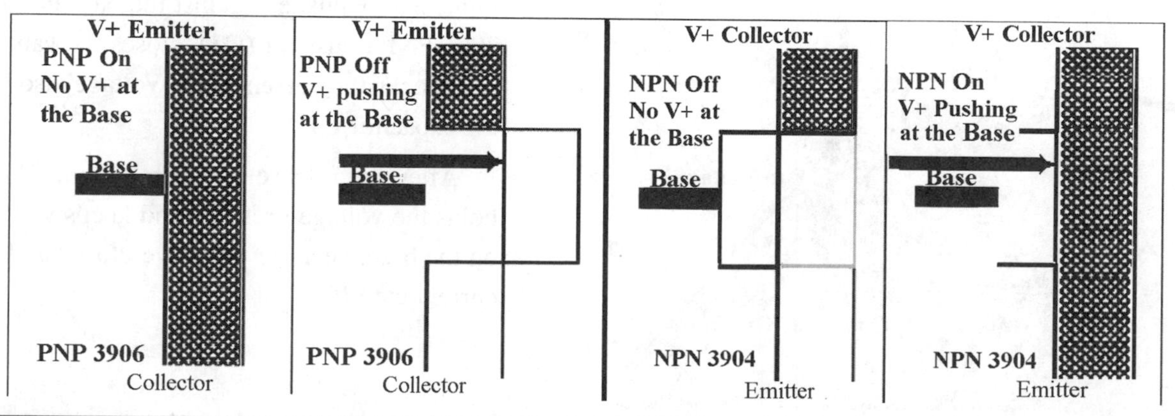

Figure L10-5

Breadboard the PNP Transistor Demonstration Circuit

The capacitor is powering the transistor. But remember for this PNP transistor that when the capacitor is charged, it is going to put pressure on the base of the transistor that will stop the flow.

Note the similarity between this schematic in Figure L10-6 and what the schematic was for the NPN transistor. Also, notice that the transistor in Figure L10-7 is physically reversed compared to the NPN transistor in the previous lesson.

PARTS BIN
■ Q1—PNP 3906
■ R1—100 kΩ
■ R2—22 kΩ
■ R3—470 Ω
■ C1—10 μF
■ PB—N.O.
■ LED—5 mm round

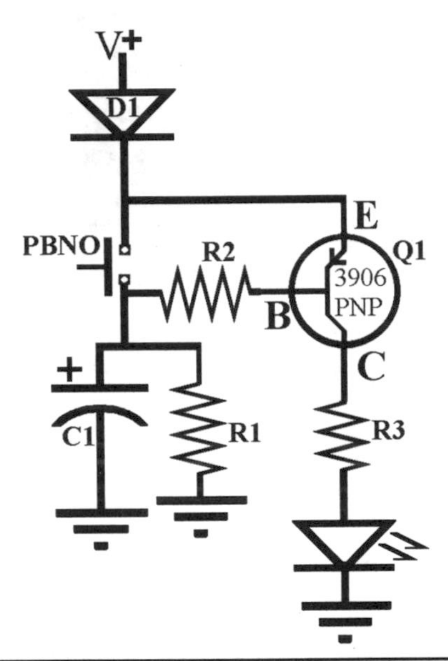

Figure L10-6

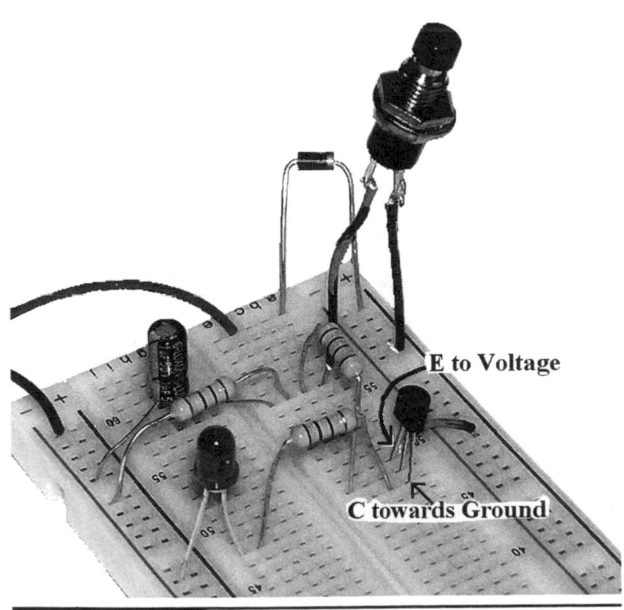

Figure L10-7

What to Expect

The LED turns on as soon as you attach your battery.

1. Push and release the push button.
2. The LED will turn off immediately. It will slowly turn back on.

How It Works

1. When you first attach your battery, the LED turns on immediately because there is no voltage pressure pushing at the base, so the valve is in the opened position, allowing the current to flow from emitter to collector.
2. When you push the plunger down, the voltage immediately pushes against the base of the 3906 PNP transistor (Q1), closes the valve, and blocks the current flow. Voltage also fills the capacitor C1.

   After you release the push button, C1 holds the voltage pressure and keeps voltage on the base, keeping the valve closed and the current cut off.

3. As the voltage drains from C1 through R1, the voltage pressure against the base is released. The transistor starts passing current and voltage again slowly. The LED turns back on.
4. Why the extra resistor (R1)? (a) Before the push button is closed, both C1 and the base of the 3906 PNP transistor have no voltage. Because there is no voltage pressure on the base of Q1, the valve is open and current flows from emitter to collector; (b) when the voltage in the capacitor is high, Q1's valve stays shut; (c) the path for current to escape from C1 through the transistor is blocked because the valve is closed; (d) so, R1 is necessary to drain the charge from the capacitor. This allows Q1's valve to open again.

The capacitor is unable to drain and the transistor stays off because the voltage from the capacitor keeps the pressure on the base of the transistor, keeping the valve closed. The capacitor cannot drain through the base of the PNP transistor like it did in the previous 3904 NPN circuit. The extra resistor allows the cap to slowly drain, decreasing the voltage pressure on the base of the PNP transistor, allowing the valve to reopen and let current flow again.

Exercise: The PNP Transistor

1. In the schematic, Q stands for what component? Q represents the ____________ ____________________________.
2. The arrow in the transistor symbol represents what action?
   a. Direction of current flow
   b. Direction of the collector
   c. Direction of the base
   d. Direction of the emitter
3. The arrow is always on the side of which leg in the schematic?
   a. Voltage
   b. Emitter
   c. Base
   d. Collector
4. What would happen if R3 were not in the circuit and the LED was connected directly to the collector of the 3906 transistor?
   a. LED would burn out.
   a. LED would be bright.
   a. LED would not work.
   a. LED would flash.

   Explain your answer to the above.

   ____________________________.
5. Replace C1 with the 100-microfarad capacitor. Describe what happens.

   ____________________________.

   Why does changing the capacitor affect the circuit this way?

   ____________________________.
6. Change C1 back to 10 microfarad. Now change R1 to 10 megohm (brown-black-blue). Describe what happens.

   ____________________________.
7. Think of the capacitor as a sink, holding water. Think of the resistor as the drainpipe. Which of the following statements best explains how changing to a higher resistance has the same effect as changing to a larger capacitor?
   a. The drain is bigger and empties the water faster.
   b. The drain is smaller and empties the water slower.
   c. The volume of water is bigger and takes longer to drain.

   d. The volume of water is smaller and drains faster.

8. Play a little.
   a. Replace R3 and the LED with the buzzer. Make sure the buzzer's red wire is getting voltage from the 3906's collector and the black wire is connected to ground.
   b. Push button and release. What happens to the sound as the capacitor discharges?

      ______________________________.

9. Carefully describe in your own words how this circuit works.

   ______________________________.

Lesson 11
The Phototransistor: Shedding Some Light on Another Component

All transistors are light-sensitive. Shortly after transistors were invented, engineers recognized that their new toys gave different results as lighting conditions changed. Lots of work went into shielding the guts of the early transistors. But at the same time, lots of effort was put into understanding and finding uses for this phenomenon.

The results of this research? The phototransistor. Figure L11-1 shows both the phototransistor (darkened glass) and its corresponding infrared (IR) LED (clear glass).

Many of the available phototransistors have been developed to react to specific wavelengths of light. This is important, because the intent is to control the transistor, leaving nothing to chance.

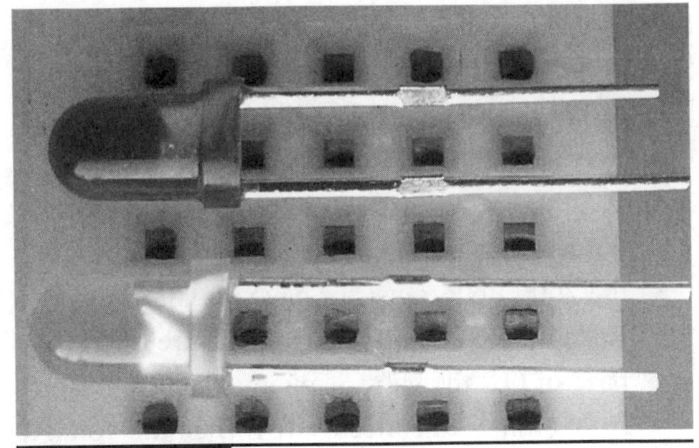

Figure L11-1

So how do we use this "tuned" phototransistor to our advantage? Here, I'm using a phototransistor that reacts specifically to IR light with a wavelength of 940 nanometers. Both sunlight and filament light bulbs create IR that will affect our phototransistor. Fluorescent lights don't have an infrared component in their spectrum.

So what to use as a light source? Hmmm? What color LEDs are readily available? Red, yellow, blue, green, and orange. Table L11-1 shows the wavelength of each color. Our eyes don't respond to the wavelengths above 750 nm, but digital cameras do respond to low-end IR.

LEDs that emit IR are also available. We use one that is specifically matched to the phototransistor we'll be using. Even the part numbers are the same, except for one added letter.

Let's start playing. The Parts Bin has a full parts list.

TABLE L11-1 We Need an Infrared LED for an Infrared Phototransistor

400nm	475nm	510nm	570nm	590nm	650nm	750 to 2,500nm
Violet	Blue	Green	Yellow	Orange	Red	Infrared

PARTS BIN

- R1, R2, R3, R4—470 Ω
- LED 1—Yellow or orange work best
- LED 2, 3, and 4—Whatever colors you want
- Infrared LED 3mm (clear glass)—LTE 4206/IR 3mm 940nM
- LDR—1 MΩ dark
- NPN phototransistor 3mm (darkened glass)—LTE 4206E/IR 3mm 940nM

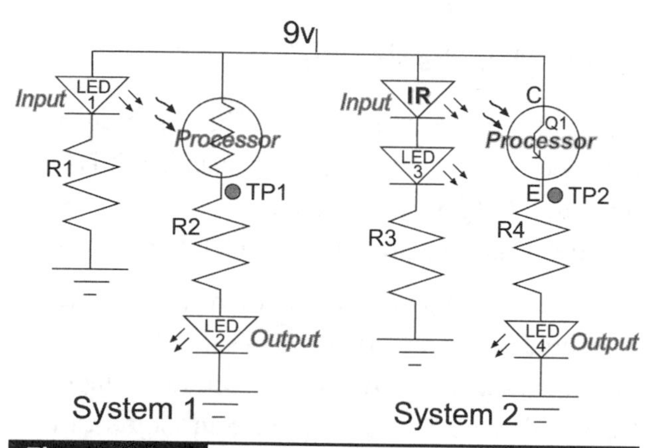

Figure 11-2

Figure L11-2 presents the basic schematic. Yes, there really are two separate circuits next to each other. Well, how else can they be compared? Each system has its own input, processor, and output. Build the two circuits. The second system has the two new components. The LDR responds best to a orange or yellow LED. The IR LED is paired with the phototransistor. Figure L11-3 shows the setup on the breadboard.

Each pair represents a separate system.

- LED 1 is the input to the LDR. The LDR is the processor. LED 2 is the output of the first system.
- IR LED is the input to the phototransistor. The phototransistor is the processor. LED 4 is the output of the second system.

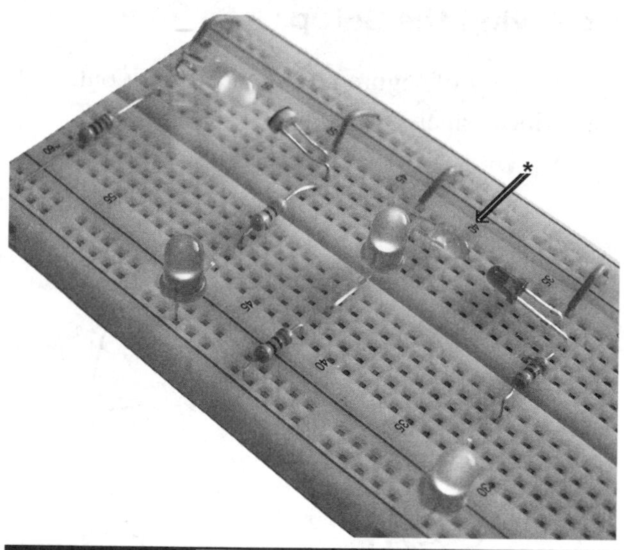

Figure L11-3

What to Expect

Do your work away from sunlit windows. Daylight provides a full spectrum that affects both the LDR and the phototransistor.

As you slide a piece of thick paper between the input and processor of each system, the outputs turn on and off. Does the phototransistor act like an NPN or PNP transistor? The output should be off when there is no input.

No big deal. Pretty much the same. Why bother?

But they aren't the same. Let's do some measuring. Why else would I have labeled a test point on each circuit? For each system, measure and record the test point voltage for both conditions. Record your results. System 1: TP1 _____volts. System 2: TP2 _____volts.

For System 1, the LDR uses a large amount of the voltage. However, the phototransistor uses about as much as the diode did back in Lesson 3. There is significantly more voltage available at System 2's test point.

"Yeah?" you say. "So what?" So let's play. Let's see if it means anything.

Modifying the Setup

Make a copy of Figure L11-4 and paste it onto some thick paper.

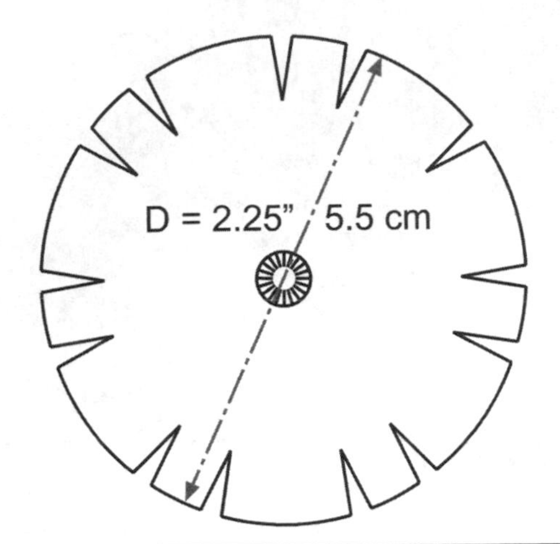

Figure 11-4

The complete setup is shown in Figure L11-5. Cut the disk and slots. Center your disk on a pencil so it spins evenly.

Now, as you spin the pencil, rate the responsiveness of each circuit on the "Mushiness Scale" (Table L11-2).

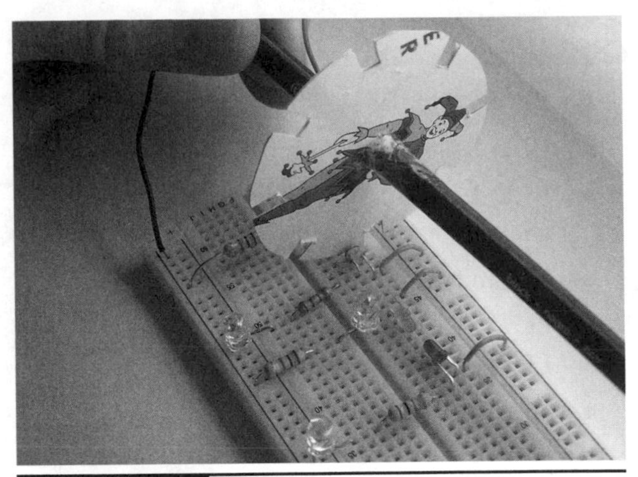

Figure 11-5

How It Works

The LDR's response is limited. The photosensitive area does not instantly change resistance. It is measured in Ω per millisecond. Also, you've seen that the resistance does not go much below 1 kΩ.

However, the phototransistor's response time is measured in microseconds. That's 1,000 times faster than the LDR. It doesn't need much input to "saturate" the base, opening the transistor's valve quickly and completely. There is little resistance, providing a clean, fast, and accurate response.

TABLE L11-2 Does the LDR or Phototransistor Work Better, Even at Different Speeds?

Input through slot	Processor	Output	Slow									Very crisp
LED 1	LDR	LED2	1	2	3	4	5	6	7	8	9	10
IR LED	Photo Transistor	LED4	1	2	3	4	5	6	7	8	9	10

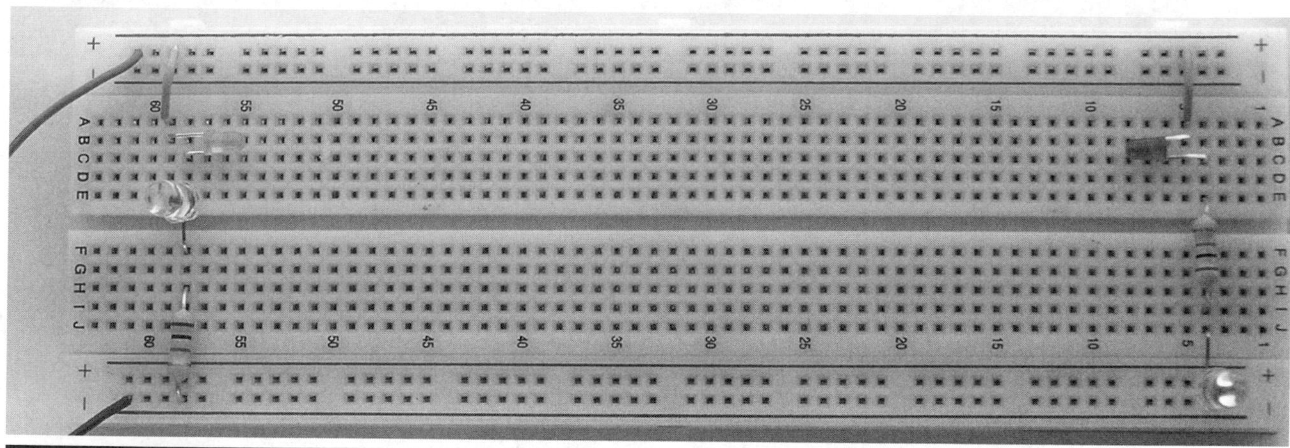

Figure 11-6

Modify It Some More

Take a moment. Compare the system response when the light source is a bit farther away. In Figure L11-6, the input half of the circuits has been moved to the opposite end of the breadboard. There will be a noticeable difference between the two systems.

Did you know that your TV remote uses infrared? See if your phototransistor responds to your remote control. Press a key. Any key will do. It should respond.

SECTION 4

Two Projects and Then Some More

Enough of this playing around. Let's do something.

Lesson 12

Your First Project: The Automatic Night Light

We combine our knowledge and the components to make an automatic night light. It dims in the light and brightens in the dark. Using the LDR, build the circuit shown in Figure L12-1 on your breadboard. The photograph of the circuit appears in Figure L12-2 (see also the Parts Bin).

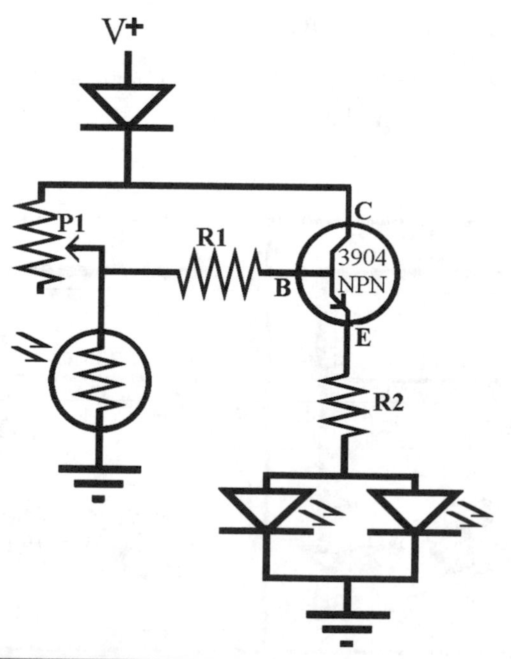

Figure L12-1

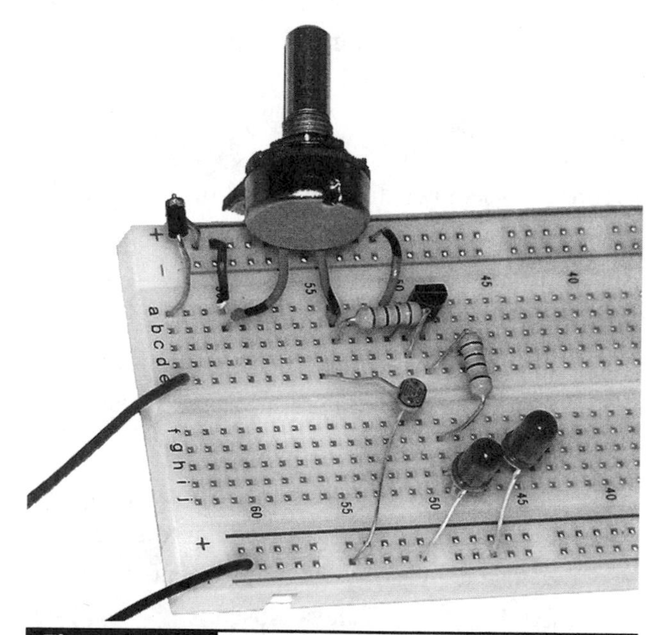

Figure L12-2

PARTS BIN
■ D1—1N400X diode
■ P1—100 kΩ pot
■ R1—22 kΩ
■ R2—470 Ω
■ LDR—1 MΩ dark
■ Q1—NPN 3904
■ LEDs—5 mm round

What to Expect

1. Attach your power supply.
2. Turn the knob on the pot one direction until the light-emitting diodes (LEDs) are barely off.
3. Now darken the room or stand in a closet. The LEDs will turn on.
4. Move to a semi-lit area. The LEDs will dim as you move into the light. Adjust the pot so that the LEDs are again barely off. Any reduction of the amount of light will now turn the LEDs on. This is your automatic night light.

How It Works

Figure L12-3 shows the movement of current and voltage in the circuit as the light changes. Remember the NPN transistor needs positive voltage to its base to turn on.

1. The potentiometer adjusts the amount of voltage shared by the 22-kilo-ohm resistor and the LDR.
2. In light, the LDR has low resistance, allowing all of the voltage to flow through to ground. Because the base of Q1 gets no voltage, the valve from C to E stays closed.
3. The resistance in the LDR increases as it gets darker, providing more voltage to the base of the transistor, pushing the valve open.
4. As the voltage flows through the transistor, the LEDs turn on.

If P1 is set to a low resistance, more voltage gets through. The more voltage that gets through the potentiometer, the easier Q1 turns on, because the LDR cannot dump all of the voltage.

Exercise: Your First Project—
The Automatic Night Light

1. Set the pot so the resistance between legs A and C, the center leg, is close to 50 kilo-ohms. Figure L12-3 shows how the voltage and current move in the different conditions for this setup.
2. Now set your circuit under a fairly good light. Measure and record the voltage at the test point shown in Figure L12-4. ______ V
3. Now cover the LDR with a heavy, dark pen cap. Measure and record the voltage at the test point again. ______ V

   Did the LED output change at all?

   So here's a major question: Do you recall how much voltage is used from V+ to ground?

   - Does it depend on the amount of voltage available? NO!
   - Does it depend on the number or types of parts in the circuit? NO!

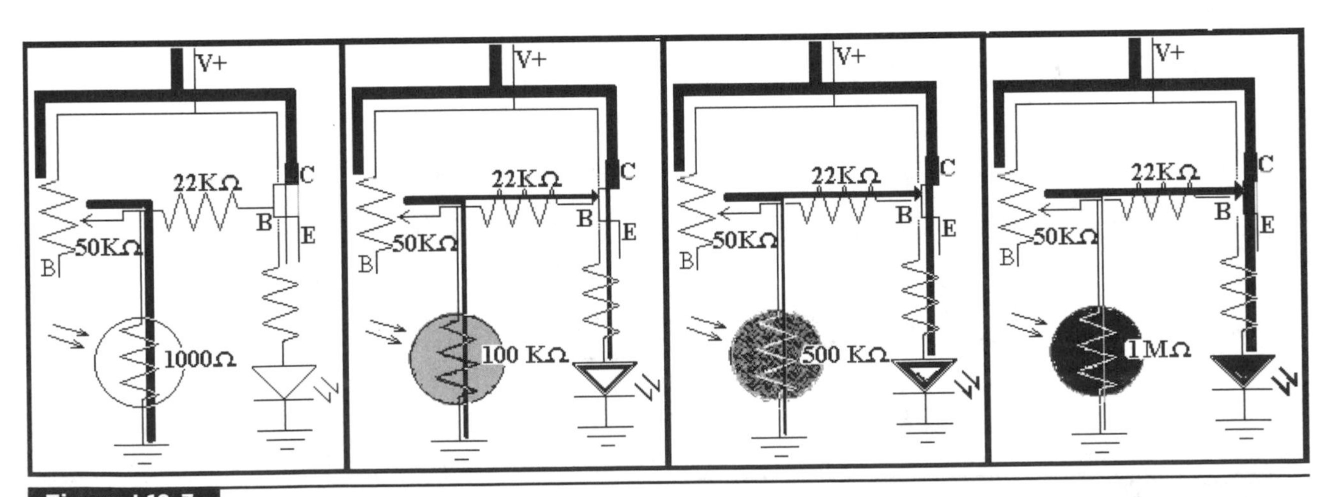

Figure L12-3

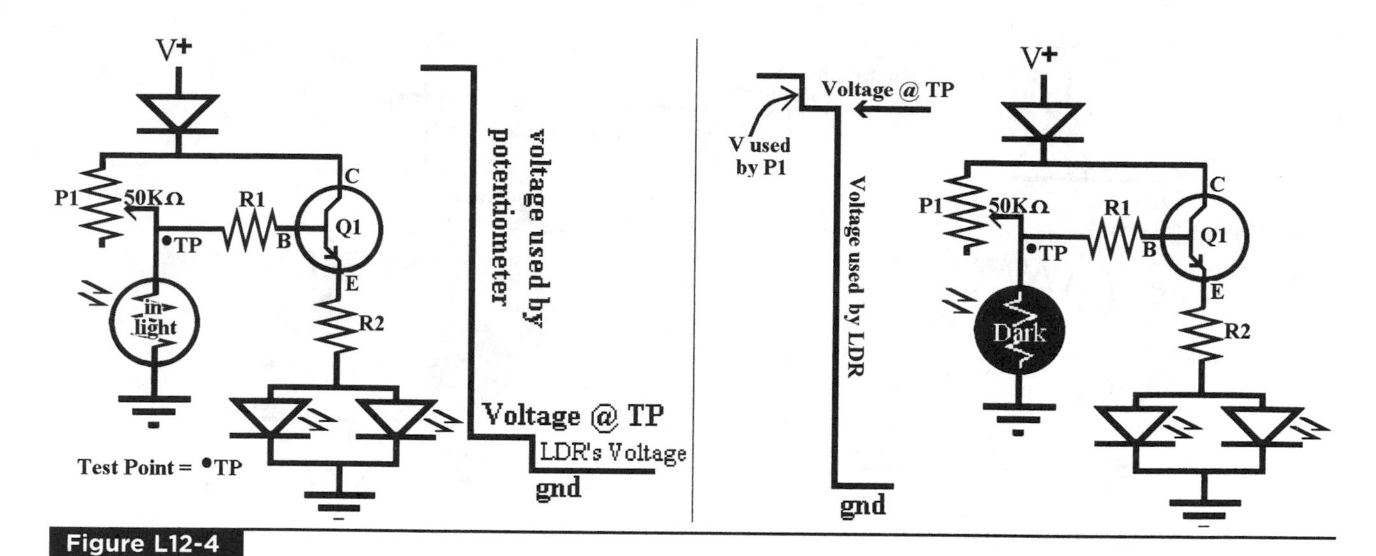

Figure L12-4

By definition, voltage used between V+ and ground doesn't relate to any circuit variables.

The question is like asking, "How much distance is there between this altitude and sea level?"

The answer is—whatever the altitude is—all the distance!

By definition, ground is 0 volts.

- How much voltage is used between V+ and ground?
- The answer will always be *All the voltage is always used between V+ and ground.*

4. Okay. Figure L12-5 will help explain how all the voltage in this circuit is used.

Remember the idea of the waterfall when we first looked at resistors. As the size of the load increased, the amount of voltage used increased *proportionately*. The same thing happens with two resistor loads in this circuit. More detail is given in Figure L12-6.

The pot uses some voltage because it is set near 50,000 ohm here.

The LDR uses a small amount of the voltage in the light because it has a small resistance. When it is in the dark, the LDR has a great deal of resistance. The base of the

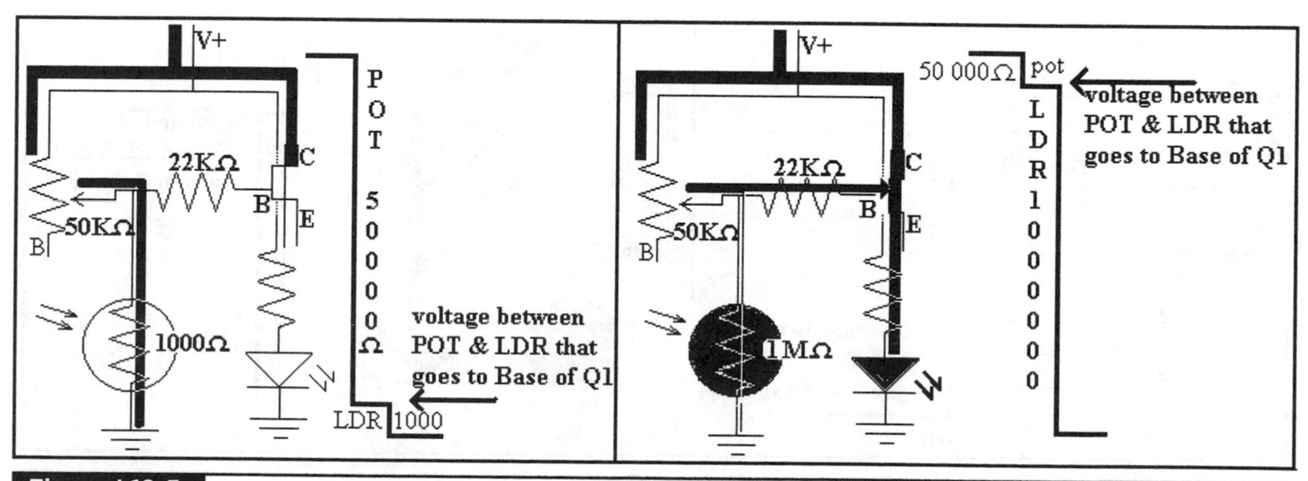

Figure L12-5

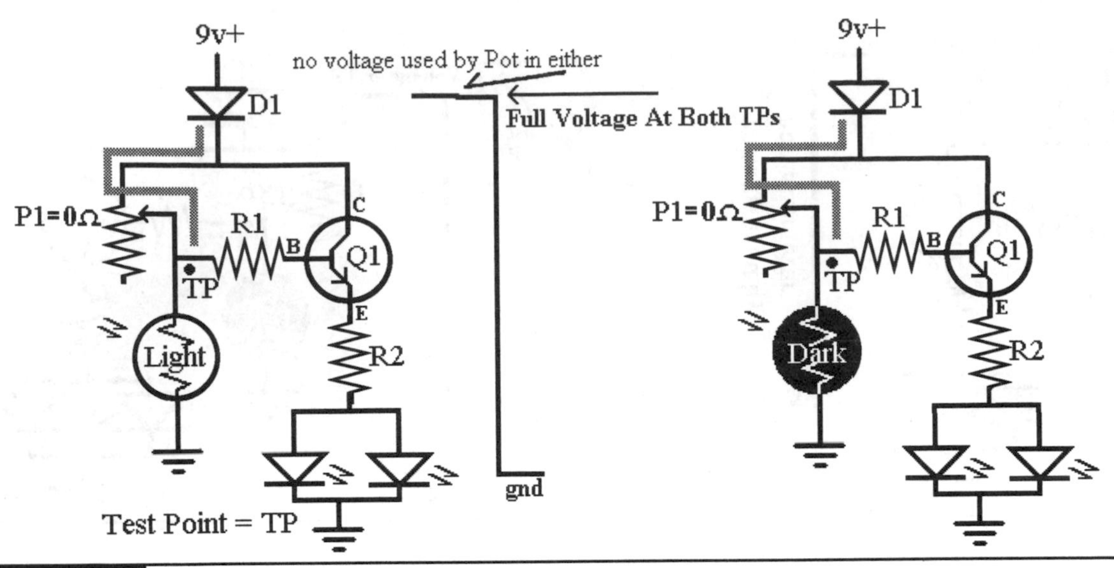

Figure L12-6

transistor reacts to the voltage available at that point where the LDR and pot connect. It becomes obvious which situation provides more voltage to the transistor's base.

5. Set the pot so the resistance between legs A and C, the center leg, is close to 0 ohms. Figure L12-6 shows how the voltage available to the transistor base is identical in both the light and the dark. The potentiometer uses none of the voltage, so the base of the transistor is exposed to nearly full voltage in both circumstances.

6. Now set your circuit under a fairly good light. Measure and record the voltage at the test point shown in Figure L12-6. ______ V

7. Now cover the LDR with a heavy, dark pen cap. Measure and record the voltage at the test point again. ______ V

   Did the LED output change at all? Did you expect the output to change?

8. Set the pot so the resistance between legs A and C, the center leg, is close to 100 kilo-ohms. Figure L12-7 shows how the voltage is responding to the changing resistance of the

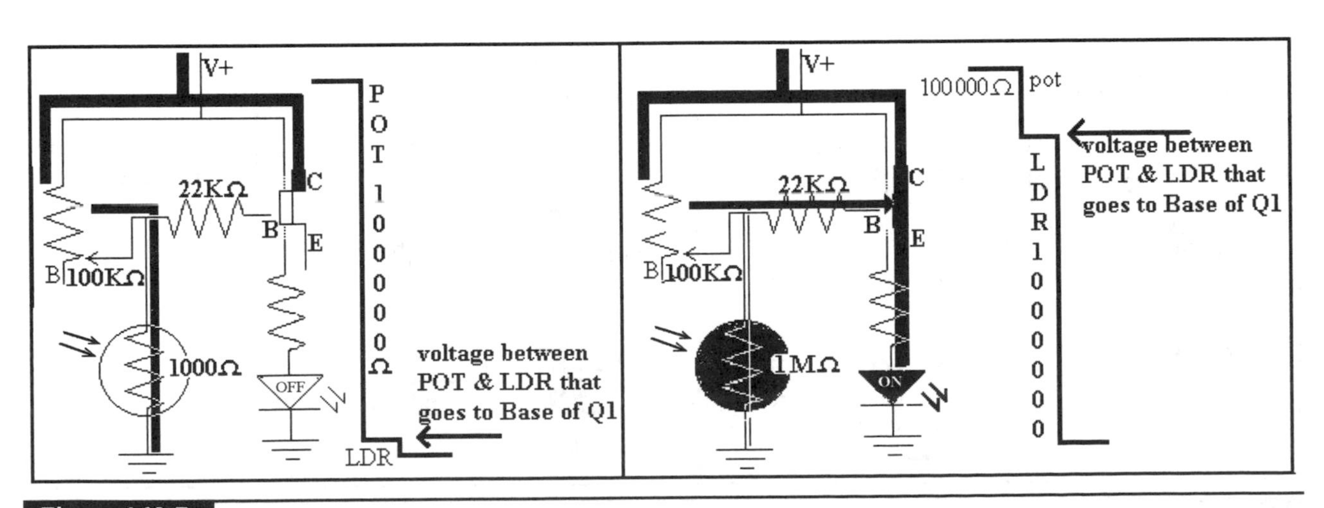

Figure L12-7

LDR, changing the voltage available to the base of the transistor. The pot here is adjusted to twice as much resistance as before, so it will use twice as much of the voltage available. The LDR must be set into nearly complete darkness or you will get muddy results.

9. Now set your circuit under a fairly good light. Measure and record the voltage at the test point. ______ V

10. Now cover the LDR with a heavy, dark pen cap. Measure and record the voltage at the test point again. ______ V

    Did the LED output change at all?

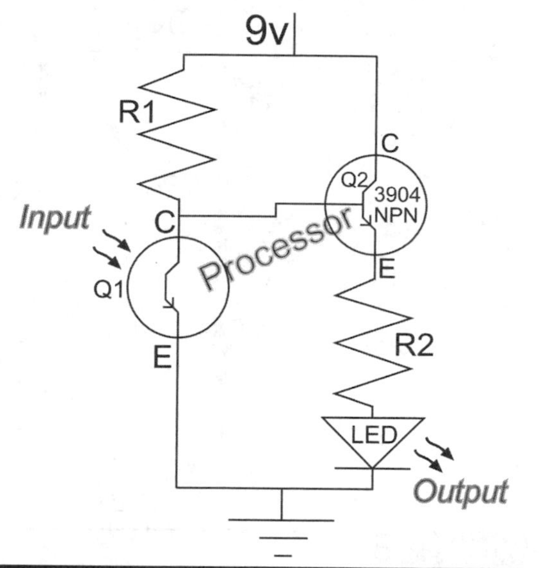

Figure L12-8

Substitute the Phototransistor for the LDR

The LDR has been the traditional component used to respond to ambient light, commonly applied in a variety of circuits from toys to streetlights. However, this simple and useful component is falling out of favor because it contains cadmium, a highly toxic substance. Don't eat them. Import to the European Union is already restricted.

Here is a substitute night light circuit that uses the phototransistor. The few components are listed in the Parts Bin. Because this is a much simpler circuit, it is not useful for explaining the concepts just introduced.

With so few components, the schematic in Figure L12-8 is very simple.

PARTS BIN

- R1—22 kΩ
- R2—470 Ω
- Q1—NPN photo transistor LTE 4206 E (darkened glass) 3mm diam
- Q2—2N3904 NPN in a TO-92 package
- LED—Standard 5mm

Make sure that the phototransistor is set in properly. The flat edge and short leg indicate the collector. This phototransitor acts like an NPN transistor. The collector is connected towards V+.

How It Works

Remember—sunlight has the complete spectrum, including infrared (IR). Incandescent light bulbs also produce some IR. Most fluorescent lights do not have an IR component, so the phototransistor night light will react to daylight and older "wire" light bulbs, but most likely will stay on whether the fluorescent lights are turned on or off.

1. Look at Figure L12-9. When there is IR input to the base of Q1, the phototransistor, voltage and current flow through it directly to ground. Nothing is left at TPA to tickle the base of Q2. So Q2 stays off. No voltage or current moves through Q2. As long as Q1 stays on, Q2 remains off. No power for the LED.
2. When there is no input to the base of Q1 (no infrared) the voltage and current cannot move through the phototransistor. Instead, they are rerouted to the base of Q2, turning on the

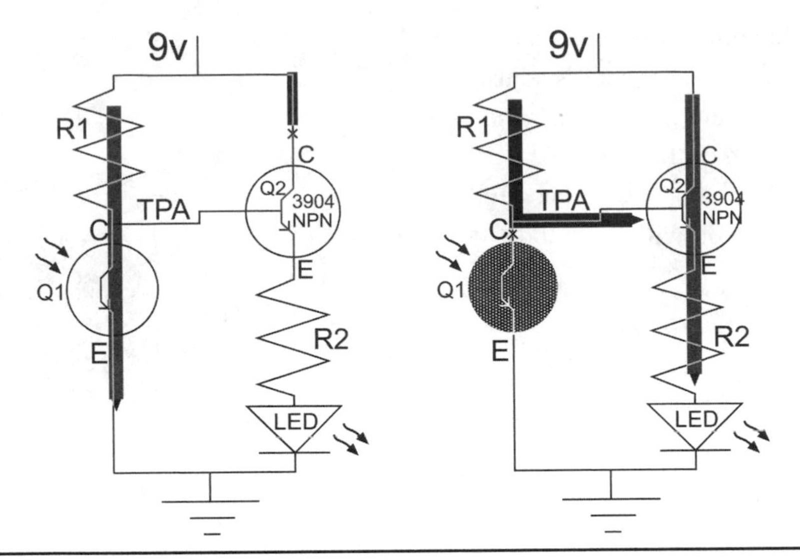

Figure L12-9

2N3904 transistor. Voltage and current move through Q2 from C to E, turning on the LED.

Like plumbing, it's all about controlling the flow. Once you understand the initial concept, it really is simple.

Building the Automatic Night Light Project

This is a top view of your *printed circuit board* (PCB). The traces on the underside are shown as gray in Figure L12-10.

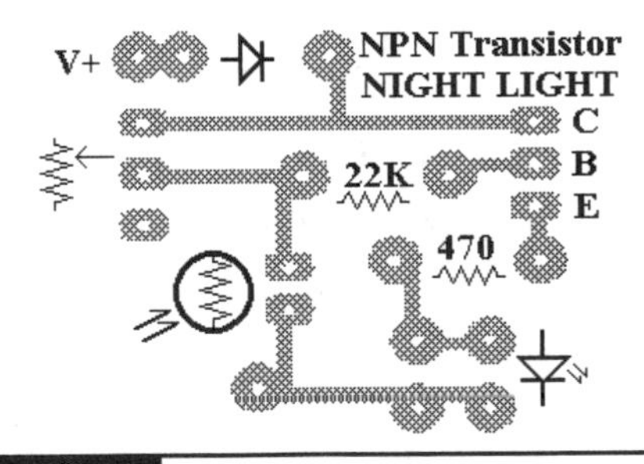

Figure L12-10

Figure L12-11 shows the view of the PCB looking directly at the bottom. The copper traces replace the wires used in the solderless breadboard.

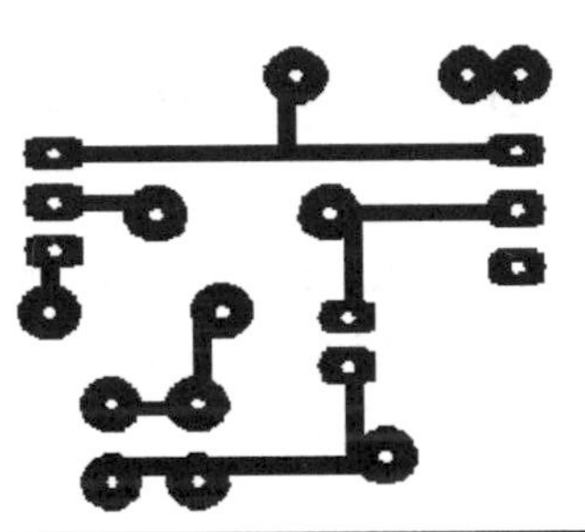

Figure L12-11

The series of frames in Figure L12-12 show how the PCB layout was developed. Creating such a layout is a relatively simple task.

Mounting Your Parts

Be certain to get your parts into the correct holes, as displayed in Figure L12-13. Your soldering technique is vital. Check out the soldering animation on the website www.mhprofessional.com/computingdownload. Each solder connection should look like a Hershey's Kiss. Just as one bad apple spoils the bunch, one bad solder can spoil your project (and your fun). The parts are mounted onto the board so that their legs stick out through the bottom. The copper traces are on the bottom. The soldering is on the bottom. This is a good job.

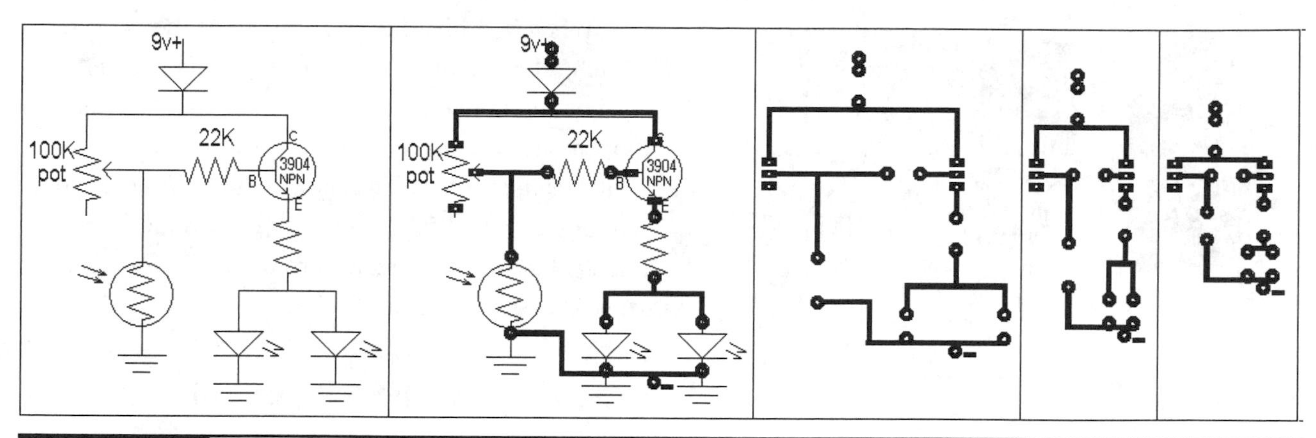

Figure L12-12

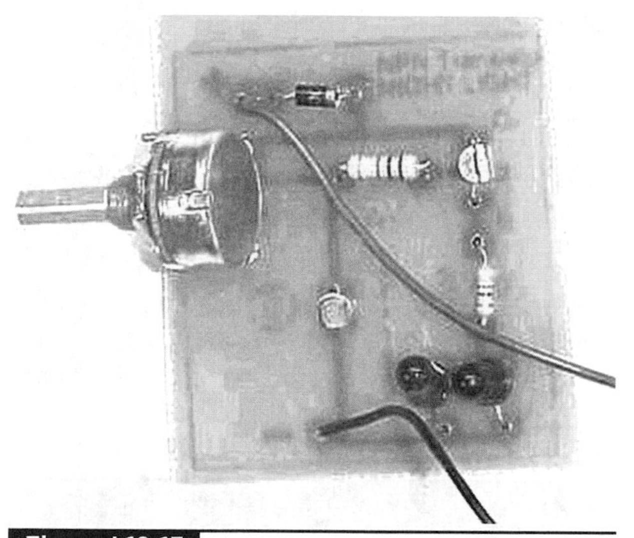

Figure L12-13

Figure L12-14

I got such a laugh when the circuit in Figure L12-14 was presented to me for troubleshooting. This person did not follow directions. They managed to mount the parts on the wrong side of the printed circuit board. They made a mess. It didn't work. There is never enough time to do it right, but there is always time to do it over.

Finishing Up

I would recommend using a touch of hot glue on each corner of the PCB and then pressing the circuit onto a thick piece of cardboard. You can mount it on anything that is an insulating material. Anything metallic would short-circuit your project, and probably destroy the transistor.

Lesson 13
Specialized Transistors—The SCR

There are many highly specialized types of transistors. Here we use a latching switch called a *silicon-controlled rectifier* (SCR). This component is also referred to as a *triode for alternating current* (TRIAC). The SCR acts like a "trapdoor." Once triggered, it stays latched open. You are familiar with this because it is the basic component

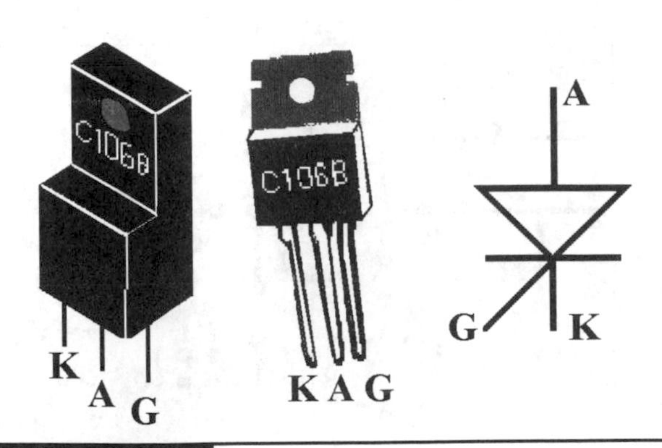

Figure L13-1

in fire alarms and burglar alarms. Once triggered, it stays on. This is your second project, a professional-quality alarm circuit.

Electronics is all about using an electrical pulse to control things or pass information. A commonly used solid-state switch (no moving parts) is the SCR. It is provided in a variety of packages, the most common ones are pictured next to their symbol in Figure L13-1.

Remember, not everything in this type of package is an SCR.

- "A" leg is anode, the positive side.
- "G" leg is gate; not to be confused with ground, "gnd."
- "K" leg is cathode, referring to the grounded side. "C" is already used for capacitor, for example, C1, C2.

The SCR is often used in alarm systems; once it is triggered, it stays on. Its action is best depicted in the series of frames shown here in Figure L13-2.

When V+ hits the G (gate) leg of the SCR, a latch releases, opening a "trapdoor" between A (anode) and K (cathode). This trapdoor remains open until the power is removed. In other words, it latches itself open. That is why an SCR is called a *latching circuit*. The only way to turn the SCR off is to shut off the power. Turn the power on again, and the SCR is reset.

Breadboard the SCR Circuit

There are four stages to building this professional-style alarm circuit. Each one will be considered individually.

Stage 1: The Basic System

Build the basic SCR circuit by carefully following the schematic presented in Figure L13-3. The parts layout is shown in the photograph in Figure L13-4 (see also the Parts Bin). Don't get too comfortable using the photograph. You will have to depend on schematics from now on to help you develop your alarm circuit.

PARTS BIN

- D1—1N400X
- R1—100 kΩ
- R2—470 Ω
- SCR—TRIAC C106B
- Buzzer—9v
- LED—5 mm round

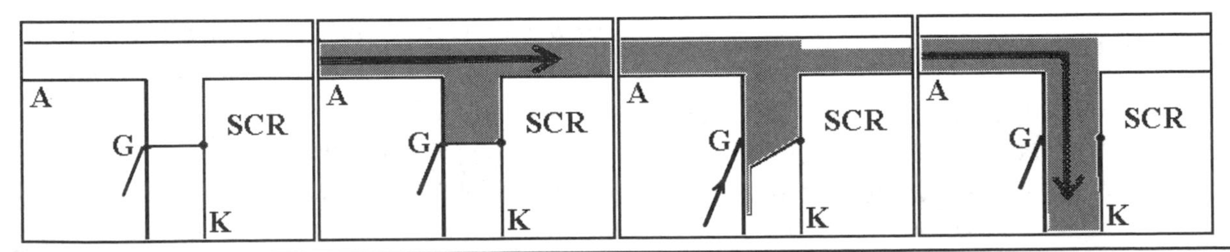

Figure L13-2

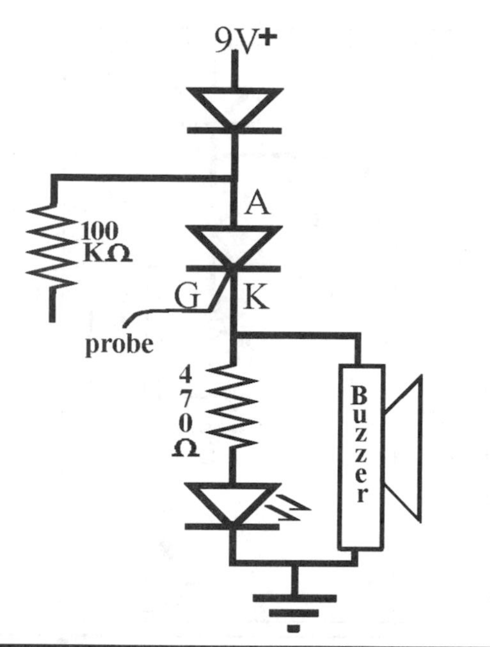

Figure L13-3

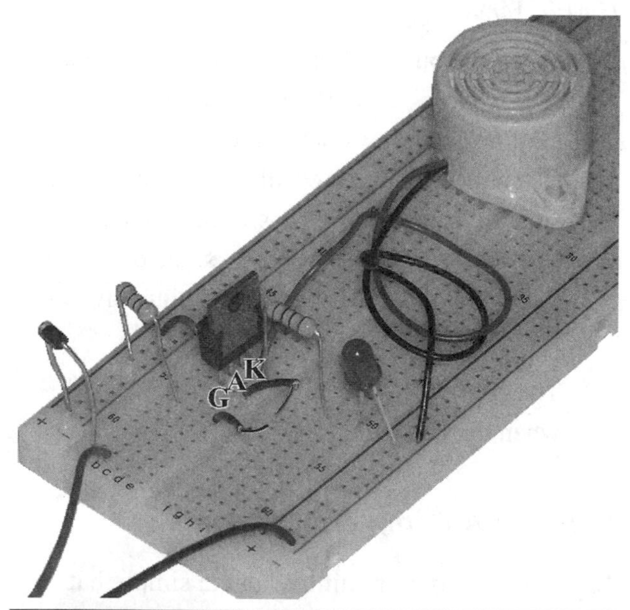

Figure L13-4

What to Expect

- When you attach the battery, it should be quiet.
- Touch the sensor probe to the end of R1. The buzzer should turn on.
- You need to disconnect the battery to reset the SCR.

How It Works

1. When you attach the battery, the LED is off and the buzzer is quiet. Voltage is not available to the LED and buzzer because the SCR has not been activated. The circuit path between A and K is not available until voltage is put to the gate.
2. When you touch the end of the probe to the bottom of the 100-kilo-ohm resistor, voltage is fed to the gate (G) leg.
3. The voltage activates the latch and opens the circuit path from A (V+) to K (gnd). Current moves on this path from A to K, through the SCR, providing voltage and power to the LED and the buzzer.
4. The buzzer and LED should turn on and stay on until you disconnect the power. When you reconnect the power, the LED should not light and the buzzer should be quiet.

Stage 2: Make Life Easier

It is a hassle to disconnect the battery each time just to reset the SCR. Wouldn't it be so much easier just to use a push button to reset the circuit? So go ahead and use the normally closed push button to do just that.

It does not matter where the voltage to the SCR is cut off. As you can see in Figure L13-5, any of the following suggestions work. Each one disconnects the voltage to the circuit and resets the SCR.

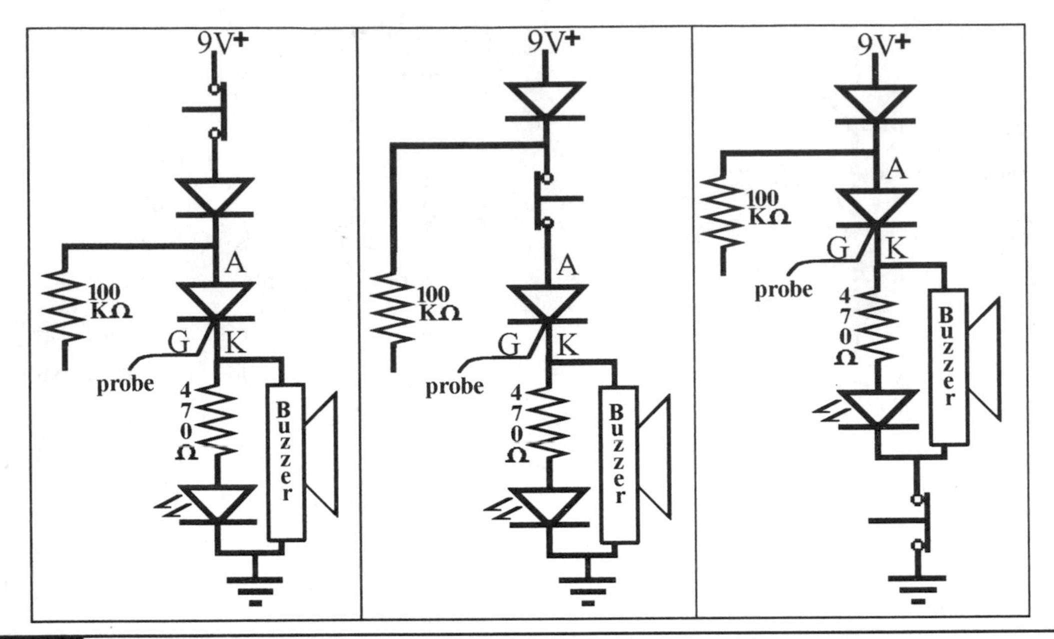

Figure L13-5

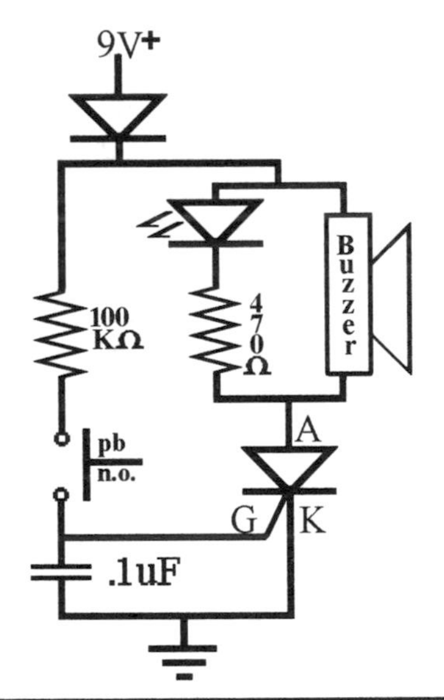

Figure L13-6

Stage 3: Avoid Static Buildup

Modify the SCR circuit with a few changes, shown in Figure L13-6.

1. Shift the SCR below the LED and buzzer. In most circuits, you can be flexible. It really doesn't matter which component comes first.
2. Add the 0.1-microfarad capacitor.
3. Add the PBNO for a trigger.

How It Works

1. When the push button is pushed, the capacitor fills "nearly" instantly because the 100,000-ohm resistor slows down the current. The capacitor acts like a cushion and dampens any small but annoying jumps in the voltage to the gate that might be caused by static electricity. It avoids false alarms by preventing any accidental triggering of the SCR.
2. The gate (G) of the SCR senses the signal when the capacitor is filled.

Stage 4: A Complete System

Now modify your circuit to make a simple but professional-quality alarm system by adding three more parts to your working breadboard, as shown in Figure L13-7.

PARTS BIN
■ D2—1N400X
■ R1—47 kΩ
■ PBNC—Normally closed push button

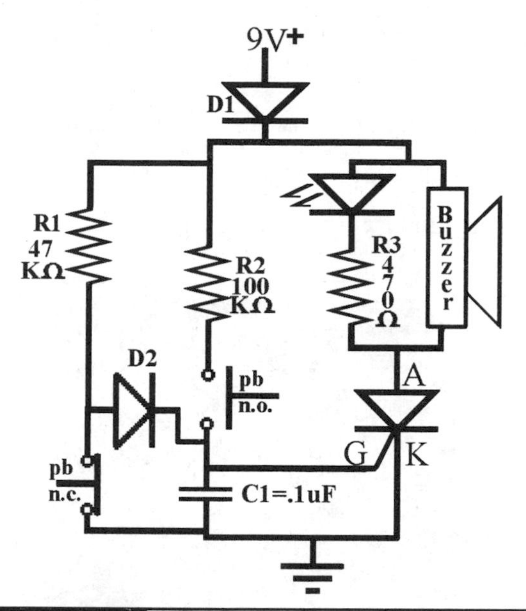

Figure L13-7

What to Expect

- Attach your power supply. The LED should be off and the buzzer quiet.
- Push the plunger on the PBNC. The buzzer will start, and the LED light will turn on.
- To reset, disconnect the power—count ten seconds—and then reconnect.
- Now push the plunger on PBNO. The buzzer and LED will turn on.

How It Works

As long as the PBNO stays closed, the voltage takes the "path of least resistance" and goes directly to ground. No voltage reaches G, the gate of the SCR.

When the PBNO is pushed, the connection to ground is broken and the voltage has to move through D2, activating G, and turning the SCR on.

As long as the PBNC stays open, the voltage cannot move across the gap. Air is a great insulator. No voltage reaches G. When the PBNC is pushed, the voltage travels through to the gate leg, turning the SCR on.

Exercise: Specialized Transistors—The SCR

1. If a component looks like your SCR, it is an SCR. True/false. Support your answer.
2. How can you tell if it is an SCR without putting it onto your breadboard?
3. Many alarm systems use an SCR. For example, a fire alarm, once triggered, keeps going. Describe what needs to be done to reset the SCR in the alarm.
4. Use Figure L13-7 to follow the current's path when the PBNC is in its normal position.
5. How does the path change when you push the plunger on the PBNC?
6. Follow the current's path when the PBNO is in its normal position.
7. How does the path change when you push the plunger on the PBNO?
8. Have you ever seen metal foil tape on the edge of store windows? There is a small current running through this tape. If the window breaks, the foil is torn. What component does the foil tape replace in your burglar alarm system?
9. Describe how a normally open type of switch would be used in a burglar alarm?

Assembling the Alarm

Figure L13-8 represents the SCR's printed circuit board when viewed from the bottom. Figure L13-9 shows lines that appear faded. This is the view of the PCB when you look at the traces through the fiberglass backing.

Consider this: If you mount your simple push buttons onto the PCB directly, you have a very cute demonstration circuit. If you mount the push buttons on a long wire, you still need to wait for someone to push one of the buttons. That still isn't much of an alarm. What you really need to do is create your own contact switches. When you do

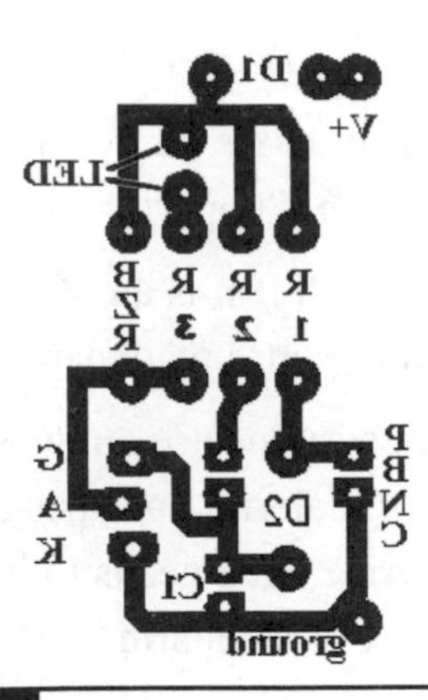

Figure L13-8

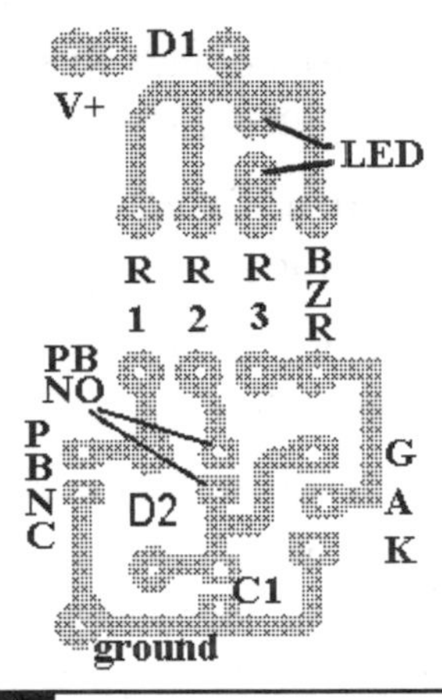

Figure L13-9

this, you can apply this as a real alarm system. The best setup for your power supply is to have both a wall adapter and a 9-volt battery. This way, the circuit still works even if your house has no power.

Lesson 14
The Regulated Power Supply

Have your batteries been feeling tired lately? Not what they used to be? All used up? Just don't have that push they used to have? There's a better way.

Here, we build a power supply that provides a stable 9v DC that you can always use as you continue to develop your skills. We'll use another specialized transistor, the voltage regulator, in conjunction with a basic wall adapter. Any 9v DC wall adapter will do—even the one provided in the Evil Genius Kit.

The few important things are highlighted in Figure L14-1.

1. Voltage input (120v AC) and output (9v DC)
2. Amount of output current (200mA)
3. The output plug wiring as a graphic

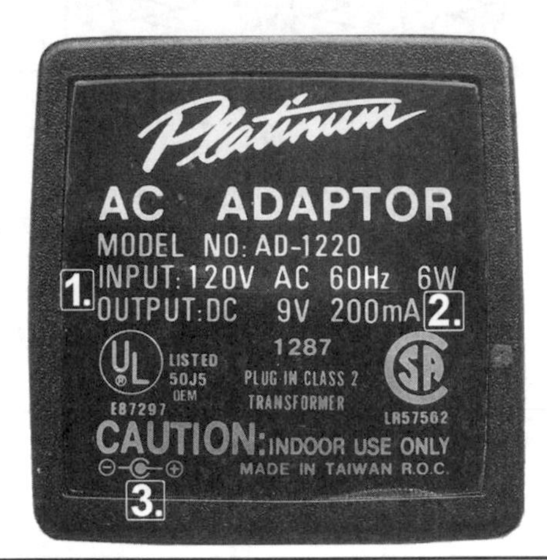

Figure L14-1

Now that you've looked at it, plug it into the wall. Don't worry about what you are about to do next. Any wall adapter in this range actually puts out less power than most 9-volt batteries. Set your DMM to vDC. Hold the red lead to one part of the output plug and the black line to the other. See the display in Figure L14-2.

Figure L14-2

Why does it show 13 vDC? Because it is unregulated. It will give more than 9 volts, depending on how hard it's working. What we're going to do is build a regulated power supply that will provide us with 9 volts all the time.

Let's Get to Work

The following directions will get you to Figure L14-3:

1. Remove the adapter from the wall (duh!).
2. Clip the wires as close as possible to the DC output plug. Separate and strip the output wires.
3. Plug the adapter back in. Use the DMM again to determine which line is V+.
4. Remove the sheaths from the alligator clips. Slide the red sheath onto the V+ line and the black onto the ground.
5. Double-check that you have the lines properly identified.
6. Unplug it while you solder the clips onto each line. Figure L14-3 shows the finished product.

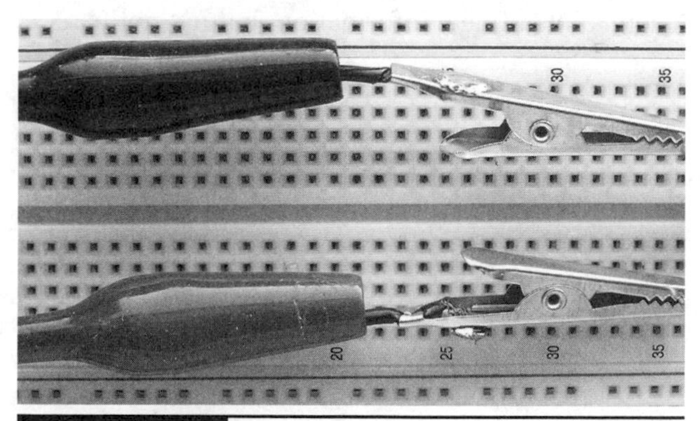

Figure L14-3

All that, and now we can get to the real workhorse of this circuit, the 7809 voltage regulator, proudly displayed in Figure L14-4.

The 7809 is only one of many in the 78XX and 79XX voltage regulator series. Table L14-1 shows what is available in the full series.

Figure L14-4

TABLE L14-1 Other Voltage Regulators Are Available, but the 78XX Series Is Most Popular

Positive Regulators		Negative Regulators	
7803	+3vDC	7903	-3vDC
7805	+5vDC	7905	-5vDC
7809	+9vDC	7909	-9vDC
7812	+12vDC	7912	-12vDC
7815	+15vDC	7915	-15vDC

Respectively, place a red hookup wire into hole E-60 of your SBB and a black hookup wire into the ground line on the bottom. The schematic in Figure L14-5 shows how to set up the circuit, and the photo in Figure L14-6 shows the layout.

The capacitor actually is part of the working circuit. The LED isn't necessary, but conveniently tells us when the circuit is powered up.

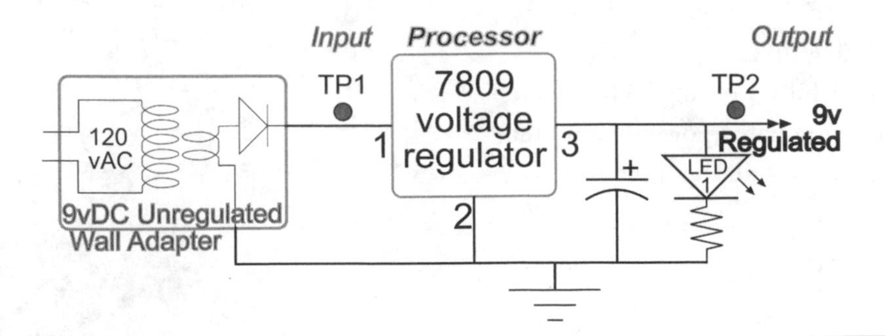

Figure L14-5

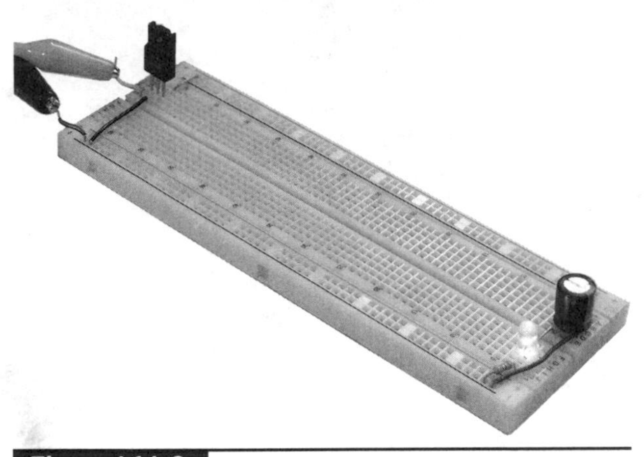

Figure L14-6

1. The V+ input line from the wall adapter should be wired directly to the 7809's input leg.
2. The adapter's ground wire is connected directly to the blue ground line across the bottom of your SBB.
3. The output leg is connected directly to the red V+ line that runs across the top of the breadboard.

What to Expect

Plug it in. First off, the LED should come on. If the LED does not turn on, unplug the wall adapter immediately and check your wiring.

Now, measure the voltage across the capacitor, because conveniently, it has exposed wires.

It should have a readout of 9v DC. This 7809 voltage regulator has a tolerance of 4 percent. So its maximum would be 9.36v and minimum 8.64v.

Okay, that's it for your quick, regulated power supply. Use this as your power supply for the rest of the book.

Like I've said before: Electronics is not hard. There's lots of new information, but it's not hard.

PART TWO

Introduction to Digital Electronics

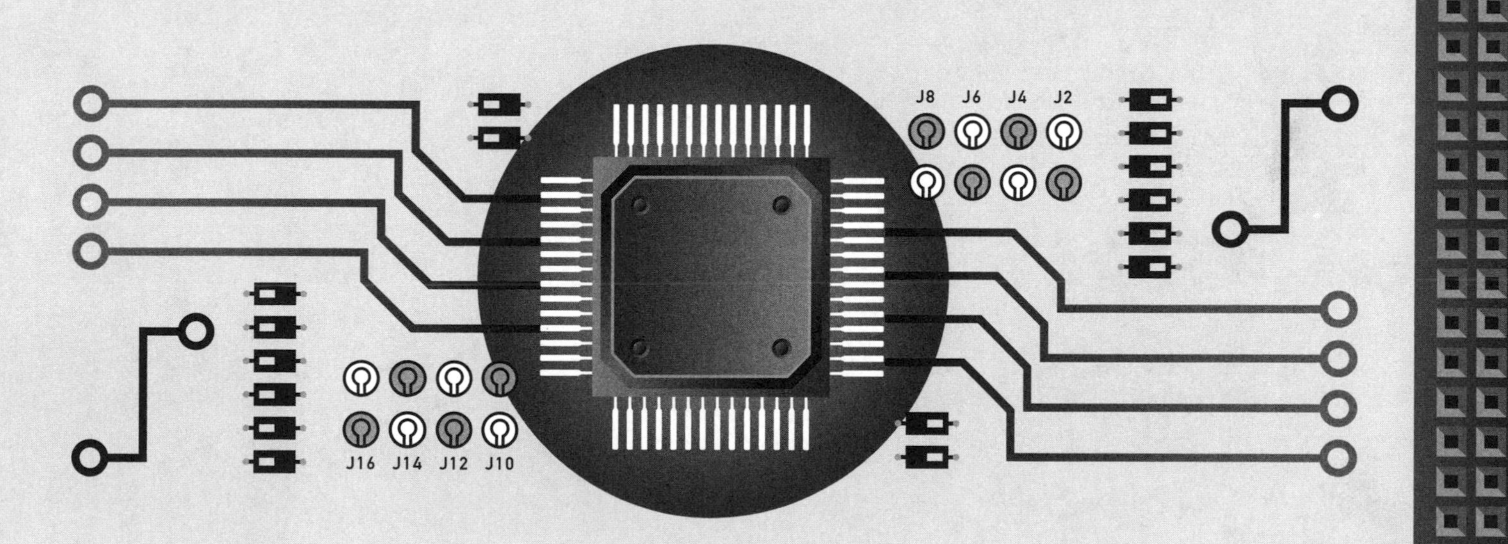

The Start of the Digital Revolution

Actually, the idea of "digital" was first introduced by the textile industry. It all started with an attachment to an existing loom. The Jacquard head attached to the large Dobby loom (figure below). A set of punched cards was connected to a chain. Once the cards were set, the up/down motion of a single lever moved them forward, advancing the pattern. Thus, digital as a functional technology was introduced over 200 years ago.

Digital technology circa 1801.

PARTS BIN FOR PART TWO

Description	Type	Quantity
1N4148	Signal diode	1
2N-3906 PNP transistor	TO-92 case	2
2N-3904 NPN transistor	TO-92 case	2
Phototransistor	LTE 4206 E (darkened glass) 3mm diameter: tuned to 940nm	1
Infrared LED	LTE 4206 (clear glass) 3mm diameter: tuned to 940nm	1
LED	5 mm	2
100 Ω	Resistor	1
1,000 Ω	Resistor	2
10,000 Ω	Resistor	1
22,000 Ω	Resistor	1
39,000 Ω	Resistor	1
100,000 Ω	Resistor	2
220,000 Ω	Resistor	1
470,000 Ω	Resistor	2
1,000,000 Ω	Resistor	1
2,200,000 Ω	Resistor	1
4,700,000 Ω	Resistor	1
10,000,000 Ω	Resistor	2
20,000,000 Ω	Resistor	1
100 kΩ trim pot	Potentiometer	1
Light-dependent resistor	LDR	1
10 μF	Cap	
100 μF	Cap	
.01 μF disk	Cap	1
.1 μF film	Cap	1
1 μF radial	Cap	1
4011 quad NAND gate	IC	1
1″ × 1/8″ heat shrink	Hardware	4
1/8″ male plug	Hardware	1
14-pin DIP socket	Hardware	1
2″ × 1/4″ heat shrink	Hardware	1
Alligator clips (red and black)	Hardware	1 each
Battery clip	Hardware	1
PBNC	Hardware	2
PBNO	Hardware	2
Speaker 8 Ω	Hardware	1
24 gauge wire	Hookup wire	Various colors
Speaker wire 3' 20 gauge	Hardware	1
Printed circuit board	PCB	1

- Not all components will be consumed by project work.

SECTION 5

Digital Logic

THIS COURSE DEALS WITH a vast amount of new information. Remember that electronics is not hard; there's just lots of new information to learn.

Lesson 15
A Spoiled Billionaire

As we begin the "digital" electronics unit, this lesson shows us a system that we can use to be certain that all of the information is transferred all the time, perfectly. You must be able to count to 255 and know the difference between "on" and "off."

With apologies to Bill Gates, but inspired by his explanation in his book, *The Road Ahead*: There is an eccentric billionaire who lives near Seattle. He is particular about his lighting in different rooms of his homes, especially his den. He likes it set at exactly 187 watts. And when you're rich, you get what you like. But here's the real problem. His wife also uses the den and she "likes" 160 watts.

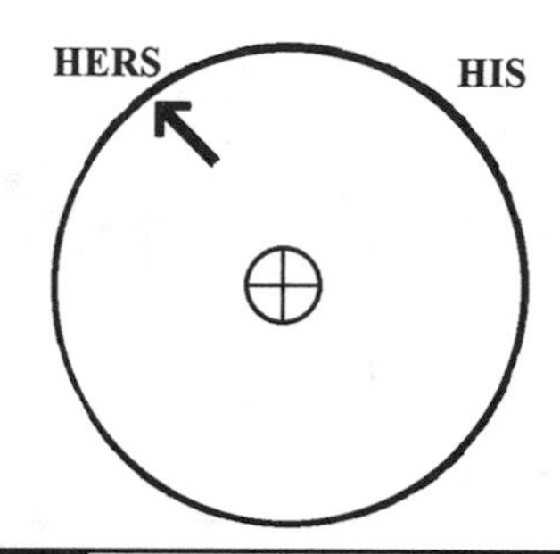

Figure L15-1

They asked their groundskeeper to come up with a solution they could use in all their homes around the world. He first installed a dimmer switch and put a mark next to the spot that represented their preferences, which is pictured in Figure L15-1.

On inspection, the groundskeeper was told that his solution was not acceptable. It was not exact enough. So he thought further and had an idea. Actually, eight ideas.

His idea is shown in Figure L15-2: a light bar with eight separate specific value lights. Each light

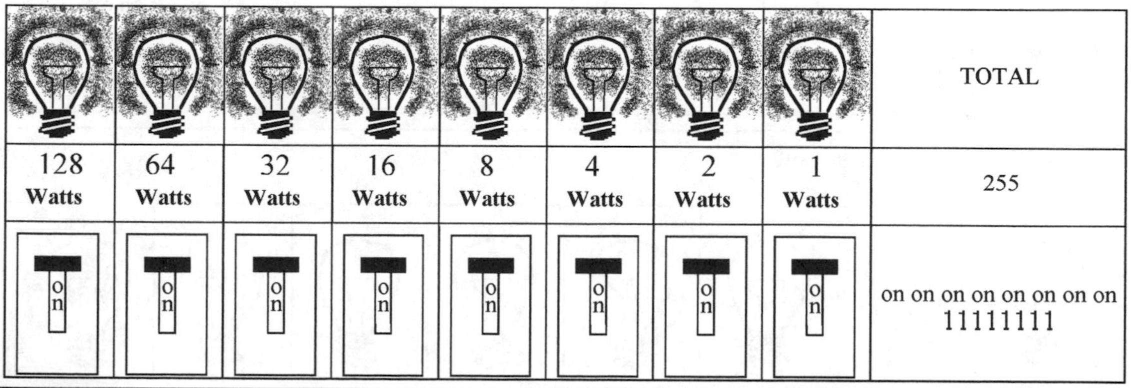

								TOTAL
128 **Watts**	64 **Watts**	32 **Watts**	16 **Watts**	8 **Watts**	4 **Watts**	2 **Watts**	1 **Watts**	255
on	on	on	on	on	on	on	on	on on on on on on on on 11111111

Figure L15-2

would have a different wattage rating, and each would have its own switch. To adjust the lighting for his own needs, the billionaire would only have to turn on a set of specific switches. Those are displayed specifically in Figure L15-3.

And to make the softer lighting situation for Mrs. Billionaire, she flips the switches shown in Figure L15-4.

So all the groundskeeper had to do was write a note under the set of switches in the den.

His = 10111011
Hers = 10100000

(*The Road Ahead*, Bill Gates, pg. 25)

By turning these switches on and off, you can adjust the lighting level in watt increments from 0 watts (all switches off) to 255 watts (all switches on). This gives you 256 possibilities.

If you want 1 watt of light, you turn on only the rightmost switch, which turns on the 1-watt bulb (shown in Figure L15-5).

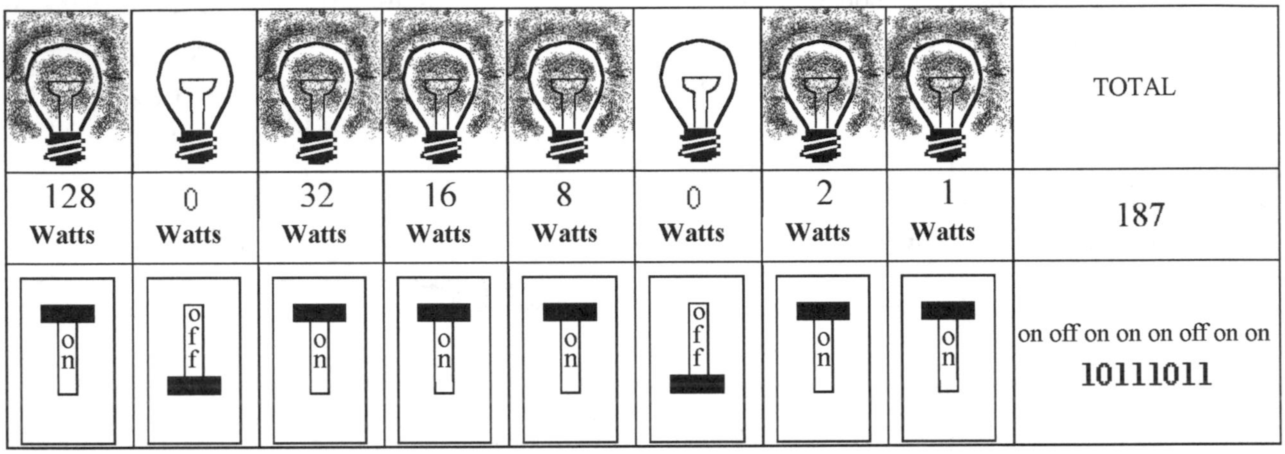

								TOTAL
128 Watts	0 Watts	32 Watts	16 Watts	8 Watts	0 Watts	2 Watts	1 Watts	187
on	off	on	on	on	off	on	on	on off on on on off on on **10111011**

Figure L15-3

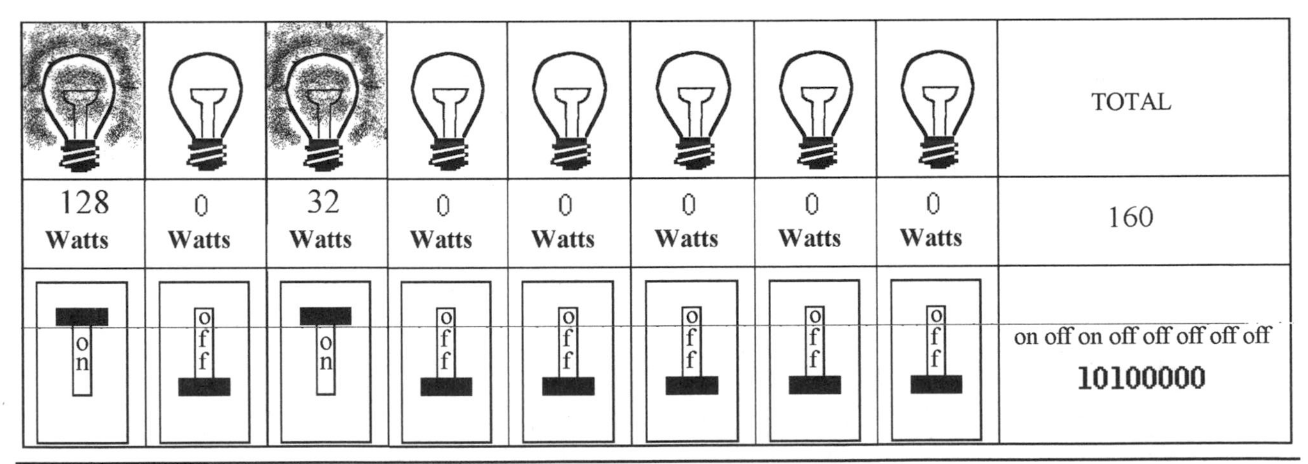

Figure L15-4

Figure L15-5

Figure L15-6

Figure L15-7

If you want 2 watts, you turn on only the 2-watt bulb (shown in Figure L15-6).

If you want 3 watts of light, you turn on both the 1-watt and 2-watt bulbs, because 1 plus 2 equals the desired 3 watts (shown in Figure L15-7).

If you want 4 watts of light, you turn on the 4-watt bulb.

If you want 5 watts, you turn on just the 4-watt and 1-watt bulbs (shown in Figure L15-8).

If you want 250 watts of light, you turn on all but the 4-watt and 1-watt bulbs (shown in Figure L15-9).

If you have decided the ideal illumination level for dining is 137 watts, you turn on the 128-, 8-, and 1-watt bulbs, like this (shown in Figure L15-10).

Figure L15-8

Figure L15-9

Figure L15-10

This system makes it easy to record an exact lighting level for later use or to communicate it to others who have the same light switch setup. Because the way we record binary information is universal—low number to the right, high number to the left, always doubling—you don't have to write the values of the bulbs. You simply record the pattern of switches: on, off, off, off, on, off, off, on. With that information, a friend can faithfully reproduce the 137 watts of light in your room. In fact, as long as everyone involved double-checks the accuracy of what they do, the message can be passed through a million hands; at the end, every person will have the same information and be able to achieve exactly 137 watts of light.

To shorten the notation further, you can record each "off" as 0 and each "on" as 1. This means that instead of writing down "on, off, off, off, on, off, off, on," meaning turn on the first, the fourth, and the eighth of eight bulbs and leave the others off, you write the same information as 1,0,0,0,1,0,0,1, or 10001001, a binary number. In this case, it's 137. You call your friend and say: "I've got the perfect lighting level! It's 10001001. Try it." Your friend gets it exactly right by flipping a switch on for each 1 and off for each 0.

This may seem like a complicated way to describe the brightness of a light source, but it is an example of the theory behind binary expression, the basis of all modern computers.

The simplest computers use an 8-bit system like the eight light switches. Each bit is a bit of information. The binary word of 8 bits makes 1 byte, as shown in the explanations earlier.

From the earliest days of computing, the alphabet and numerals have been assigned specific values (Table L15-1).

TABLE L15-1 The Binary Alphabet: The ASCII Table

Space = 20 = 00010100	
A = 65 = 01000001	a = 97 = 01100001
B = 66 = 01000010	b = 98 = 01100010
C = 67 = 01000011	c = 99 = 01100011
D = 68 = 01000100	d = 100 = 01100100
E = 69 = 01000101	e = 101 = 01100101
F = 70 = 01000110	f = 102 = 01100110
G = 71 = 01000111	g = 103 = 01100111
H = 72 = 01001000	h = 104 = 01101000
I = 73 = 01001001	i = 105 = 01101001
J = 74 = 01001100	j = 106 = 01101010
K = 75 = 01001101	k = 107 = 01101011
L = 76 = 01001110	l = 108 = 01101100
M = 77 = 01001101	m = 109 = 01101101
N = 78 = 01001110	n = 110 = 01101110
O = 79 = 01001111	o = 111 = 01101111
P = 80 = 01010000	p = 112 = 01110000
Q = 81 = 01010001	q = 113 = 01110001
R = 82 = 01010010	r = 114 = 01110010
S = 83 = 01010011	s = 115 = 01110011
T = 84 = 01010100	t = 116 = 01110100
U = 85 = 01010101	u = 117 = 01110101
V = 86 = 01010110	v = 118 = 01110110
W = 87 = 01010111	w = 119 = 01110111
X = 88 = 01011000	x = 120 = 01111000
Y = 89 = 01011001	y = 121 = 01111001
Z = 90 = 01011100	z = 122 = 01111010

The real question for the rest of us becomes—Why use binary? It seems so confusing!

The answer is actually very simple. We are dealing with machines. The easiest thing for a machine to sense is whether something is on or off.

So we are forced to use a system that can count in on's and off's.

But still you ask, why bother?

Well, look at it (Table L15-2).

TABLE L15-2 Compare Analog to Digital

	Analog	Digital
Advantages	1. Varying voltages. 2. Easy to record. 3. Easy to play back.	1. Precise transfer of information. 2. No generational loss. 3. Footprint of a bit can be done at the molecular level. 4. Common transfer rate is a billion bits per second. 5. Any material can be used for storing data: 0s and 1s.
Disadvantages	1. Not precise. 2. Signal loss with each generation recorded. Have you ever watched a copy of a copy of a videotape? Ugh! 3. Takes up large "recording" space. Compare the size of a videotape to a DVD or a memory stick. 4. Limited transfer time—how long does it take to record a video, compared to transfer of a CD or MP3 file? 5. Specific media used for storage doesn't translate from one media to another—for example, video to film. 6. Signal fades as media ages.	1. Needs special equipment to transfer, record, and read information.

Exercise: A Spoiled Billionaire

I am writing in binary code!

0100100100010100011000010110110100010100 01

1101110111001001101001011001110110100001 11

0100011010010110111001101110001010001101 00

1011011100001010001100010011010010110111 00

1100001011100100111001000101000110001101 1

0111101100100011 0010

1. What do you think is the most important advantage of using digital information?

   Use Table L15-3 to help translate from binary to decimal, and back.

TABLE L15-3 Binary to Decimal and Back

Bit number	Bit 8	Bit 7	Bit 6	Bit 5	Bit 4	Bit 3	Bit 2	Bit 1
Value	128	64	32	16	8	4	2	1

2. Translate the following 8-bit binary codes to the decimal equivalent.

Binary	Decimal
10101100	
01100110	
10010011	
00110001	

3. Translate the following decimal numbers to the binary codes.

Decimal	Binary
241	
27	
191	
192	

4. Consider this. If 8 bits count up to 255 decimal, what is the maximum that a 9-bit binary word can count?

Lesson 16
The Basic Digital Logic Gates

You build prototypes of the five main types of logic gates in this lesson. Each gate can be built with individual transistors that act like normally open or normally closed push buttons. All these switches do is redirect the output to voltage (V) or ground.

The AND gate and NOT AND gate are perfectly opposite in their outputs. The OR gate and NOT OR gate are also perfectly opposite in their output.

Voltage States	
V+	Ground
1	0
High	Low

In digital electronics, the *voltage state* is named in three different ways:

- Each term for voltage has an alternate name for ground.
- These terms are generally interchangeable.
- The terms are usually paired. Terms like V+ and ground are used together just like the term *high* is used with *low*, and 1 with 0.
- Digital gives an *output* of high or low, but we often have to use a real-world *analog input.*

All gates have at least one input, but all gates have only one output.

Inputs are the analog "sensors." They compare the voltage they feel to the voltage of the chip. They sense if the input is to be seen as high or low.

Output is the result of the logic function, whether the gate provides a full V+ or ground at the output.

- The real world analog input usually never happens as a convenient high or low.
- So digital gates have been designed to compare the input against their own power source of V+.
- Anything above half of V+ is seen as a high input.
- Anything below half of V+ is seen as a low input.

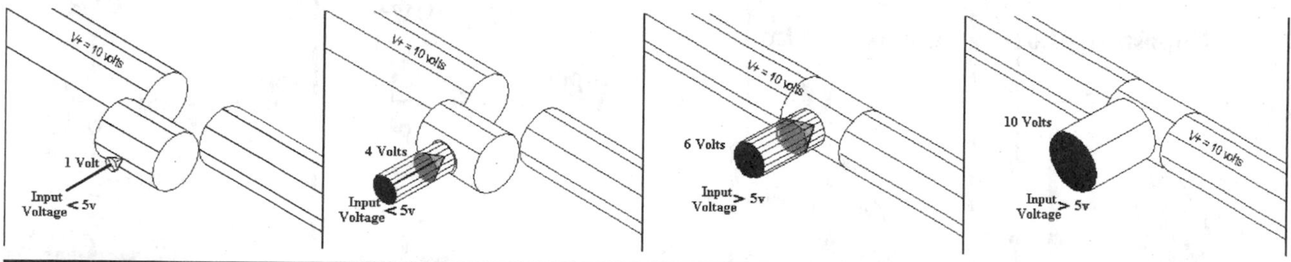

Figure L16-1

Inputs Are the Analog Sensors

- For example, as shown in Figure L16-1, the gate is powered by 10 volts.
- Anything connected to the input that is over 5 volts is seen as a high input.
- Alternatively, any input voltage that is below 5 volts is seen as a low input.

The chip contains inputs, processors, and outputs.

Each of these needs power. They are all powered from the same power source. The voltage source is shown as the line coming into the symbol marked with the V+.

The NOT Logic Gate

The input given is NOT the output. You can see this in Figure L16-2.

This is often referred to as an inverter. It "inverts" the input.

- You can see that the input is like a push button. It controls the voltage flow through the transistor switch inside the chip.
- The transistor processor inside is like the push button demonstration circuits. The transistor responds to the input, controlling voltage to the output.

For the purpose of this exercise, refer to Figure L16-3.

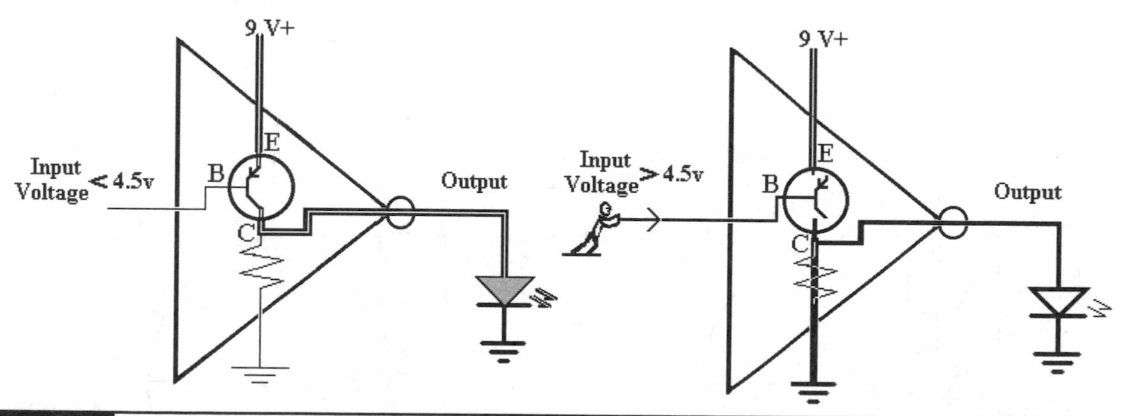

Figure L16-2

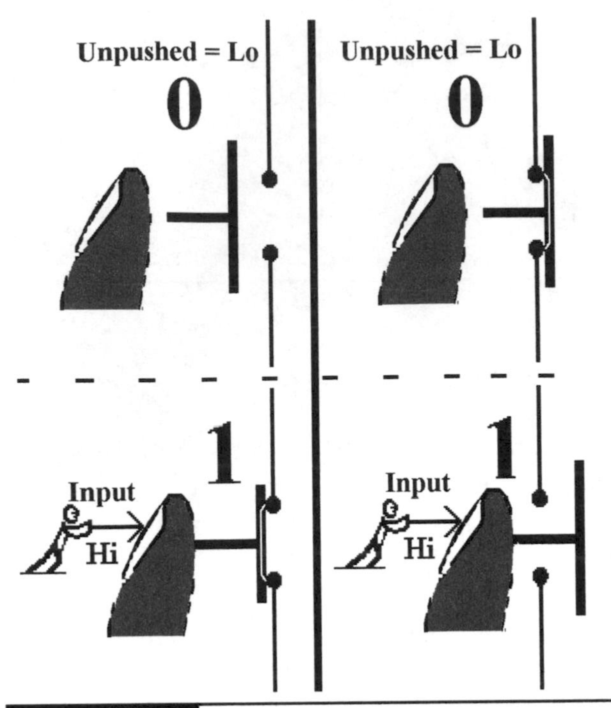

Figure L16-3

- The pushed button is high.
- The unpushed button is low.

Breadboard the NOT Gate Simulation Circuit

This circuit quickly demonstrates the action of the NOT gate (see Figure L16-4).

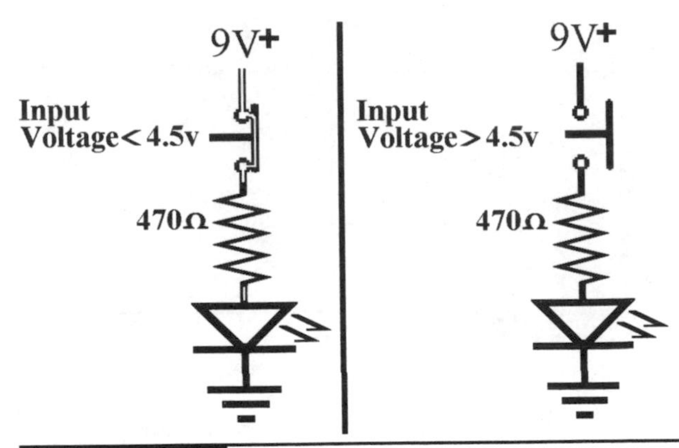

Figure L16-4

The input is the force of your finger. The binary processor is the push button's position (up = 0, down = 1). The output is the LED. On is high, and off is low (Table 16-1).

TABLE L16-1 NOT Gate: Complete the Logic Table for This Gate

Input	Output
High	______
Low	______

The AND Logic Gate

As shown in Figure L16-5, both input A AND input B have to be Hi to get an Hi output.

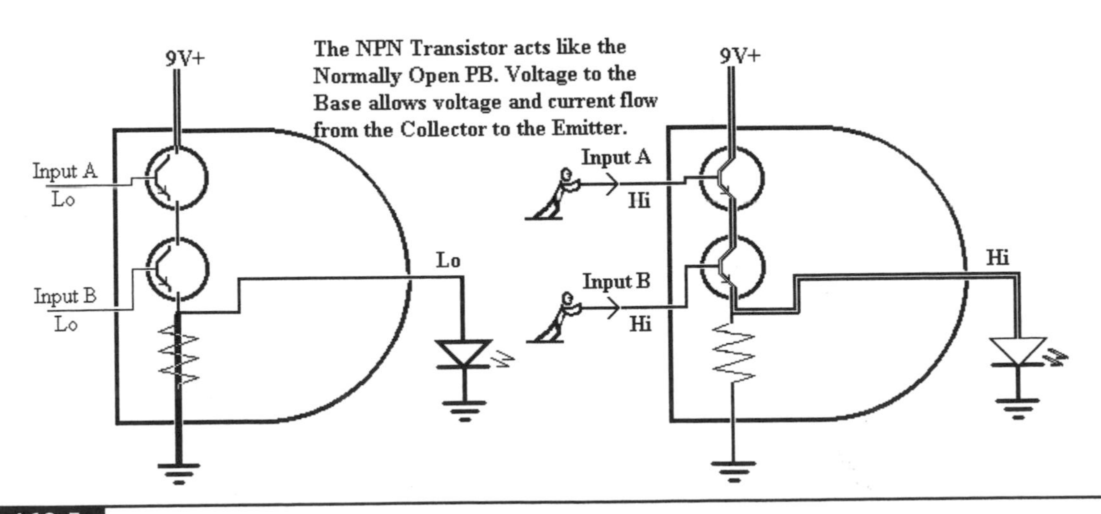

Figure L16-5

These inputs are like normally open push buttons. They control the voltage flow through the NPN transistors inside the chip. The transistor processors inside are like the push button demonstration circuits shown here. The transistors respond to the inputs, controlling voltage to the output.

Breadboard the AND Gate Simulation Circuit

Build the simulation circuit that demonstrates the action of the AND gate as displayed in Figure L16-6. Remember that for the purpose of this exercise, as shown in Figure L16-3:

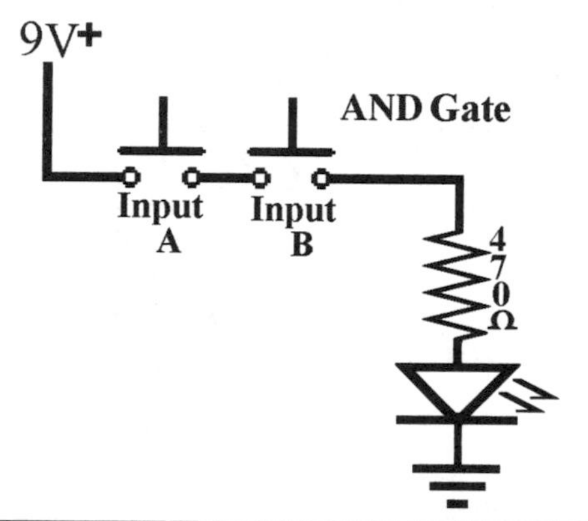

Figure L16-6

- The pushed button is high.
- The unpushed button is low.

The NPN transistors act like the normally open push buttons. Voltage at the base greater than half of V+ allows the voltage to move through (Table L16-2).

TABLE L16-2 AND Gate: Complete the Logic Table

Input A	Input B	Output
High	High	______
High	Low	______
Low	High	______
Low	Low	______

The OR Logic Gate

As shown here in Figure L16-7, input A OR input B has to be HI to get an HI output. The inputs still act like normally open push buttons. They control the voltage flow through the NPN transistors inside the chip. The transistor processors inside are like the push button demonstration circuit shown here. The transistors respond to the input, controlling the voltage to the output.

Breadboard the OR Gate Simulation Circuit

Build the simulation circuit that demonstrates the action of the OR gate as displayed in Figure L16-8.

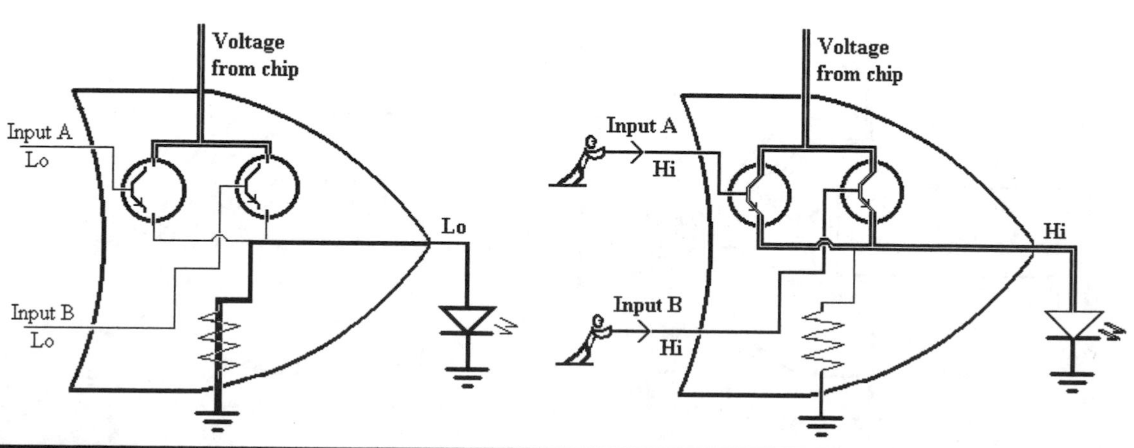

Figure L16-7

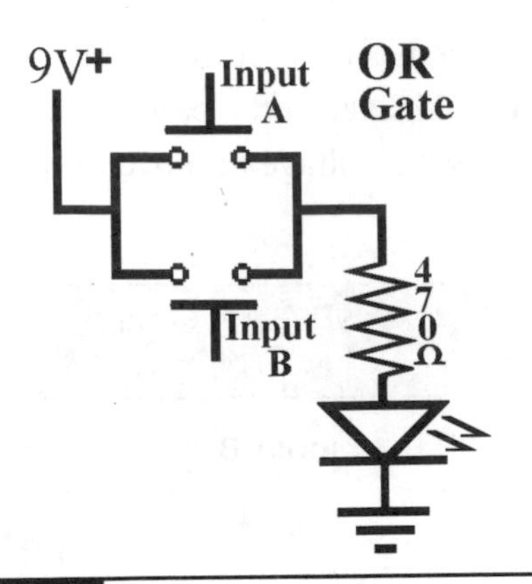

Figure L16-8

Remember that for the purpose of this exercise, as shown in Figure L16-8:

- The pushed button is high.
- The unpushed button is low.

The NPN transistors act like the normally open push buttons. Voltage at the base greater than half of V+ allows the voltage to move through (Table L16-3).

TABLE L16-3 OR Gate: Complete the Logic Table

Input A	Input B	Output
High	High	______
High	Low	______
Low	High	______
Low	Low	______

The NAND Logic Gate

As shown in Figure L16-9, the NAND gate looks a bit more complex. It really isn't. It is designed to give the exact opposite results of an AND gate. That is why it is referred to as a NOT AND gate.

Breadboard the NAND Gate Simulation Circuit

Build the simulation circuit that demonstrates the action of the NAND gate as displayed in Figure L16-10.

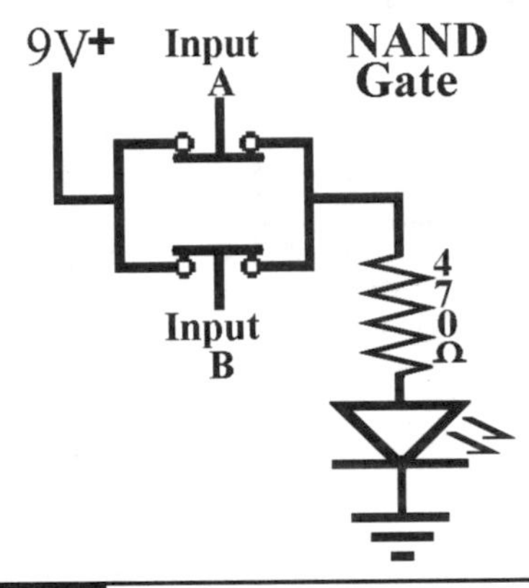

Figure L16-10

The PNP transistors act like normally closed push buttons. Voltage at the base greater than half of V+ stops the voltage from moving through (Table L16-4).

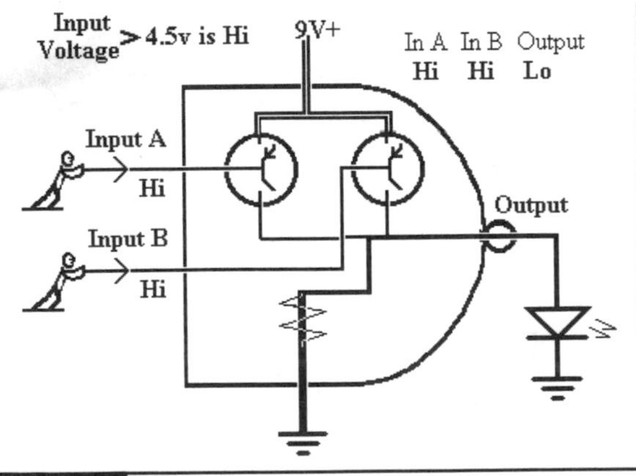

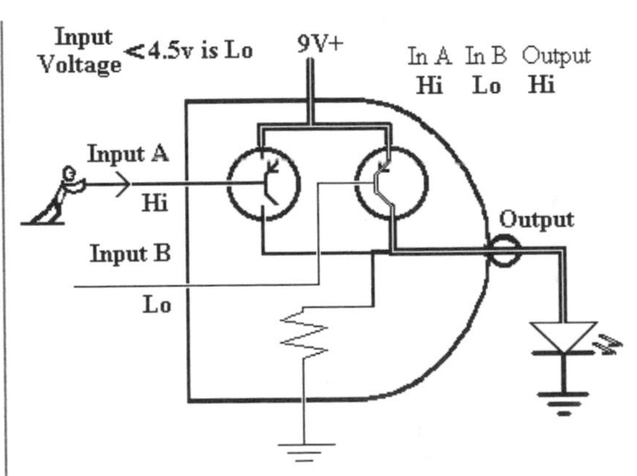

Figure L16-9

TABLE L16-4 NAND Gate: Complete the Logic Table

Input A	Input B	Output
High	High	______
High	Low	______
Low	High	______
Low	Low	______

The inputs act like normally closed push buttons. They control the voltage flow through the PNP transistors inside the chip. The transistor processors inside are like the push button demonstration circuit shown in L10-16. The transistors respond to the input, controlling the voltage to the output.

The NOR Logic Gate

The NOR gate is displayed in Figure L16-11. Just like the NAND gate, it looks overly complex. Relax. It, too, was designed to give the exact opposite results of an OR gate. That is why it is referred to as a NOT OR gate. The inputs act like normally closed push buttons. They control the voltage flow through the PNP transistors inside the chip. The transistor processors inside are like the push button demonstration circuit shown here. The transistors respond to the input, controlling the voltage to the output.

Breadboard the NOR Gate Simulation Circuit

Build the simulation circuit that demonstrates the action of the NOR gate as displayed in Figure L16-12.

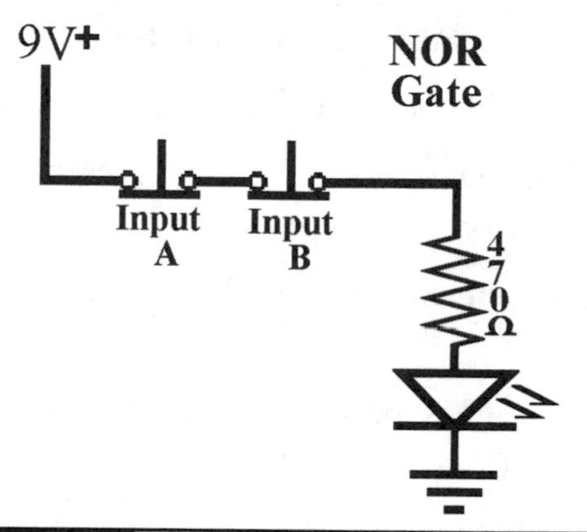

Figure L16-12

The PNP transistors act like normally closed push buttons. Voltage at the base greater than half of V+ stops the voltage from moving through (Tables L16-5 and L16-6).

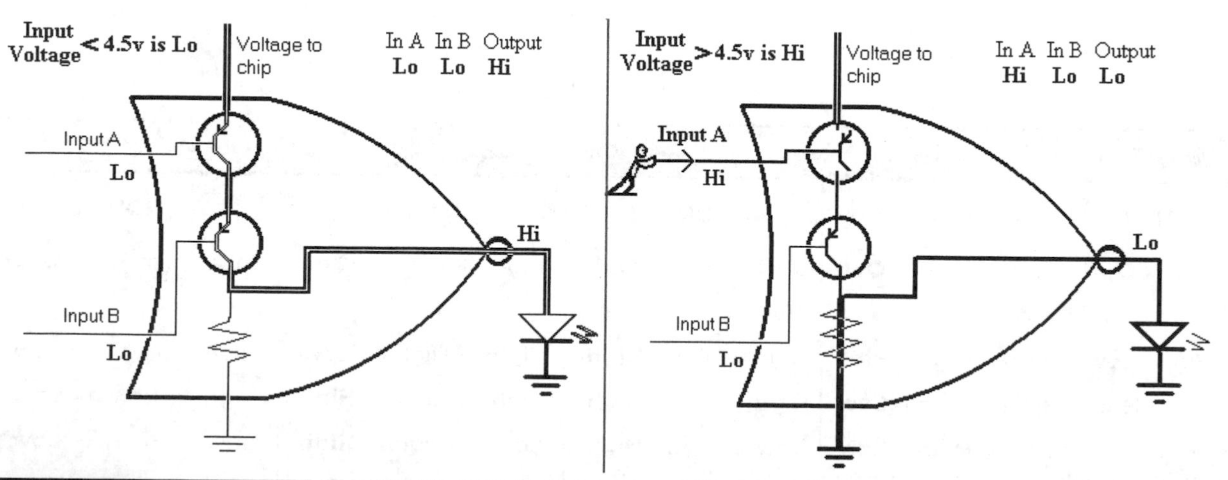

Figure L16-11

TABLE L16-5 NAND Gate: Complete the Logic Table

Input A	Input B	Output
High	High	______
High	Low	______
Low	High	______
Low	Low	______

Exercise: The Basic Digital Logic Gates

1. What did you use to represent the inputs?
2. What component is used to represent the processor?
3. What component is used to represent the output?
4. Where does the voltage powering the output come from?
5. When the push buttons were *un*pushed, that represented a state of high or low at the input?
6. When the push buttons were pushed, that represented a state of high or low at the input? Look again at the graphics of the logic gates.
7. What components are actually used in the integrated circuit (IC) chip as the processors?
8. The input is actually similar to which part of the transistor?

Lesson 17
Integrated Circuits CMOS ICs

There are thousands of *integrated circuits* (ICs). The 4000 series of complementary metal oxide semiconductor (CMOS) ICs are popular because they are inexpensive and work with as little as 3 volts and as much as 18 volts. With mishandling, however, they are easily destroyed by static electricity. So they go from packing material to your solderless breadboard. Some basic vocabulary is developed. Common layout for these chips is discussed. Handling instructions are given. And mentioned again. Did I already say that these things are static sensitive? Don't rub them in your hair. They won't stick to the ceiling like balloons afterward. They just won't work.

You need this information, because in the next chapter, you will:

- Build a prototype of a digital alarm system
- Learn to use a variety of events that can start the system
- Learn how to determine the system output
- Learn how to determine how long the system stays on after it is triggered.

The IC that you will use is the 4011 CMOS. The 4011 chip is set into a 14-pin *dual inline package* (DIP). Figure L17-1 shows an 18-pin chip in a DIP format.

TABLE L16-6 Comparing the Gates

NOT		AND			OR			NAND			NOR		
In	**Out**	In A	In B	**Out**	In A	In B	**Out**	In A	In B	**Out**	In A	In B	**Out**
	Put		**Put**		**Put**		**Put**		**Put**				
High	**Low**	High	High	**High**	High	High	**High**	High	High	**Low**	High	High	**Low**
Low	**High**	High	Low	**Low**	High	Low	**High**	High	Low	**High**	High	Low	**Low**
		Low	High	**Low**	Low	High	**High**	Low	High	**High**	Low	High	**Low**
		Low	Low	**Low**	Low	Low	**Low**	Low	Low	**High**	Low	Low	**High**

Figure L17-1

Precautions

The 4000 series CMOS IC has been used in electronics since the 1970s. They are versatile and widely used. They are used here because they work on a range from 3 to 18 volts. They are inexpensive, too! **But they are static sensitive.** Yes, I know I'm repeating myself. You know. Shuffle across the carpet and zap your friend. Even the smallest zap can toast a CMOS chip (see Figure L17-2).

Figure L17-2

CAUTION Ignore these precautions at your own risk!

1. Always store the IC in a carrying tube or in "static" foam until it is placed into the circuit.
2. Remove static from your fingers. Touch some type of large metal object to remove any static electricity from your fingers before you handle the CMOS chips.
3. Don't walk across the room with a CMOS chip in hand. Walking across linoleum or a rug in a dry room will build up a static charge.
4. Always check that the chip is set in properly. A backward chip is a dead chip ($$).
5. Always tie any unused inputs to ground. I'll note where we have done that in this circuit.
6. Having an unused input pin unconnected is not the same as tying it to ground. If an input is not connected, the small voltage changes in the air around us can affect the input.

Take a Moment and Look at the Partial List of CMOS Series

For a more complete idea of the chips available, visit www.abra-electronics.com.

- 4000 Dual 3-input NOR gate plus inverter
- 4001 Quad 2-input NOR gate
- 4002 Dual 4-input NOR gate (same as 74HC4002)
- 4006 18-stage shift register, serial-in/serial-out
- 4007 Dual CMOS pair plus inverter
- 4008 4-bit, full-adder arithmetic unit
- 4009 Hex inverter—OBSOLETE, use 4049 instead
- 4010 Hex buffer—OBSOLETE, use 4050 instead
- 4011 Quad 2-input NAND gate
- 4012 Dual 4-input NAND gate
- 4013 Dual Type D Flip-Flop
- 4014 8-stage shift register, parallel-in/serial-out
- 4015 Dual 4-stage shift register, serial-in/parallel-out (same as 74HC4015)
- 4016 Quad bilateral analog switch
- 4017 Decade counter, synchronous 1-of-10 outputs
- 4018 Programmable counter, walking ring
- 4019 4-pole, double-throw data selector
- 4020 14-stage binary ripple counter (same as 74HC4020)
- 4021 8-stage shift register, parallel-in/serial-out

- 4022 Octal counter, synchronous 1-of-8 outputs
- 4023 Triple 3-input NAND gate
- 4024 7-stage binary ripple counter (same as 74HC4024)
- 4025 Triple 3-input NOR gate
- 4026 Decade counter and 7-segment decoder with enable
- 4027 Dual JK Flip-Flop with preset and clear
- 4028 1-of-10 decoder
- 4029 Up-down synchronous counter, decade or hexadecimal
- 4030 Quake EXCLUSIVE-OR gate—OBSOLETE, use 4077 or 4507
- 4031 64-stage shift register, serial-in/serial-out
- 4032 Triple-adder, positive-logic arithmetic unit
- 4033 Decade counter and 7-segment decoder with blanking
- 4034 8-bit bidirectional storage register
- 4035 4-stage shift register, parallel-in/parallel-out
- 4038 Triple-adder, negative-logic arithemtic unit
- 4040 12-stage binary ripple counter (same as 74HC4040)
- 4041 Quad inverting/noninverting buffer
- 4042 Quad latch storage register
- 4043 Quad Flip-Flop, R/S NOR logic
- 4044 Quad Flip-Flop, R/S NAND logic
- 4046 Phase-locked loop, special device
- 4047 Astable and monostable multivibrator
- 4049 Hex inverter/translator (same as 74HC4049)
- 4050 Hex buffer/translator (same as 74HC4050)
- 4051 1-of-8 analog switch (same as 74HC4051)
- 4052 Dual 1-of-4 analog switch (same as 74HC4052)
- 4053 Triple 1-of-2 analog switch (same as 74HC4053)
- 4060 14-stage binary ripple counter with oscillator (same as 74HC4060)
- 4063 4-bit magnitude comparator arithmetic unit
- 4066 Quad analog switch, low-impedance (same as 74HC4066)
- 4067 1-of-16 analog switch
- 4068 8-input NAND gate
- 4069 Hex inverter
- 4070 Quad EXCLUSIVE-OR gate
- 4071 Quad 2-input OR gate
- 4072 Dual 4-input OR gate
- 4073 Triple 3-input AND gate
- 4075 Triple 3-input OR gate (same as 74HC4075)
- 4076 4-stage tri-state storage register
- 4077 Quad 2-input EXCLUSIVE-NOR gate
- 4078 8-input NOR gate (same as 74HC4078)
- 4081 Quad 2-input AND gate
- 4082 Dual 4-input AND gate
- 4089 Binary rate multiplier, special device
- 4093 Quad 2-input NAND Schmitt Trigger
- 4097 Dual 1-of-8 analog switch

The 4011 Dual Input Quad NAND Gate

The 4011 IC is a semiconductor that looks like Figure L17-3.

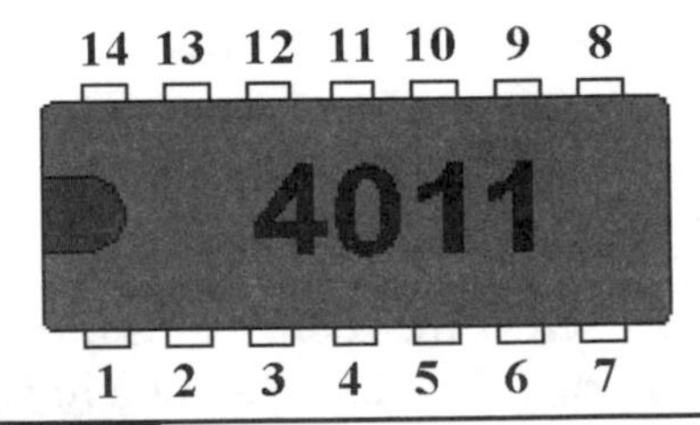

Figure L17-3

Notice the method of numbering the pins. All DIPs use this system.

Looking at the IC from the top, the reference notch should be to the left as shown. Then, pin 1 is on the bottom left. The numbering starts there and moves counterclockwise.

What happens when you flip the chip upside down? Where is pin 1 now?

Study Figure L17-4. The 4011 IC contains four separate NAND gates, each able to work independently. It has 14 pins. These pins are the connecting points.

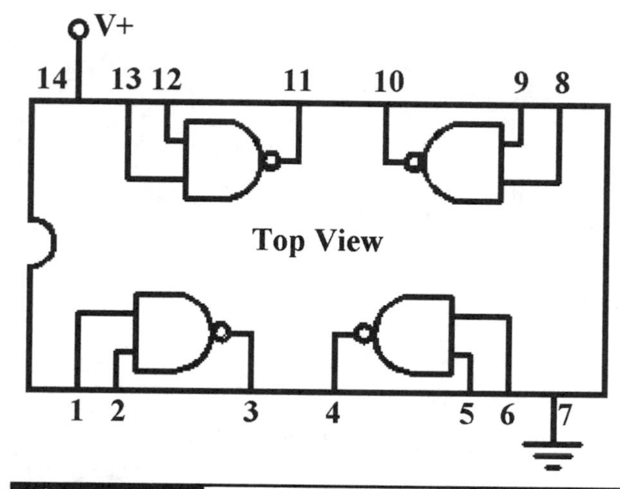

Figure L17-4

- Here is the pin-out diagram for the 4011.
- Each pin has a specific function for each IC. It is important to connect these pins correctly.
- Pins 1 and 2 are inputs to the NAND gate that has its output at pin 3.
- Power to the chip is provided through pin 14.
- The chip's connection to ground is through pin 7.
- The inputs redirect the output of each gate to voltage (pin 14) or to ground (pin 7).

Exercise: Integrated Circuits, CMOS ICs

1. What is a DIP?
2. State the functional voltage range of a CMOS IC. _____ to _____ V
3. Briefly state the six major precautions regarding the proper care and feeding of an IC.
   a. ______________________________
   b. ______________________________
   c. ______________________________
   d. ______________________________
   e. ______________________________
   f. ______________________________
4. Draw a picture of your 4011 chip.
5. Indicate on your drawing all writing on the chip.
6. Include the marker notch on your drawing.
7. Label pins 1 through 14.
8. From the diagram sheet, which pin powers the 4011 chip? ________
9. Which pin connects the chip to ground?

   ____
10. Describe clearly how to identify pin 1 on any IC. ________________
11. What would happen if a wire connection of the 4011 was made to the wrong pin?

    ____

    ____
12. Would your answer for the previous question be true for any IC? Look at the list of some CMOS ICs, mostly the 4000 series.
13. Look at the list of some of the CMOS ICs of the 4000 series. How many ICs shown are dedicated logic gate chips?

SECTION 6

The First NAND Gate Circuit

THIS IS GOING TO LOOK very messy and complicated at first. As you get used to it, you will see that it really is very simple. In this chapter, we will do the following:

- We will build and become familiar with a basic digital circuit.
- We will learn how an analog signal is translated into a digital output, using different input devices.
- We will be introduced to control timing in a resistor-capacitor circuit.

PARTS BIN

- R1—100 kΩ
- R2—10 MΩ
- C1—1μF axial or radial
- LED—5 mm round
- D1—Signal diode (skinny golden)*
- IC 1—4011 Quad NAND gate

* Don't use the fat golden diode. You can substitute the black power diode.

Lesson 18
Building the First NAND Gate Circuit

Enough talk, already. Back to the fun stuff.

Breadboard the circuit shown in Figures L18-1 and L18-2 (see the Parts Bin).

Even though they don't look the same, match the pin numbers in the drawing with the pin numbers on the schematic. Refer back to the 4011 pin-out diagram of Figure L17-4 to help you with this.

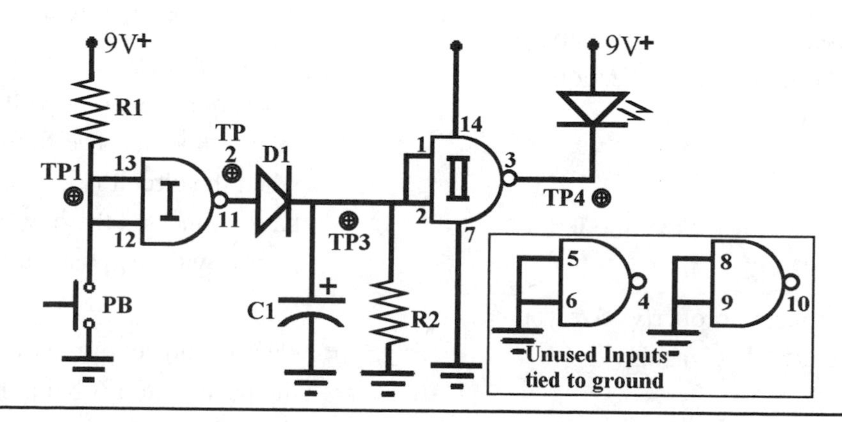

Figure L18-1

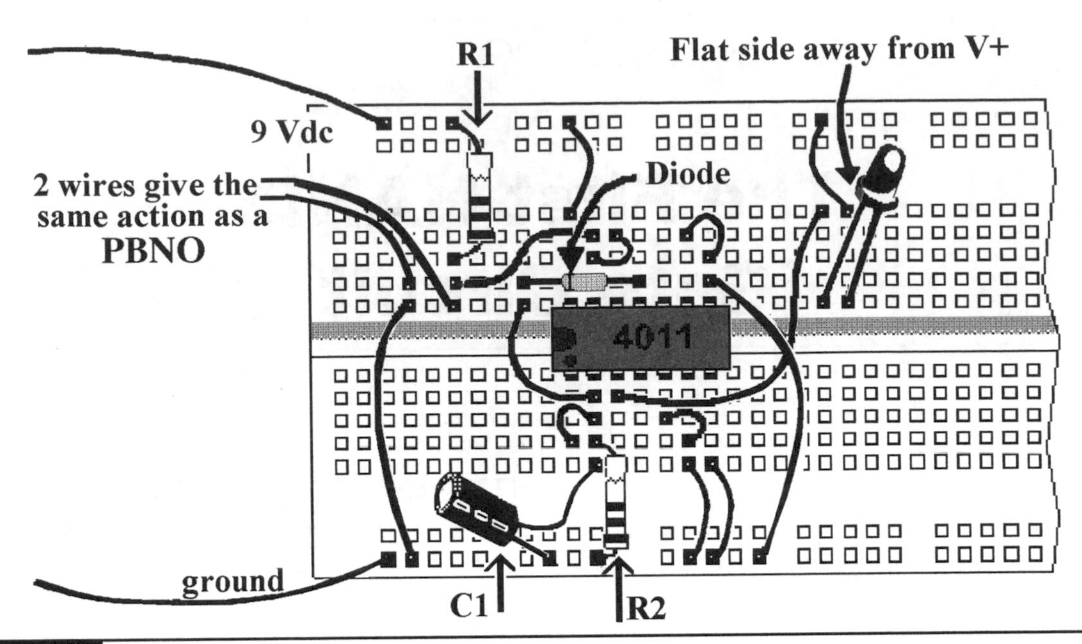

Figure L18-2

What to Expect

When you attach the power supply, the LED should stay off.

The circuit is working when you momentarily close the PBNO and the LED turns on for about eight seconds, and then automatically turns off again.

If the LED turns on as soon as you attach the power supply, immediately disconnect the power. Something is wrong. If the power remains connected, you could burn out your chip. Also, if all you wanted to do was turn an LED on, what's all the other stuff doing in the circuit? Just use an LED and resistor for that.

If the circuit refuses to work immediately, you need to refer to the Troubleshooting section.

Troubleshooting

Some general questions to ask as you look for errors:

1. Is the power connected properly? Are you positive about that? Check again!
2. Are your breadboard connections done properly? All connections need to be in the small rows of five dots. Look quickly for any side-by-side connections.
3. Are your parts in the right way?
   - Is your chip in the right way?
   - Is your diode going the right way, as shown in the schematic?
   - A backward capacitor won't help either.
   - A backward LED won't turn on, even if everything else is working.
4. Now examine Figure L18-3.
   - Set your DMM to voltage DC. Connect the black probe to ground.
   - Use the red probe to take two measurements of thc voltage at each of the 19 checkpoints. The first measurement is when the circuit is at rest. That is, you have connected the power supply but not played with the push button. Record your results.
   - The second set of measurements is when the push button is being held down and pins 12 and 13 are being connected to

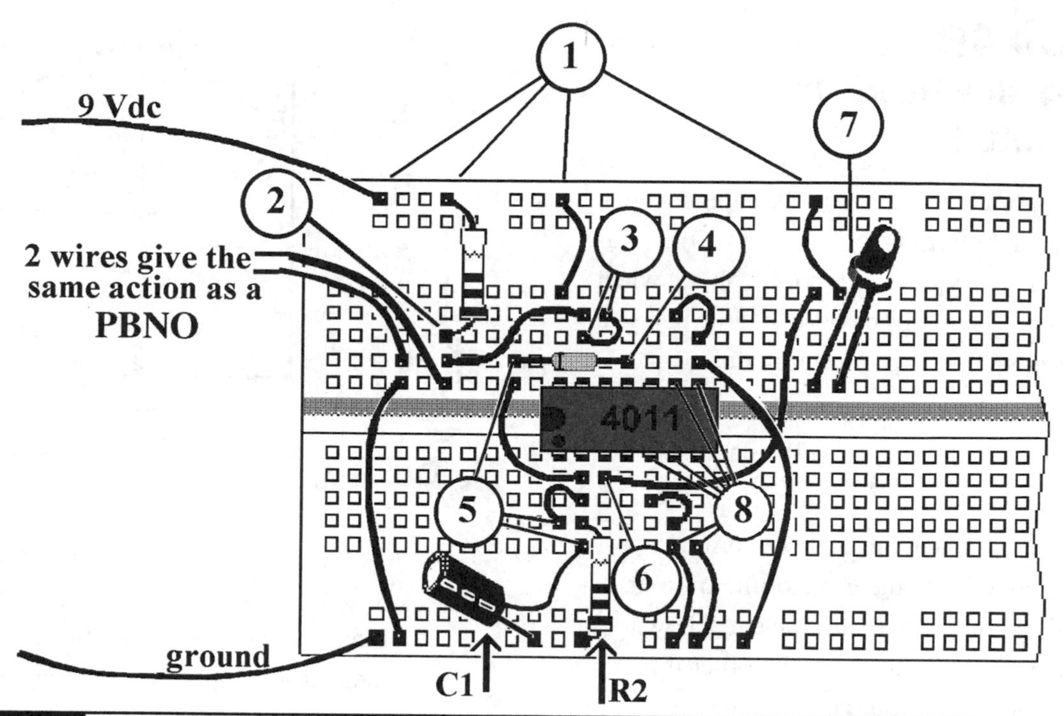

Figure L18-3

ground through the push button. This is to guarantee that you are measuring the circuit in its active state. See Table L18-1.

Focus on the area that does not match the expected results.

TABLE L18-1 Measurements for Figure L18-3

At Rest	Active *Pins 12 and 13 Connected to Ground*
1. V+ These points are connected directly to voltage.	V+ These points are still connected directly to voltage.
2. This is a bit less than V+.	0 volts because when the push button is closed, it is connected directly to ground.
3. This is the same as 2.	This is the same as 2.
4. Reading should be 0.0 V.	Reading should be V+.
5. 0 volts or close to it.	Reading should be V+.
6. The readout from pin 3 should read V+.	This should be V+ minus 2 V (LED uses 2 V). Remove LED and reading should change to V+.
7. LED flat side toward pin 3 LED should be off.	LED should be on and remain on for eight or so seconds after you release the push button.
8. Pins 5, 6, 8, and 9 should have 0.0 V.	These are connected directly to ground and should still have 0.0 V.

Lesson 19
Testing the Input at Test Point 1

You will measure the voltage at the inputs when the input (push button) is high and low. If the push button works, you will measure the voltage when it is not pushed. There will be no voltage when you push it, because you are connecting directly to ground when it is pushed.

Record all of your results on the data table at the end of Section 6.

Wrap a piece of wire around the common probe (black) and attach it to the ground line of your breadboard. Figure L19-1 shows clearly how to do it. This makes it easier to do all the other things you need to do as you use the red probe to measure the voltage at different test points.

Attach the power supply and measure the voltage at the test point 1 (TP1) shown in Figure L19-1.

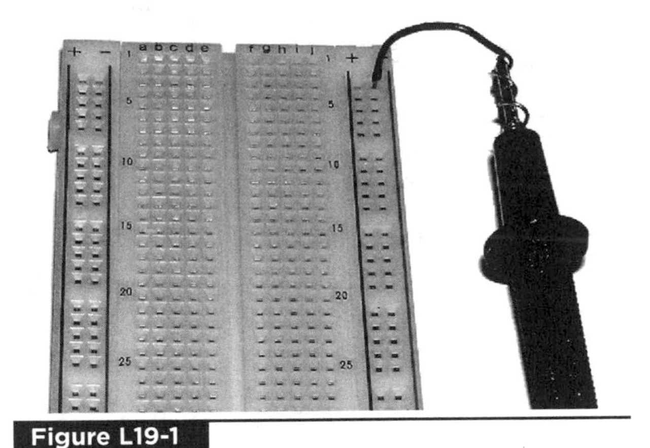

Figure L19-1

- Figure L19-2 shows that the voltage through R1 is putting pressure on pins 12 and 13, the connected pair of inputs to one of the 4011 NAND gates. The input voltage is seen as a high.

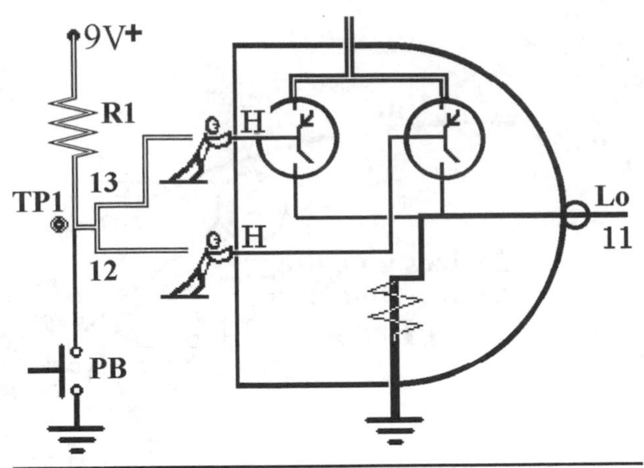

Figure L19-2

- When you measure the voltage at TP1 without pressing the normally open push button, you are measuring the actual voltage at the input to the NAND gate.

  Is that voltage greater than half of the power source voltage? _____

  It should be much greater than half of the voltage.

  Describe the *state* of the inputs to the first NAND gate when the system is at rest. ______________________________

With the power supply still connected, push the plunger of the push button down. Record the voltage.

What Is Happening Here?

- Figure L19-3 shows that when you close the push button, the voltage through R1 flows directly to ground. Like water, the voltage and current always takes the easiest path.
- The connected pair of inputs 12 and 13 no longer have any voltage pressure on them. They are all connected to ground. Ground at 0.0 volts is definitely less than half of the voltage supplied to the chip. Remember that ground is just another word for a low

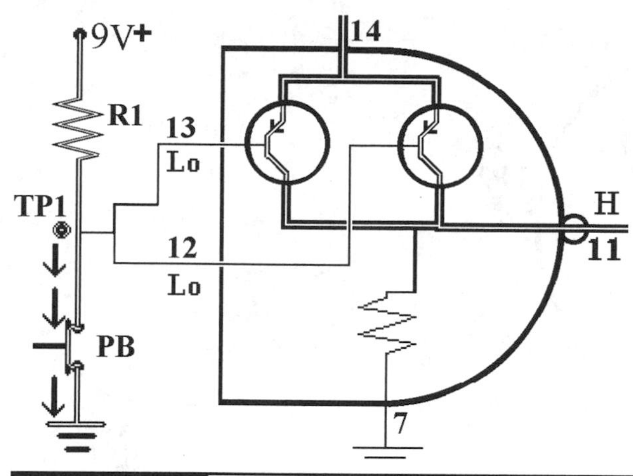

Figure L19-3

state. Is that voltage greater than half of the power source voltage?___________

It should read 0.0 volts. With the plunger down, it should be connected directly to ground.

Describe the *state* of the inputs to the first NAND gate when the push button is closed.

Describe what happens to the voltage on the DMM when you release the push button.

The other instrument reading shown in Figures L19-4 and L19-5 is an oscilloscope. Each horizontal line equals 2 volts. The line representing voltage is at 4.5 lines up. 4.5 3 2 represents 9 volts.

Don't worry if you don't have a proper oscilloscope available. Your computer can be used as a cheap "scope" with freeware available online. Soundcard Scope was developed by Proffessor Christain Zeitnitz.

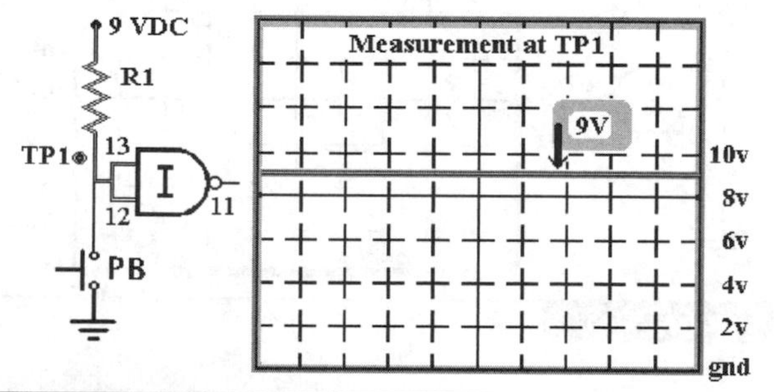

Figure L19-4

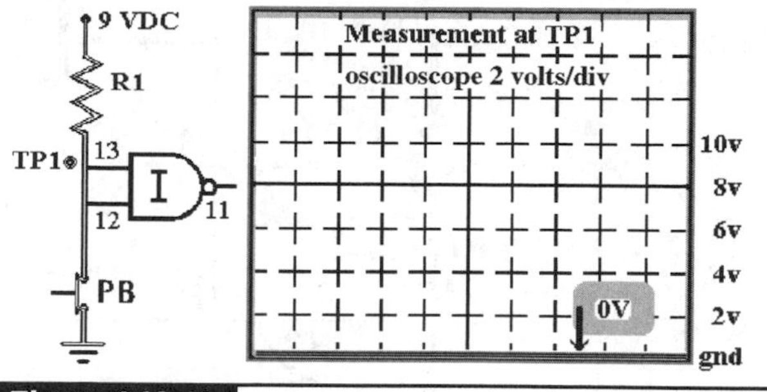

Figure L19-5

It is a wonderful tool and available at www.zeitnitz.de/christian/scope_en. The only failing is that it is limited to measuring rapidly changing signals. We'll be doing that soon enough. It will not measure a stable DC input.

Lesson 20
Test Point 2—The NAND Gate Processor at Work

When the inputs are connected to ground, the output will be internally connected to voltage. Try it out!

For the NAND gate, if the inputs are connected to voltage, the output should be connected to ground. Conversely, if the inputs are connected to voltage, the output is internally connected to ground. The results happen instantly. Take a minute and understand the NAND gate.

What to Expect

With the system at rest and the push button untouched, measure the voltage at the output of the first NAND gate at pin 11 (Table L20-1).

TABLE L20-1 NAND Gate Logic Table

Input A	Input B	Output
High	High	Low
High	Low	High
Low	High	High
Low	Low	High

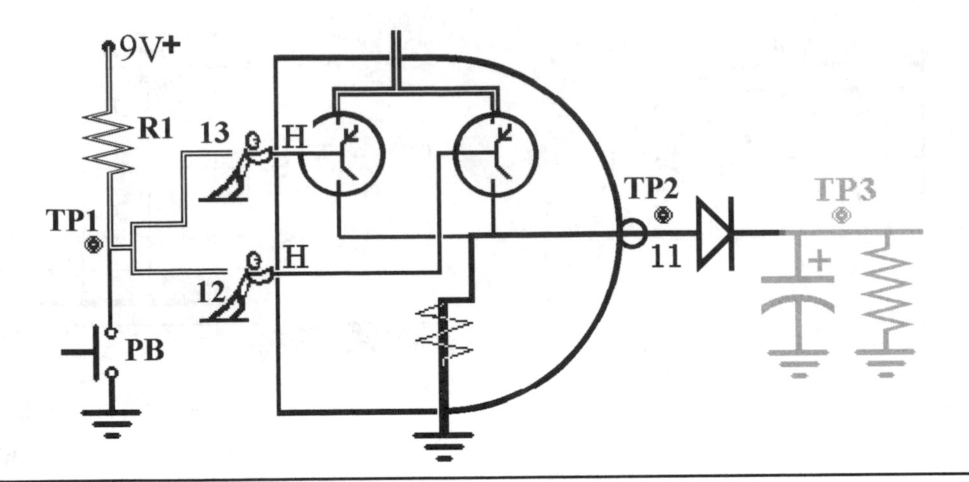

Figure 20-1

The measurement for TP2 is at pin 11. Referring to the schematic in Figure L20-1, this is right at the output of the first gate, but before the diode.

Record all of your results on the data table at the end of Section 6.

1. Measure the voltage at TP2 with the push button open and record your results.
   - Because pins 12 and 13, the inputs to the NAND gate, are high, the output at pin 11 is low.
   - This would be the expected value of the output when the circuit is "at rest."
2. Measure the voltage at TP2 with the push button held closed and record your results.
   - Both inputs of the NAND gate are now connected directly to ground; that is to say, they are *tied to ground*, as shown in Figure L20-2. Remember that in the NAND gate, if either input receives a low, it produces a high output.
   - This would be the expected value of the output when the circuit is just started, or "activated."

The output at pin 11 is normally low when the circuit is at rest. Remember low equals ground.

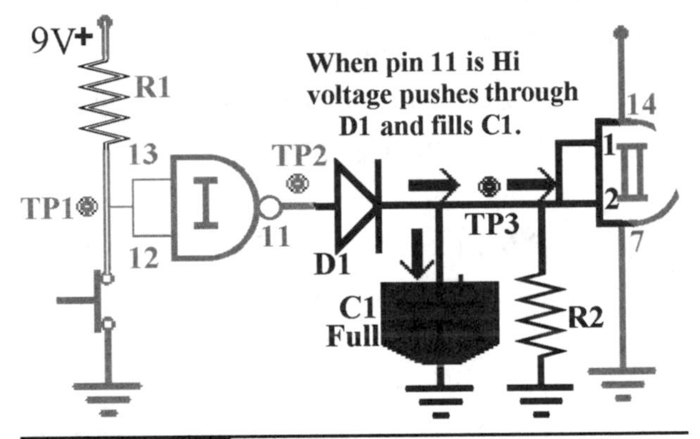

Figure L20-2

This is how the output at TP2 would look on an oscilloscope. Figure L20-3 shows where the voltage would be when the circuit is at rest. Figure L20-4 shows where the voltage would be when the plunger is held down and pin 11 is connected to ground.

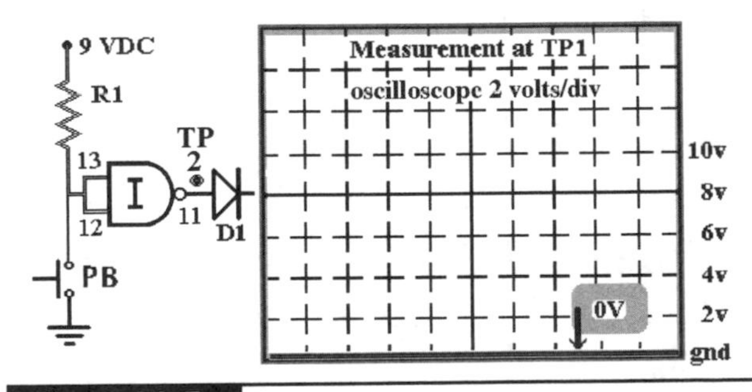

Figure 20-3

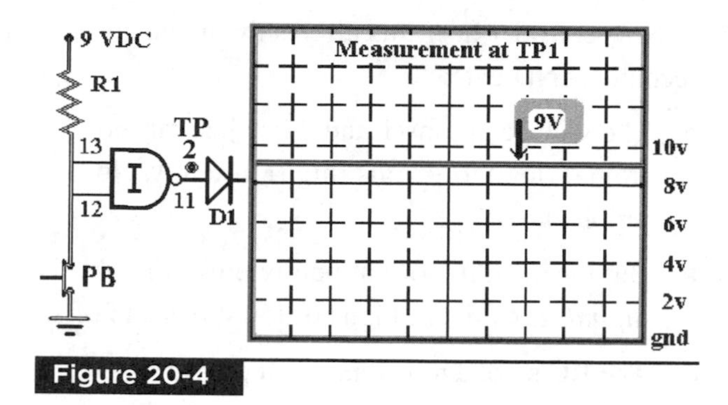

Figure 20-4

The voltage produced at pin 11 pushes through the diode and gets trapped on the other side. Referring to the schematic in Figure L20-4, it would be okay to say that the voltage and current are trapped on the right side of the diode and not able to drain to ground through pin 11. It reaches ground another way.

Lesson 21
Test Point 3—Introducing the Resistor/Capacitor Circuit

You are introduced to a resistor/capacitor working as a pair to control timing. You saw this briefly when you were introduced to capacitors in Part One. This is called a *resistor/capacitor* (RC) circuit. This is actually one of the major subsystems in electronics used to control timing. Here it is used to control how quickly the voltage to the second NAND gate drains, which controls how long the LED stays on. In an RC timing circuit, imagine this: The capacitor is like a sink and holds the charge. The resistor is like a drain pipe. It allows the charge in the sink to flow out.

Note that TP3 can be considered anywhere along the common connection of D1, R2, and C1 and pins 1 and 2. They are all connected together as they are shown in Figure L21-1.

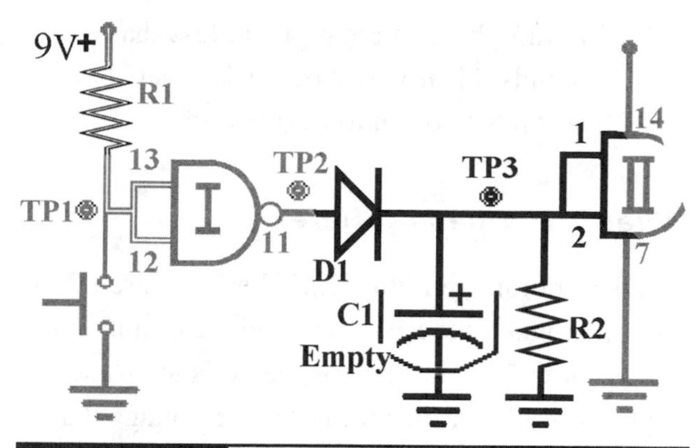

Figure L21-1

Record all of your results on the data table at the end of Section 6.

1. Before you take any measurements, allow the circuit to remain unused for at least a minute. Measure the voltage at TP3 with the push button plunger untouched and record your results.
2. Measure the voltage at TP3 with the push button held closed and record your results. Figure L21-2 shows that as long as the NAND gate's output at pin 11 is high, the voltage and current push through D1 and do two things. They influence the inputs of the second NAND gate and fill the capacitor.

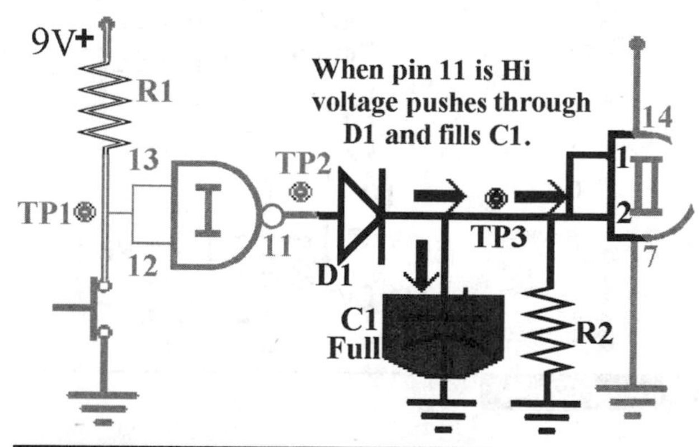

Figure L21-2

3. Now keep the probe at TP3. Release the push button (PB) and watch the DMM. What happens here is completely different from TP1 or TP2. The voltage slowly decreases. If you are using a poor-quality DMM, the voltage

will drain through the probe in less than 2 seconds. Ideally, it should take over 20 seconds to drain to near 0 volts.

What Is Happening Here?

The illustrations in Figure L21-3 and Figure L21-4 show that when the PB is released the output at pin 11 goes low. The diode traps the voltage on the "right" side. The capacitor holds the voltage that influences the inputs of the second NAND gate at pins 1 and 2. Meanwhile, the voltage is draining from C1 through R2.

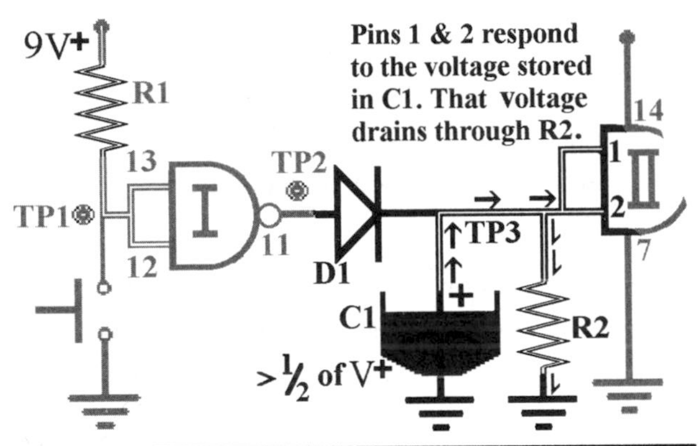

Figure L21-3

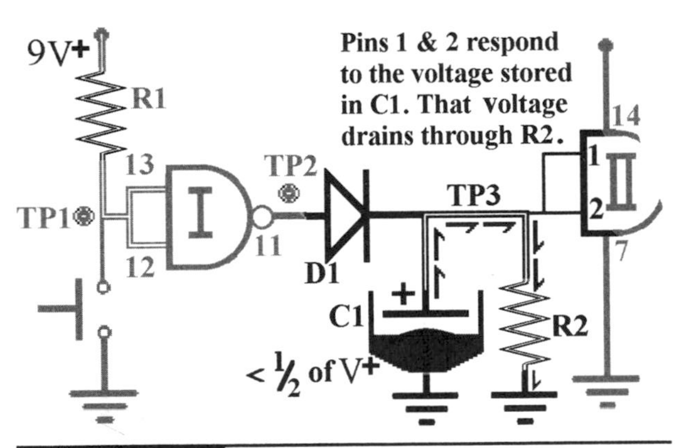

Figure L21-4

You can see the voltage draining on your DMM, too.

When a resistor and capacitor are used together like this for timing, this is referred to as an *RC circuit*.

Now consider how this affects the inputs to the second NAND gate:

- The voltage at pins 1 and 2 has just moved from 0 (low) to 9 volts (high) instantly. No problem.
- But the most important point here is that this is *digital*. It works in highs or lows, ons and offs.
- The RC's voltage is *analog*. Inputs are designed to respond to analog inputs. The input for the second NAND gate is sliding downward, controlled by the RC circuit. Where along that sinking voltage path do the second gate's inputs at pins 1 and 2 switch from sensing the voltage as high to sensing that input as low?

Start the circuit again and carefully watch the DMM. What is the voltage when the LED turns off?

What's that you say? It looks pretty close to 4.5 volts.

That's right, but only if you use precisely 9 volts here.

Note that Figure L21-5 shows how the falling voltage would be shown on an oscilloscope. The vertical lines are used to measure time. The time units here are very large for electronics, representing half-second units.

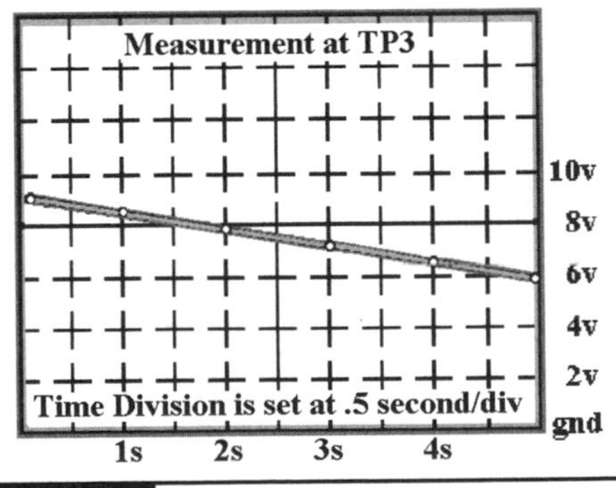

Figure L21-5

Remember that *digital inputs* are designed to:

- Sense anything above 1/2 of V+ as high.
- Sense anything below 1/2 of V+ as low.
- The state of the input changes at half of the voltage supplied to pin 14.
- If the voltage powering the chip through pin 14 is 9 volts, the state of the inputs to the gate changes from high to low at 4.5 volts.
- If the supply voltage is 12 volts, the inputs to the gate will change from high to low input at 6 volts.

Exercise: Test Point 3 and Introducing the RC Circuit

1. Record the "at rest" voltage of TP3 on the data sheet at the end of Section 6.
2. Push the button's plunger and hold it closed while you check and record on the data sheet the voltage at TP3.
3. How much voltage is used up by the diode (tp2@HI = tp3@HI = V used)? Diode voltage = ____ V
4. Describe what happens to the voltage on the DMM when you release the push button.
5. How long did the LED stay on for your circuit? ______ s
6. What was the voltage on your DMM for TP3 when the LED was turned off? ______ V
7. How much time does it take for the voltage to drain to 1 volt? ______ s
8. What is the capacitor in the RC circuit being compared to? A ____________
9. What is the resistor in the RC circuit compared to? The ____________
10. Replace R2 with a 20-megohm resistor. How much time does the LED stay on now? ______ s
11. For R2 = 20 MΩ, what is the voltage when the LED turns off? ______ V. Is the LED going off at almost the exact point when the capacitor is half drained? Yes or No? It should be very nearly the same as before. Why would you expect this? ______________________
12. With R2 = 20 MΩ, how much time does it take the voltage to drain to 1 volt? ______ s How does that relate to the time it took to drain when it was the 10-megohm resistor?

    ____
13. Reset R2 back to 10 megohms. Replace the capacitor with the value of 10 μF. What is the time for the LED to stay lit with the capacitor 10 times larger? ______ s

    Was that predictable?
14. Use this as a rough formula for RC timers. It gives an *estimate* of the time it takes an RC circuit to drain from being filled to when it is near the voltage that affects the digital inputs.

    R*C = T

    R is in ohms.

    C is in farads.

    T is time measured in seconds.

    Here

    $C = 1\ \mu F = 0.000001F = 1 \times 10^{-6}F$

    Also

    $R = 10\ M\Omega = 10{,}000{,}000\ \Omega$

    $R \times C = T$

    $(1 \times 10^{7}\ \Omega) \times (10^{-6}F) = 10^{1}s$

    See Table L21-1.

TABLE L21-1 Component Values

	Capacitor C1	Resistor R2	Expected Time On	Real Time
1	1 μF	10 MΩ	10 S	_____
2	1 μF	20 MΩ	20 S	_____
3	10 μF	1 MΩ	_____	_____
4	10 μF	2.2 MΩ	_____	_____
5	10 μF	4.7 MΩ	_____	_____

15. Now check each RC timer in your circuit.

    See if your predictions are close. They should be in the ballpark.

    This is not a precise timer. RC circuits are as accurate as the components that make them. Consider what affects their accuracy.

    - The resistors supplied have a tolerance of 5 percent.
    - Aluminum electrolytic caps generally have a tolerance of 20 percent.

16. Look at your predictions and results. There should be an obvious pattern. Describe the pattern you see developing.

Lesson 22
Test Point 4—The Inputs Are Switches

You now get a close-up view of the output of this circuit. The LED is removed so you can get "clean" voltage measurements. Also, you start to play with the circuit's output.

Remove the LED as shown in the schematic in Figure L22-1 for measuring the voltage at TP4. Remember to record all of your results on the data table at the end of Section 6.

Record the voltage at TP4 while the circuit is at rest.

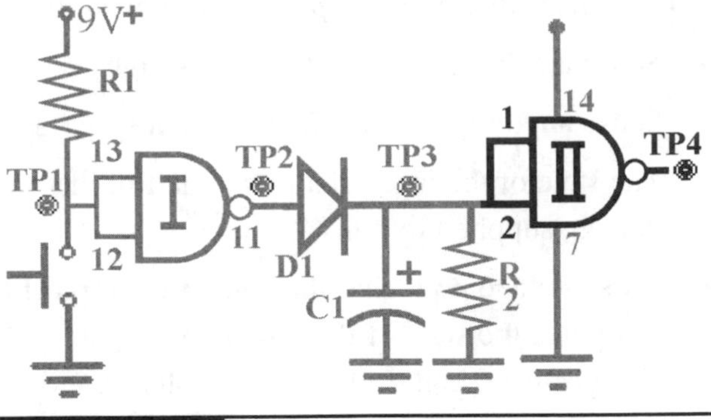

Figure L22-1

Now push the plunger of the push button. It isn't necessary to continue holding it down. Record the voltage while the circuit is active.

What Is Happening Here?

Figure L22-2 shows clearly what is occurring in the circuit when it is at rest. There is no voltage stored in C1. The inputs to the second NAND gate at pins 1 and 2 are low.

When the push button's plunger is pushed, the output of the first NAND gate goes high, filling C1 and providing a voltage nearly equal to V+ to the inputs of the second NAND gate. This is shown in Figure L22-3.

When the push button is released, as shown in Figure L22-4, the capacitor starts to drain through R2. But as long as the inputs at pins 1 and 2 sense a voltage greater than half of the voltage provided to the system, they continue to see this as a high input.

When the voltage stored in C1 drops below the halfway mark, as shown in Figure L22-5, the inputs now see the analog voltage as low. The internal circuit reacts appropriately. The switches reroute the voltage supplied to the chip to the output of the NAND gate. Pin 3 goes high.

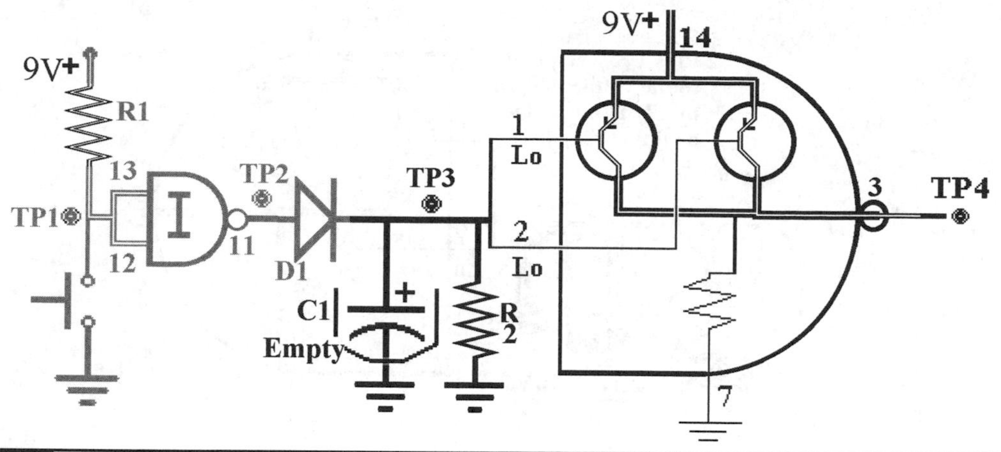

Figure L22-2

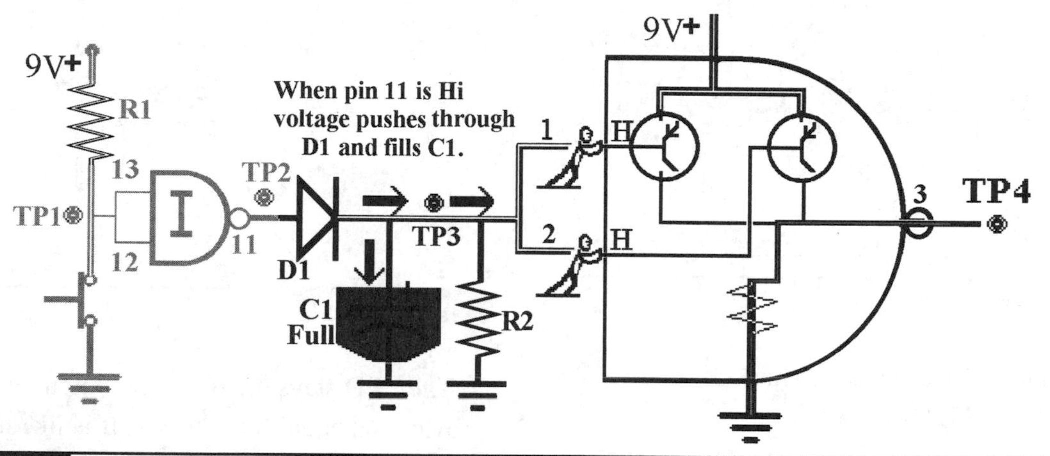

Figure L22-3

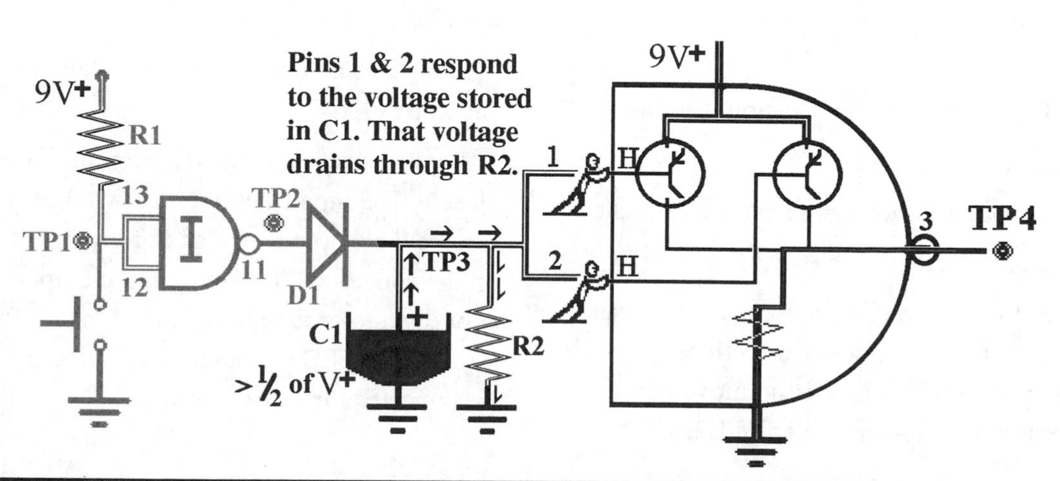

Figure L22-4

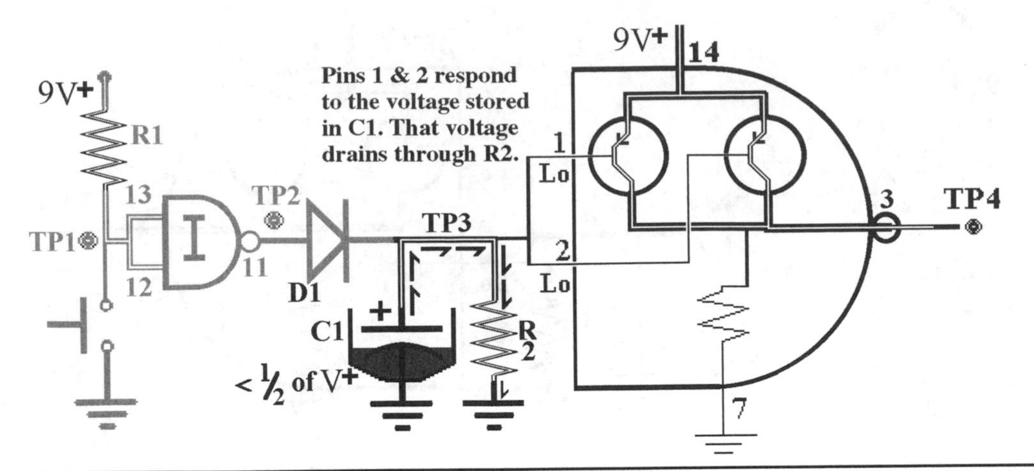

Figure L22-5

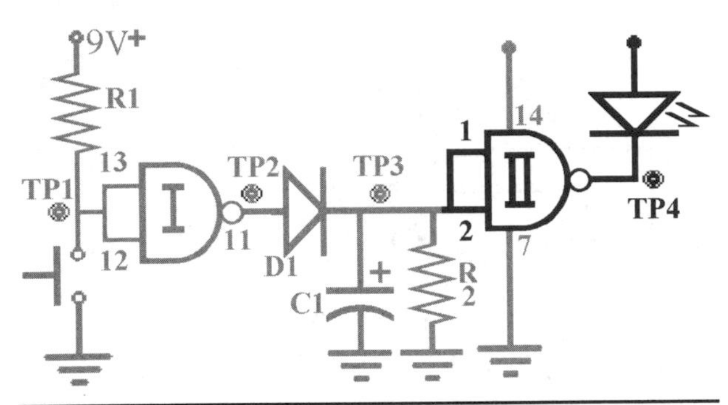

Figure L22-6

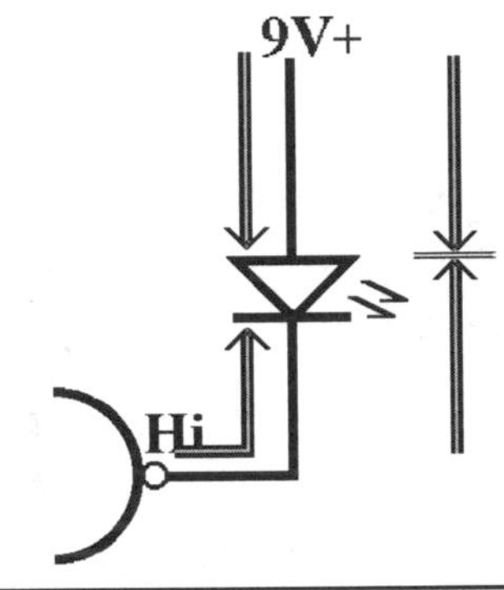

Figure L22-7

Now put the LED back into the circuit, as shown in Figure L22-6.

There is some further testing to be done, but first, a puzzling question in two parts.

Part 1: Why does the LED stay off when the output from the second NAND gate at pin 3 goes high?

In an effort to answer this question, try this:

1. Stand up.
2. Put your palms together in front of your chest.
3. Push them together with equal force.

Why don't they move? Because each hand has equal force pushing in opposite directions. One force cancels the other. There is no movement. This concept is also shown in Figure L22-7.

The LED stays off when the output at pin 3 provides an equal force to V+. It is like having both legs of the LED connected to voltage at the same time. Nothing is going to happen.

Part 2: Remember, there were two parts to the question.

How does a "low" output of 0 volts turn the LED on?

Look at the schematic diagram in Figure L22-4 closely. The inputs act only as switches. They reroute the output to voltage or ground, depending on the conditions. Here, when the output is connected to ground, the voltage moves from V+, through the LED, through the IC, to pin 7, which is connected directly to ground.

Think about the beauty of it.

- When a digital output is Hi, it can be used as a voltage source.
- And when the output of a digital system is low, it can be used as ground. Figure L22-8 gives pause for thought.

Figure L22-8

Exercise: TP4—The Inputs Are Switches

Table L20-1 is a detailed outline of the system at rest. Make a detailed outline for the active system.

1. Make a detailed outline for the ACTIVE system.

   Input

   Processor

   Output

2. Connect the LED as shown in Figure L22-9.
   a. Describe what the circuit does now when it is at rest.
   b. Describe what the circuit does now when it is active.
   c. Explain what is happening. Keep in mind the fact that pin 3 is the output of the second NAND gate.

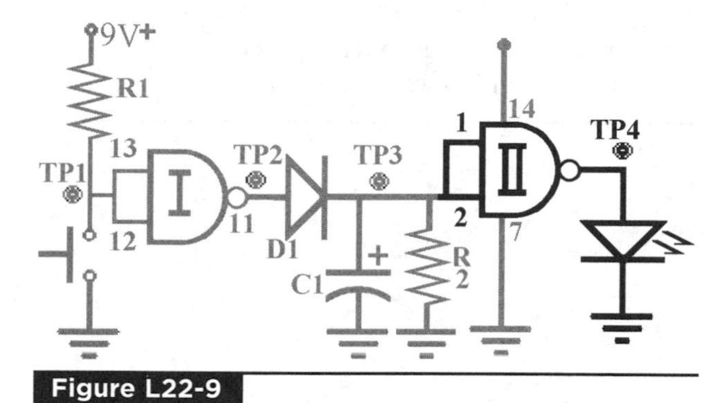

Figure L22-9

Section 6 Data Sheet

Record all of your results from Section 6 in the data table in Table L22-2. There appears to be a huge amount of information. This single table can be used for your convenient reference and review.

TABLE L22-1 Outline of System at Rest

Input	At rest, the inputs to the first NAND gate at pins 12 and 13 are tied to V+ through R1.
Processor	1. Because the inputs to the first NAND gate are both high, the output is low. 2. The capacitor has little or no voltage because it has drained through R2. 3. The inputs to the second NAND gate are both low. They are tied to ground through R2. 4. Because the inputs to the second gate are both low, the output from the second gate is high. The second gate acts as a source for voltage.
Output	Because there is no voltage difference between V+ and the high output of the second NAND gate, there is no pressure to push the current. The LED remains off.

TABLE L22-2 Data Table

	System at Rest	System Active	
	PB unpushed for 1 minute	**PB pushed**	**Immediately upon release**
TP1			
TP2			
TP3*			
TP4 A (no LED)			
TP4 B (LED In)			

*What was the voltage to the inputs of the second NAND gate when they sensed a change from a high input to low input and the LED was turned off?

SECTION 7

Analog Switches for Digital Circuits

IS THERE REALLY GOLD at the end of a rainbow? That's *only wishing for power*. But the knowledge of how to use voltage dividers? THAT is POWER! There is real power in this knowledge! Knowing this gives you the power to control. Electronics is about control.

Lesson 23
Understanding Voltage Dividers

While playing with a potentiometer and one fixed resistor, we can adjust the pot's resistance to give us any voltage at the midpoint. We can use this flexibility to our advantage, but first, we have to understand what's happening so we can control it.

You know that voltage is used as it passes through resistor loads. The higher the value of the resistor, the greater the amount of voltage used. You already know, as well, that all the voltage is used from V+ to ground. So far, so good. Here, we use two resistors to give us any voltage we want at the midpoint. We divide the voltage. There is even a simple math formula to predict the outcome.

Here we will apply our knowledge to build switches that control the circuit. Some have moving parts, but others don't. The projects you will be building can use many types of switches. Here are some examples.

Figure L23-1 shows a simple motion detector. You can make many physical switches that work like a push button.

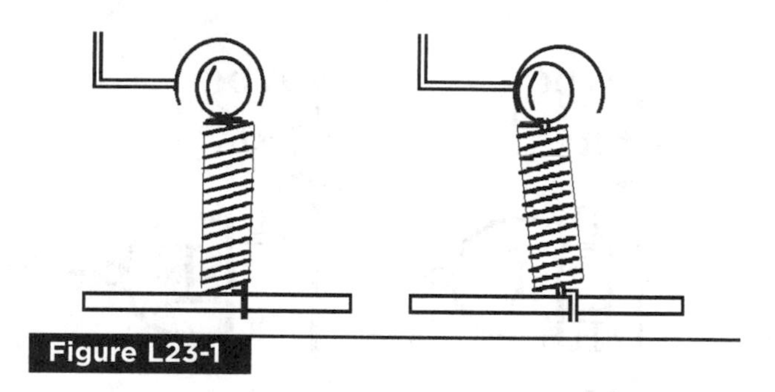

Figure L23-1

- It can be made small enough to be hidden in small boxes or cans.
- It can be made sensitive enough to trigger when a person walks by on a wooden floor.

Figure L23-2 uses what is referred to as a *break beam*. The sensor acts like a dark detector. It requires a source of light to keep the light-dependent resistor at a low resistance. If the beam of light is interrupted, the resistance increases. The changing voltage level triggers the inputs.

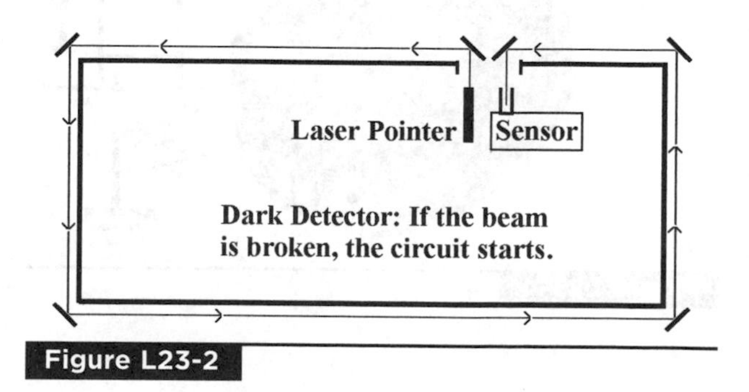

Figure L23-2

By trading positions of the resistor and LDR, you create a light detector. Such a device can be used to alert you when a car turns into your driveway. The headlights would trigger this unit.

At what voltage do digital inputs sense the change from high to low? Ideally, it would be right at half of the voltage supplied to the system.

That means that if V+ is 9 volts, the inputs sense the high-to-low change as V+ falls below 4.5 volts.

How do we get the inputs to change from high to low at the first gate? Right now, as you can see in Figure L23-3, you have the normally open PB to connect those inputs directly to ground when you push that plunger.

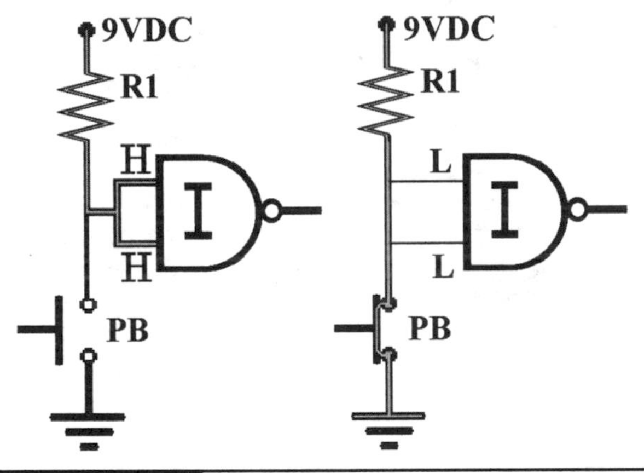

Figure 23-3

Modify the Circuit

Remember always to detach your power when making changes.

Make the three changes on your solderless breadboard that are shown in Figure L23-4.

1. Replace the PB with the trimming potentiometer.
2. Replace R1 with a 39-kilo-ohm resistor.
3. Remove C1. You remove the capacitor so that the circuit reacts instantly.

Now put the trim pot back into the circuit. Turn the trim pot backward until the LED goes off. Remove it again and measure the resistance at that point. Actually, it will be much less than 39 kilo-ohms. But this is an introduction to voltage dividers. Let's keep it simple.

What to Expect

1. Turn the trim potentiometer to full resistance.
2. Measure resistance between A and center. (B is unconnected.)
3. It should be near 100,000 ohms.
4. Attach the power.
5. The LED should be off.

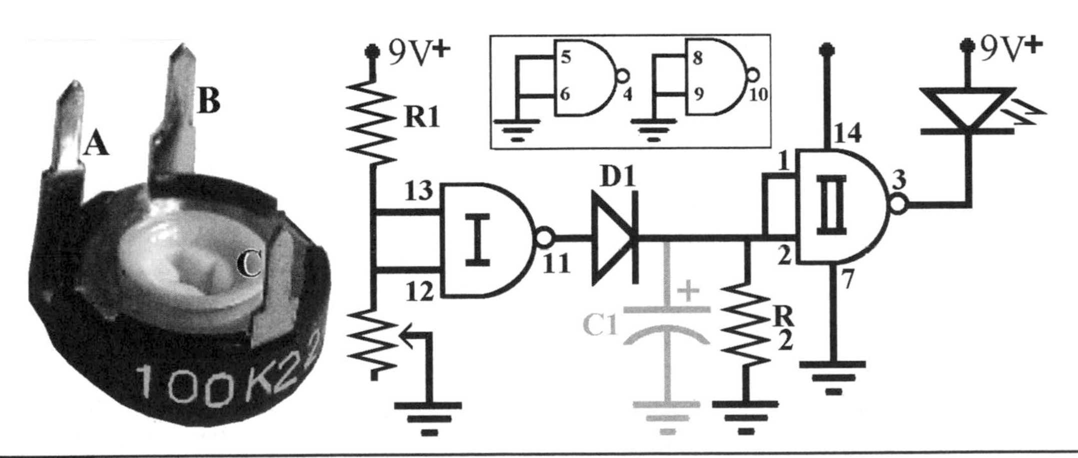

Figure L23-4

6. Adjust the trim pot until the LED goes on. It should stay on.
7. Disconnect the power; remove the trim pot.
8. Measure and note the resistance between A and center now.

How It Works

We can use different resistors and variable resistors to create changing voltages similar to what we did with the night light.

Remember the night light? This is for reference only. Don't rebuild it.

Think of how the night light worked. Use Figure L23-5 as a reference.

- The NPN transistor needed positive voltage to its base to turn on.
- The potentiometer adjusted the amount of voltage shared by the 22-kilo-ohm resistor and the LDR.
- In light, the LDR had a low resistance, allowing all of the voltage to flow through to ground. Because the base of Q1 got no voltage, the valve from C to E stayed closed.
- The resistance in the LDR increased as it got darker, providing more voltage to the base of the transistor, pushing the valve open.
- As the voltage flowed through the transistor, the LEDs turned on.

We can apply the same idea to the digital circuit inputs, as we can see in Figure L23-6.

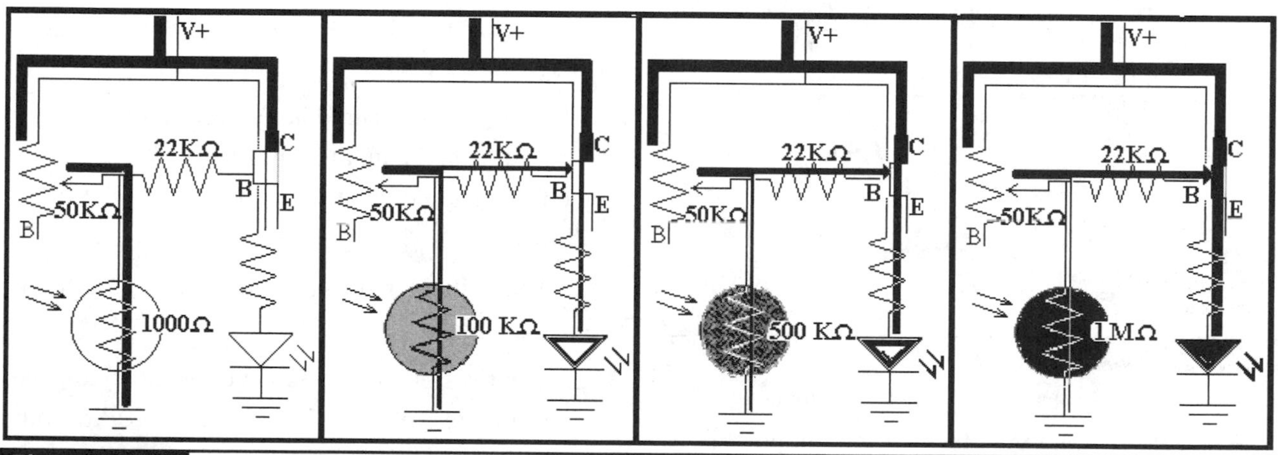

Figure L23-5

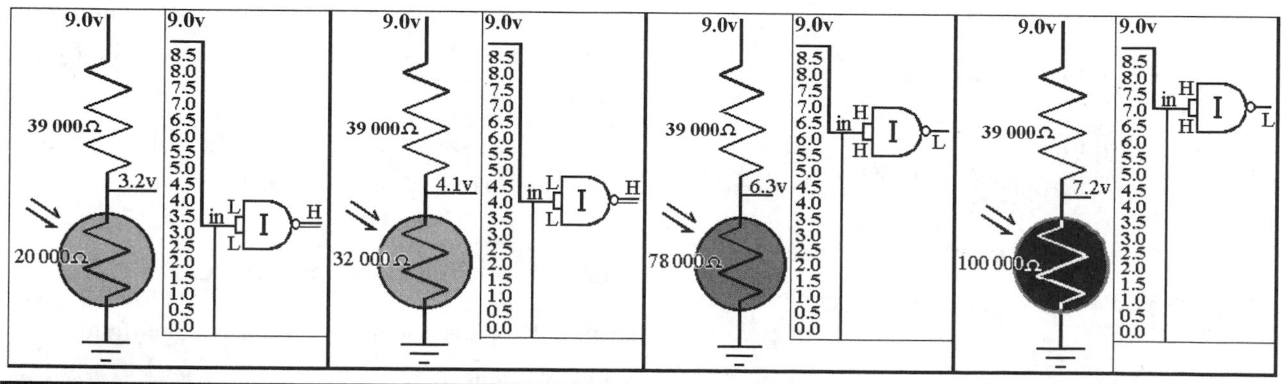

Figure L23-6

Remember: By simple definition, a circuit uses all of the voltage between the source and ground.

1. Two resistors set between voltage and ground use all of the voltage.
2. The first resistor uses some of the voltage, and the second uses the rest.
3. If you know the value of each resistor, you can figure the voltage used by each one using simple ratios. You compare the partial load to the whole load.

Look at Figure L23-7 as an example. Here R1 = R2.

R1 = 10 kΩ

R2 = 10 kΩ

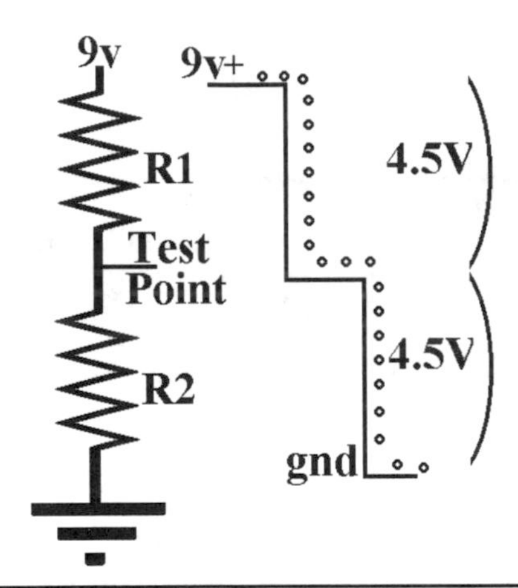

Figure L23-7

When R1 = R2, the voltage at the midpoint is exactly half of V+ because each resistor uses exactly half of the voltage.

$$V\left(\frac{R1}{R1 + R2}\right) = \text{voltage used}$$

$$9v\left(\frac{10\ k\Omega}{10\ k\Omega + 10\ k\Omega}\right) = 4.5v$$

This gives us a number of 4.5 volts used.

Why does the voltage split like this?

Simply because the resistor is a load. The larger the individual load when compared to the total load, the more voltage is used up. If there are two loads of equal value, they both use the same amount of voltage.

Build on Your Breadboard R1 = R2

Do not take apart your digital circuit.

Build the setup of two resistors on a separate spot of your solderless breadboard (SBB).

R1 = 10 kΩ

R2 = 10 kΩ

Measure and record the voltage at these points.

Voltage from V+ to ground_____ = total voltage.

Voltage across R1, from V+ to midpoint_____ = voltage used by R1.

Voltage across R2, from TP to ground_____ = voltage used by R2.

The voltage measured across R1 and R2 should be the same and equal to half of V+. It may be off by a few hundredths of a volt because of the following:

1. The voltmeter acts as a third load and affects the circuit.
2. The resistors have a range of accuracy of plus or minus 5 percent. That means a 10-kilo-ohm resistor could have a value of 9,500 ohms to 10,500 ohms.

The ideal statement for voltage at the midpoint when R1 = R2 is that the voltage is divided by half.

The load uses that portion of the voltage, in a ratio compared to the total load.

Build on Your Breadboard R1 > R2

What happens when we build a voltage divider of unequal parts?

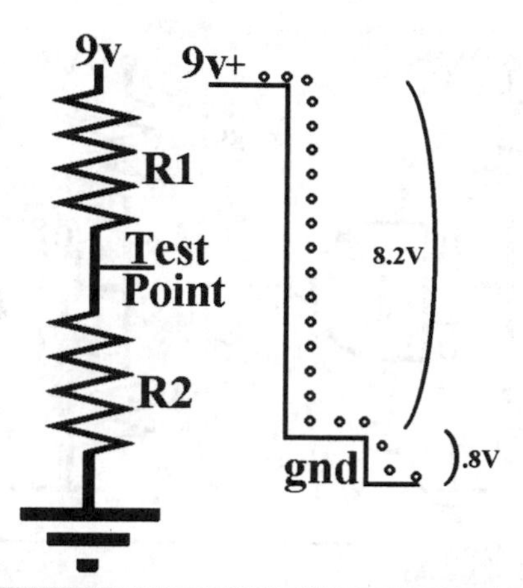

Figure L23-8

Here is what happens when R1 is 10 times the value of R2, as laid out in Figure L23-8. Replace R2 with a 1-kilo-ohm resistor.

If, for example:

V+ = 9.0 V

R1 = 10 kΩ

R2 = 1 kΩ

Again, figure the voltage used like this.

$$V_{total}\left(\frac{R1}{R1 + R2}\right) = V_{used}$$

9V (10 kΩ / 10 kΩ + 1kΩ) = 8.2V

But remember that the important information really is the voltage remaining.

R1 uses up 10/11 of the total voltage. That is 8.2 volts to be exact. The voltage left at the midpoint should be about 1/11 of the total because:

Total volts – used volts = remaining volts

11/11 – 10/11 = 1/11

9 V – 8.2 V = 0.8 V

It is important to be able to predict the voltage at the midpoint.

This allows you to trigger a digital circuit with whatever switch you create.

Exercise: Understanding Voltage Dividers

Predict the voltage in a voltage divider if you have exactly 9 volts.

Don't breadboard these voltage dividers.

Use the formula. And don't forget:

Total – Used = Remaining.

1. R1 = 1 kΩ

   R2 = 10 kΩ

   Midpoint V = ______

2. R1 = 100 Ω

   R2 = 1 kΩ

   Midpoint V = ______

3. R1 = 1 kΩ

   R2 = 100 Ω

   Midpoint V = ______

4. R1 = 39 kΩ

   R2 = 100 kΩ

   Midpoint V = ______

5. R1 = 39 kΩ

   R2 = 2.2 MΩ

   Midpoint V = ______

6. R1 = 2.2 MΩ

   R2 = 100 kΩ

   Midpoint V = ______

7. R1 = 100 kΩ

   R2 = 20 MΩ

   Midpoint V = ______

Lesson 24
Create a Light-Sensitive Switch

Here, the trim pot is replaced by an LDR to create a light-sensitive switch to trigger the NAND gate. It is a voltage divider that uses a light-dependent resistor as one of the loads.

Remove the last section's two-resistor voltage divider setup from the SBB so they are no longer in the way.

The inputs to the first gate are held high via connections through R1. The circuit is at rest. No, it is not off. The circuit is off only when the power is disconnected (Table L24-1).

TABLE L24-1 NAND Logic Table

Input A	Input B	Output
High	High	Low
High	Low	High
Low	High	High
Low	Low	High

Disconnect power when you change parts on the breadboard.

Refer to Figure L24-1 as we do a quick review of how the trim pot worked as a switch.

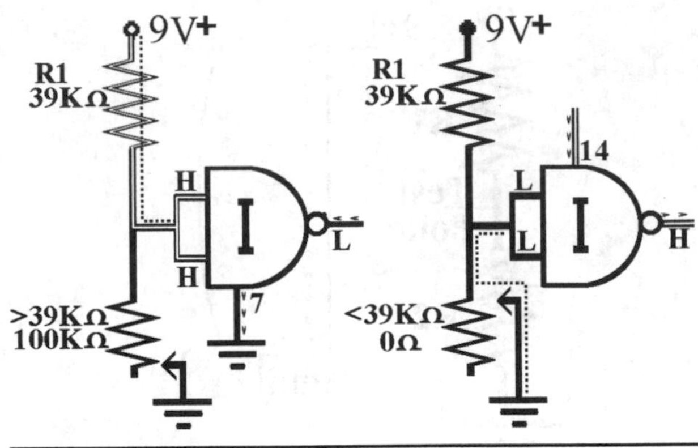

Figure L24-1

The 100-kilo-ohm trim pot replaced the normally open push button with a sliding resistance. As the trim pot changed resistance, the ratio of the voltage divider changed. Notice what happened to the output of the circuit as you adjusted the trim pot back and forth.

- As you increased the resistance, the voltage to the first NAND gate's inputs increased. It was harder for the voltage to reach ground because of the increased resistance of the trim pot.
- The inputs to the NAND gate were connected to voltage, giving a low output from the first gate.
- With the first gate inputs connected to a high, the system was at rest.

Modifying the Circuit: The Light Detector

The capacitor remains *out* (disconnected) for this lesson. Leaving it in will delay the change in output and confuse what you should see. Now try the other variable resistor, the LDR.

1. Remove the trim pot.
2. Place the LDR into your circuit as shown in Figure L24-2.

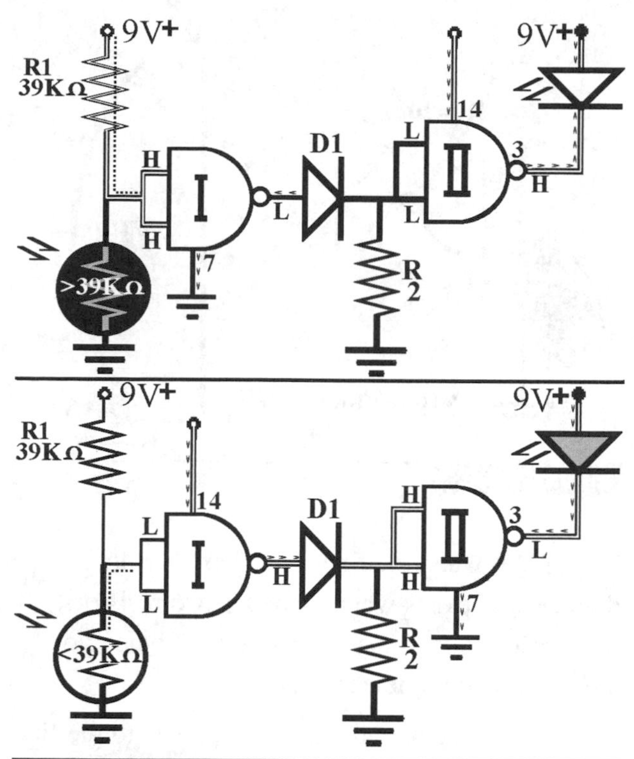

Figure 24-2

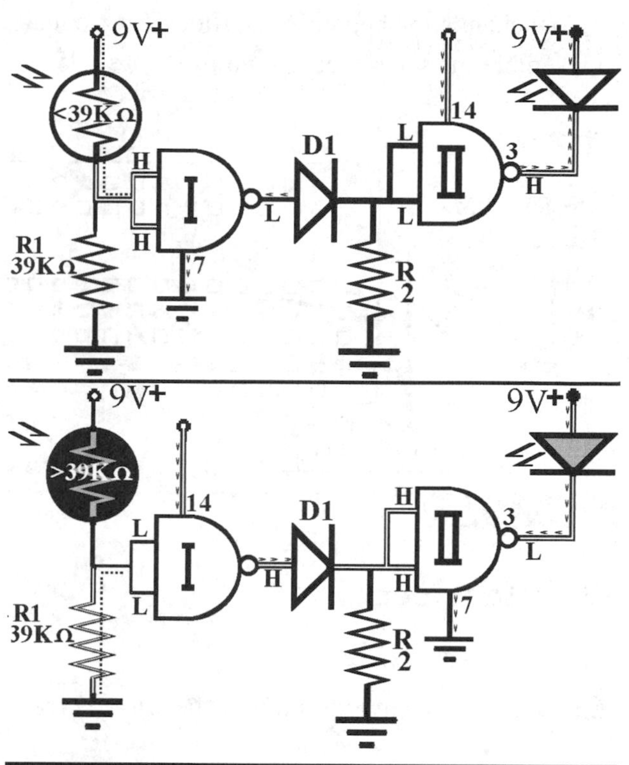

Figure 24-3

What to Expect

1. When you attach the battery, the LED should turn on instantly because this circuit is at rest in the dark.
2. Because this setup is *active* in the light, you will have to place the circuit into a dark situation for it to be at *rest*.
3. When the resistance of the LDR goes down in the light, the inputs to the first NAND gate sense that decrease in voltage. When it drops below half of the voltage from the power supply, they sense this as a low input.

Modifying the Circuit: The Dark Detector

Reverse the positions of R1 and the LDR as shown in Figure L24-3. This simple change creates a dark detector. It is at rest in the light, and becomes active if the resistance of the LDR goes above 39 kilo-ohms.

This setup needs to be in constant light to keep the circuit at rest. It will detect a break in a light source. If you put the circuit in the light and an object breaks the light source to the LDR, it will start the circuit. A common favorite is to put this switch onto a toy car. Its headlights turn on every time it goes under something.

Lesson 25
The Touch Switch

This switch depends on the fact that your skin has a resistance that is always between 100 kΩ and 2 MΩ, depending on how sweaty you are. We exploit this natural resistance in a voltage divider.

1. Set your digital multimeter to resistance.
2. Grasp a probe in each hand.

The resistance reading will constantly change, but should stay in the same range—somewhere between 100,000 ohms (100 kΩ) and 1,000,000 ohms (1 MΩ).

Now change the beginning portion of the circuit to resemble the schematic shown in Figure L25-1.

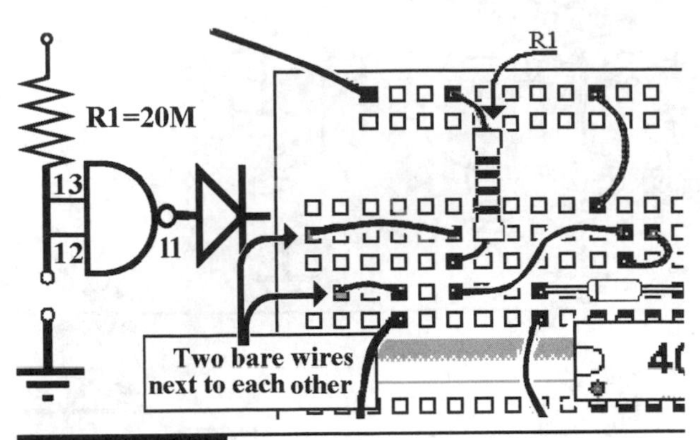

Figure L25-1

What to Expect

1. Attach your battery.
2. Touch your finger to both of the wires at the same time.

The LED turns on when your finger touches both wires that make the touch switch.

How It Works

Figure L25-2 demonstrates the effect of the finger's resistance when you become part of the circuit. The input pins 12 and 13 sense less than

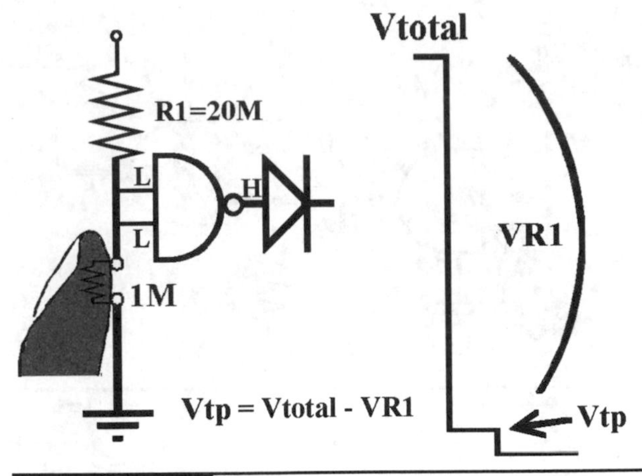

Figure 25-2

half of V+ when the finger touches. A voltage divider only exists when there are two resistors. So consider: Is there a voltage divider when you are not touching the contacts?

Figure the actual voltage at the inputs to the first NAND gate when the finger is touching. Assume you have a very dry finger and it has a resistance around 1 megohm.

$$V_{used} = V_{total} \times \left(\frac{R1}{R1 + R2} \right)$$

$$V_{total} - V_{used} = V_{midpoint}$$

SECTION 8

The NAND Gate Oscillator

HAVE YOU SEEN *THE WIZARD OF OZ*? Don't be frightened, *and don't* "ignore the man behind the curtain." Any technology that one does not understand is often interpreted as "magic!"

Knowledge, Design, Control

As you start to learn how to control digital inputs, you actually start to understand how some of the "whiz-bang" electronics around you actually work. Go out and buy a copy of a monthly electronics magazine. You will actually understand more than you expect. Remember, electronics is not hard—just lots of new information.

Lesson 26
Building the NAND Gate Oscillator

Here, just like the title says, you will incorporate the two unused NAND gates of the 4011 and build an extension onto your existing circuit. That extension will create a flashing output.

Add to your breadboarded circuit. Don't strip your breadboard.

Here you will get some dramatic changes by adding three basic components and changing some wiring to use the other two NAND gates.

Note what is needed in Figure L26-1 and the Parts Bin.

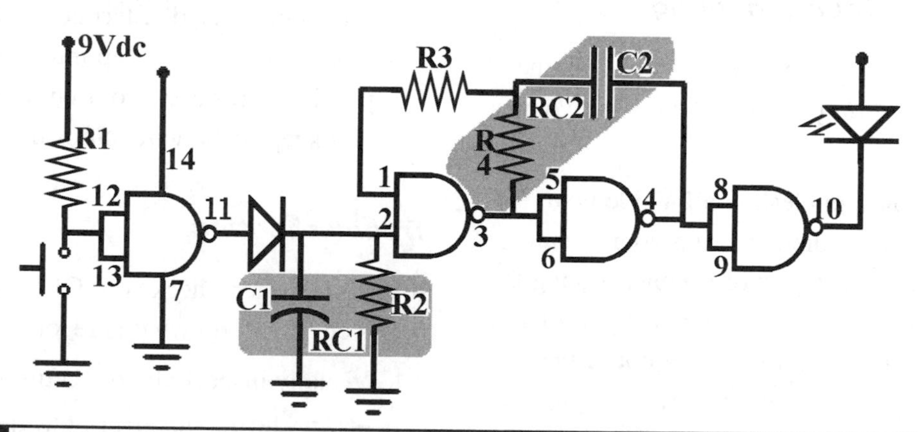

Figure L26-1 RC1 and RC2 are highlighted.

PARTS BIN

- R1—100 kΩ
- R2—10 MΩ
- R3—470 kΩ (new)
- R4—2.2 MΩ (new)
- C1—1 μF
- C2—1 μF (new)
- D1—4148 Signal diode
- LED—5 mm red
- IC1—4011 Quad NAND gate
- PB—Normally open

There are only four points connected to ground now. Make sure that inputs 5/6 and 8/9 are no longer connected to ground.

What to Expect

Press the button to make the circuit go "active." The LED should flash once a second for about eight seconds. It stops and automatically returns to its rest state.

If It Isn't Working: Problems and Troubleshooting

This troubleshooting guide will help you with the oscillator prototype on your breadboard and your finished project.

The intent of the troubleshooting guide is to help you locate the problem causing your circuit not to work. The hardest part of troubleshooting is finding the source of the problem. Once you find the cause of the problem, it's usually not difficult to fix it. I don't go into detail on how to fix it once you've located it. Upon closer inspection, that becomes self-evident.

Something to consider: If your LED is flashing faster than 24 frames per second, it looks like it is on steadily to you. *Did you know* that if the LED is blinking at 24 Hz or faster, your eyes tell you it is not blinking at all? That is why movies are shown at 26 frames per second. You are sitting in complete darkness for half of the time. You just can't notice it. Old silent movies are stuttering and jerky because they were often done at a rate just less than 24 frames per second. You can notice that.

There are usually four major problems that occur with this circuit.

1. Attach power, and the LED lights up but does not blink. Start at number 1 on the checklist below.
2. The LED blinks as soon as you attach the battery. Start at number 1 on the checklist below, but pay careful attention to the first gate. Something is triggering the inputs at pins 12 and 13.
3. The LED is off until you activate the circuit. The LED turns on but does not blink. It does time off properly. Problems at RC2. Check R3, R4, and C2 connections and values. Then start at 10. If you don't get half of V+, return to start at 1.
4. The LED is off and stays off. Is your power supply connected? Start at 1. Do not just insert a fresh 4011 chip into the circuit. If a physical error blew your chip, that same error will keep on blowing chips until you fix it.

Troubleshooting

If you find a step checks out OK, then move to the next step. If not, do what is recommended.

1. Visually inspect all connections.
   - All pins on the 4011 chip should be used. If you find an open pin, something is missing.
   - A wire left in from the previous setup could still be connecting pins 1 and 2.

- Pins 5 and 6 are connected, but the wire connecting these to ground needs to be removed.
- Pins 8 and 9 are connected, but the wire connecting these to ground needs to be removed.
- Make sure that none of the bare legs of the parts are touching at crossovers, creating a short circuit.

2. Look at all parts that have to be put in with polarity in mind. Positive must be toward V+ and negative toward ground.
   - Capacitors 1 μF and bigger
   - Chip
   - LEDs
   - Transistors (Lesson 32)
   - Diodes
   - Speaker (Lesson 29)
3. Check that the IC has power.
   - Note that V+ is being supplied to pin 14. Check that voltage is being supplied from the battery to the V+ line on the board.
   - Check that there is a wire connecting pin 14 to the V+ line.
4. Note that ground has only four connections. Check that there is continuity from the small button on the battery clip to the ground line.
   - Pin 7
   - R2
   - C1
   - Contact to the input switch
5. Here you are looking for short circuits in your wiring. This can also be caused by sloppy soldering.
   - Disconnect the power and replace the chip in the breadboard with a chip seat (empty socket).
   - Get a multimeter and start checking at points noted. Infinite or over limit means that there should be absolutely no connection between the two pins with the chip removed (see Tables L26-1 and L26-2).
   - Zero ohms means there is a direct connection.

TABLE L26-1 Measure the Resistance at Each Leg of the Chip with the Black Probe Connected to Ground

Red Probe at	Black Probe at	Expected Resistance
Pin 1	Pin 7	Infinite
Pin 2	Pin 7	Value of R2
Pin 3	Pin 7	Infinite
Pin 4	Pin 7	Infinite
Pin 5	Pin 7	Infinite
Pin 6	Pin 7	Infinite
Pin 8	Pin 7	Infinite
Pin 9	Pin 7	Infinite
Pin 10	Pin 7	Depends on the output; disconnect the output, and it should be infinite
Pin 11	Pin 7	Infinite
Pin 12	Pin 7	Infinite
Pin 13	Pin 7	Infinite
Pin 14	Pin 7	Infinite

6. Replace R1 with 100 kilo-ohms (20 megohms is too sensitive and will start the circuit).
   - With the battery connected, check that the voltage at pins 12 and 13 is well above half voltage when the switch is "open."
   - Check that the voltage at pins 12 and 13 is well below half voltage when the switch is "closed."

TABLE L26-2 Measure the Resistance from Each Pin to the Next; the Chart Assumes that RC2 Oscillator Is Installed

Red Probe at	Black Probe at	Expected Resistance
Pin 1	Pin 2	Infinite
Pin 2	Pin 3	Infinite
Pin 3	Pin 4	Infinite
Pin 4	Pin 5	Infinite
Pin 5	Pin 6	0 Ω
Pin 6	Pin 7	Infinite
Pin 7	Pin 8	Infinite
Pin 8	Pin 9	0 Ω
Pin 9	Pin 10	Infinite
Pin 10	Pin 11	Infinite
Pin 11	Pin 12	Infinite
Pin 12	Pin 13	0 Ω
Pin 13	Pin 14	Value of R1
Pin 14	Pin 1	Infinite

7. With the battery connected, check the voltage at pin 11 when the switch is "open." It should read 0.0 volts (low).
   - Close the switch and check the voltage at pin 11 when the switch is "held closed." It should be V+ (HI).
   - If pin 11 does not respond properly, either the gate is burnt out or pin 11 is accidentally connected to ground or somewhere else.
8. With the battery connected, check the voltage at pin 2 (RC1) when the switch is "open." It should be sinking toward 0 volts.
   - Check the voltage at pin 2 (RC1) when the switch is "held closed." It should be up at full voltage.
   - If RC1 does not fill, check the value of R2. Also, check if diode 1 is in the right way. Then replace D1 with a power diode 1N4005. The signal diode might have burnt out. Also, check for accidental connections to ground or somewhere else.
9. When pin 2 is low, pin 3 should be high. Conversely, when pin 2 is high, pin 3 should be oscillating.
   - Use a multimeter to check if the oscillator is working at pin 3.
   - If RC2 is set for slow pulse of 2 Hz or slower, the reading will swing from V+ to 0 volts.
   - If RC2 is set for a faster frequency, the reading will stay at half of V+.
   - For example, if V+ is 9 volts, the meter will read 4.5 volts because it will average the voltage swings between 9 and 0 volts.
10. The output at pin 3 should be directly connected to pins 5 and 6. The reading at pins 5 and 6 will be identical to the reading at pin 3.
11. The output of pins 5 and 6 is at pin 4. Check to see that the gate is working.
12. The inputs to the fourth gate, pins 8 and 9, are connected directly to pin 4. Check to see that the gate is working.
13. Is your output device working?
   - Is an LED burnt out? Test them singly in a 9-volt system with a 470-ohm resistor.
   - Perhaps your speaker is broken. Check continuity on the speaker wire.
   - Is your transistor the correct value? Maybe it is burnt out. Table L51-2 guides you through testing of transistors.

Lesson 27
Understanding the NAND Gate Oscillator

Table L27-1 describes the present system. It is an even closer look at how the NAND gate works. The NAND gate oscillator is widely used because it can be tuned easily using an RC. Watch for the new vocabulary. Master the material now or be a slave later.

TABLE L27-1 The System Diagram as It Exists Now

Input	Processors	Output
Push button	RC1-Push on/ timed off about 10 per second	LED flashing once per second
	RC2-NAND gate RCM oscillator	
	At 1 flash per second	

Recall the logic table for the NAND gate (Table L27-2).

In review, recall what happens at RC1. Gate 1 is used to start the RC1. RC1's C1 fills and drains through R2. These control the time the circuit stays on.

- Gate 1 output goes high.

TABLE L27-2 Logic Table for the NAND Gate

Input A (Pin 2)	Input B (Pin 1)	Output (Pin 3)
High	High	Low
High	Low	High
Low	High	High
Low	Low	High

- D1 traps the voltage on RC1 side.
- C1 fills.
- R2 drains the voltage.

The action of an RC circuit is always the same. The only difference is the speed that the circuit fills or drains. Figure L27-1 reviews that basic action.

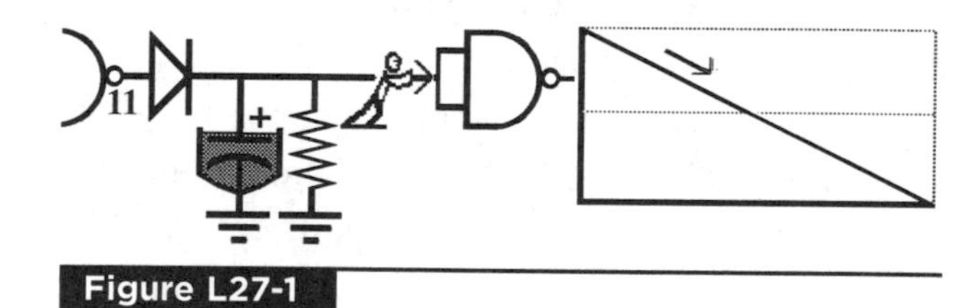

Figure L27-1

But we are interested now in RC2. RC2 is made of C2 and R4. They use gate 2 to make an oscillator. The oscillation action happens at gate 2.

Look at the setup of gate 2 shown in Figure L27-2. The high or low state of pin 3 determines

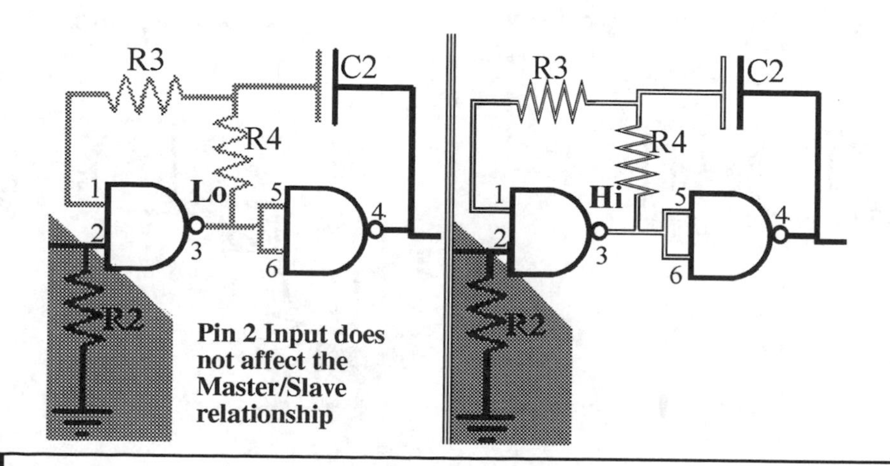

Figure L27-2

TABLE L27-3 System at Rest

Time (Seconds)	Input A at Pin 2	Input B at Pin 1	Output at Pin 3
The system remains at rest until pin 2 changes state when RC1 gets charged.		Slaved to pin 3	As long as one input is low, the output is high.
1	Low		High
2	Low	High	High
3	Low	High	High
4	Low	High	High

the state of pin 1. Pins 1 and 3 have a special relationship. Pin 3 is the master; pin 1 is the slave.

Table 27-3 shows the system at rest. There is no activity or changing voltage values happening within the circuit.

So pin 1 is a slave to pin 3. When the system is at rest, C1 has less than half of V+. Because pin 2 is low, pin 3 gives a high output. Note the NAND gate's logic in Table L27-2. In fact, if either input is low, the output is high. This is shown plainly in Figure L27-3.

But what happens when the circuit becomes active? Figure L27-4 clearly shows that when pin 2 goes high, because the capacitor is charged, pin 3 goes low.

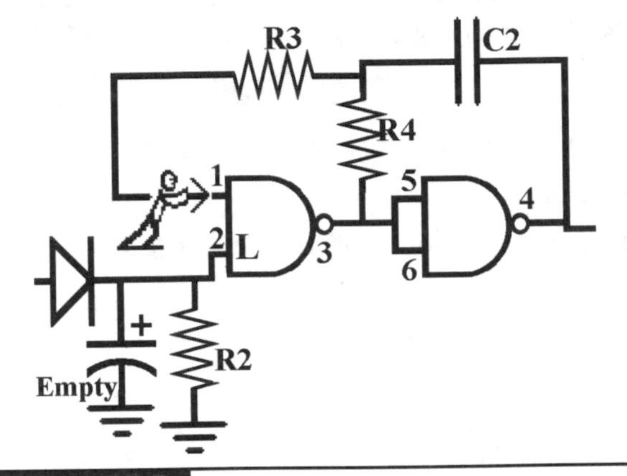

Figure L27-3

It takes a fraction of a second for the input at pin 1 to respond because C2 has to drain. Once it has drained, the voltage at pin 1 matches the output of pin 3. Wait!? Pin 3 is now low. So pin 3 makes pin 1 low, but pin 2 is high. One of the

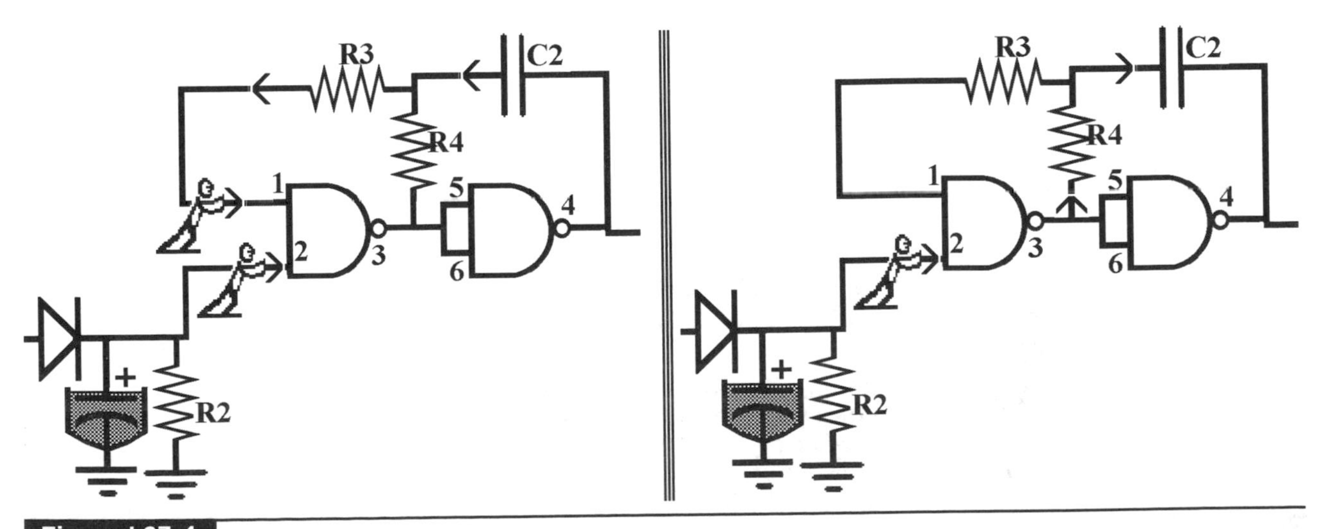

Figure L27-4

TABLE L27-4 Series of Actions

Voltage in RC1	System State	Time	Input A Pin 2	Input B Pin 1	Output Pin 3
0	Rest	0	Low		High
0	Rest	0	Low	High	High
0	Rest	0	Low	High	High
9	Active	1	High	High	Low
8.5	Active	2	High	Low	High
8.0	Active	3	High	High	Low
7.5	Active	4	High	Low	High
7.0	Active	5	High	High	Low
6.5	Active	6	High	Low	High
6.0	Active	7	High	High	Low
5.5	Active	8	High	Low	High
5.0	Active	9	High	High	Low
4.5	Active	10	High	Low	High
4.0	Rest	11	Low	High	High
3.5	Rest	12	Low	High	High
3.0	Rest	13	Low	High	High

inputs is low, which makes the output at pin 3 go high. This is better than a puppy chasing its tail.

The series of actions is actually laid out very neatly in Table L27-4.

The system starts at rest. When the system becomes active, oscillating start.

Animated version of this table is at www.mhprofessional.com/computingdownload. As you can see, the feedback loop at gate 2 creates the oscillation! All animated graphics are available online and are listed in Appendix C.

SECTION 9

How Do We Understand What We Can't See?

EVEN WHEN WE CAN'T SEE what's happening, we can predict, measure, and visualize.

Lesson 28
Controlling the Flash Rate

This lesson explains how it is done and you get to do exactly that. You learn the relationship between the RC values and the frequency output.

The values of C2 and R4 that make RC2 determines the rate of oscillation. The rate of oscillation is properly called *hertz* (Hz). Hertz is frequency per second. Another way of saying this is how many beats per second. It is a standard unit.

How It Works

First, an explanation of how the second resistor/capacitor circuit (RC2) works. Then we'll play with it. This is not an exact representation of the circuit, but it will help you understand what is really happening.

Figure L28-1 shows your original RC2 setup where C2 = .1 μF and R4 = 2.2 megohms. R4 is represented by the pipeline feeding the capacitor.

Pin 2 is low. The system is at rest and stable.

The high output from pin 3 is defined by the inputs. The capacitor C2 is fully charged. There is no place for it to empty.

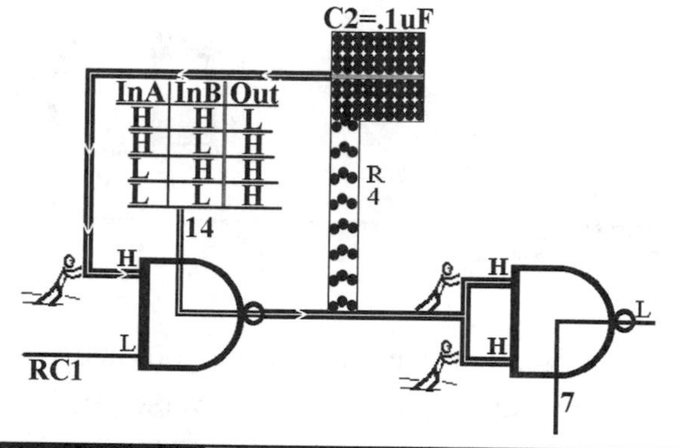

Figure L28-1

When RC1 gets charged, the system becomes active. You recognize that the high inputs at pin 1 and pin 2 create a low output. The low output at pin 1 allows the charge held in C2 to begin draining. It does so at a speed determined by the size of R4. Figure L28-2 shows the action of drainage from C2. As long as the voltage is above that magical half of the voltage mark, pin 1 sees its input as high.

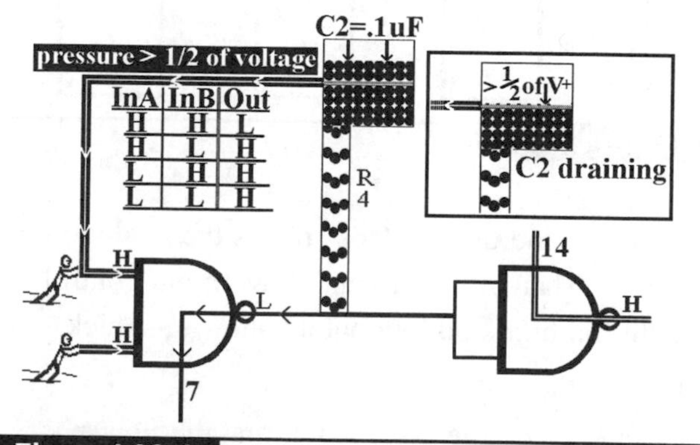

Figure L28-2

But as soon as the charge in C2 drops below a certain point, the input to pin 1 senses that input as low. HMM? Pin 2 is still high. NAND gate logic demands that the output at pin 3 become high, and C2 starts to fill (see Figure L28-3).

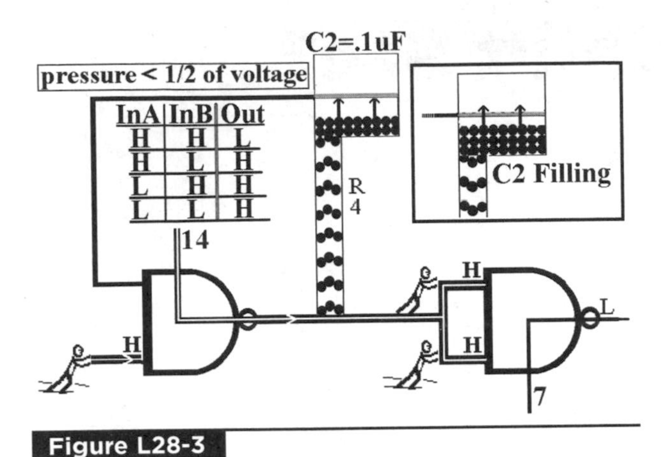

Figure L28-3

Of course, this continues until it goes above that magical marker, when the action reverses again. RC1 may be set for 10 seconds. RC2 might be set for 1 Hz. So by the time pin 2 goes low again, RC2 will have filled and drained 10 times.

RC2's rate of voltage charge/discharge is charted and shown in Figure L28-4.

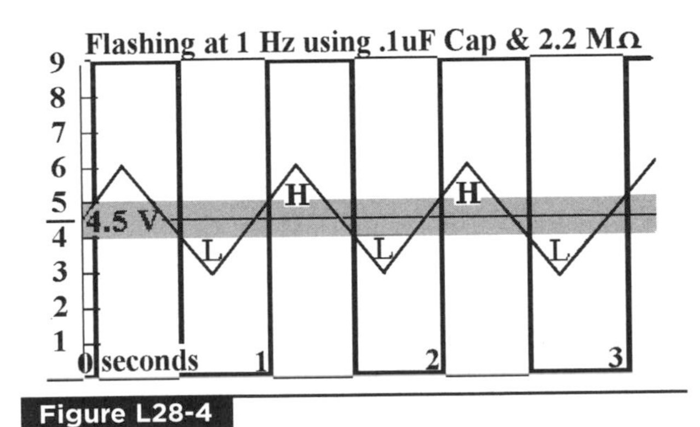

Figure L28-4

C2 fills and drains. This creates the analog input to pin 1. That sliding up and down input controls the digital high and low output shown as thick square waves.

I have made the assumption that the power supply is a convenient 9 volts. That makes the half-voltage mark 4.5 volts. Notice the grayed area around the half-voltage mark. In simplifying the explanation, I have referred to the magical point of half of V+. That's not quite true. There is a bit more of a range. If the voltage is moving upward, it has to rise above about 5 volts to be sensed as high. If the voltage is dropping, it has to drop below nearly 4 volts before it is registered by the inputs as low.

Modifying the Circuit

Make sure your power is disconnected.

Now replace C2 with a 0.01-μF capacitor. Use the DMM to check the capacitance. Ideally, the capacitor you have is marked the same way as the disk capacitor shown in Figure L28-5. There is no standard for marking capacitors. There are several generally accepted methods. You can expect to see the marking 103Z. That refers to 10 followed by three zeros. In other words, 10,000. Disk capacitors are measured in picofarads. That is a millionth of a microfarad. That is a thousandth of a nanofarad. 10,000 pF is 10 nF is 0.01 μF.

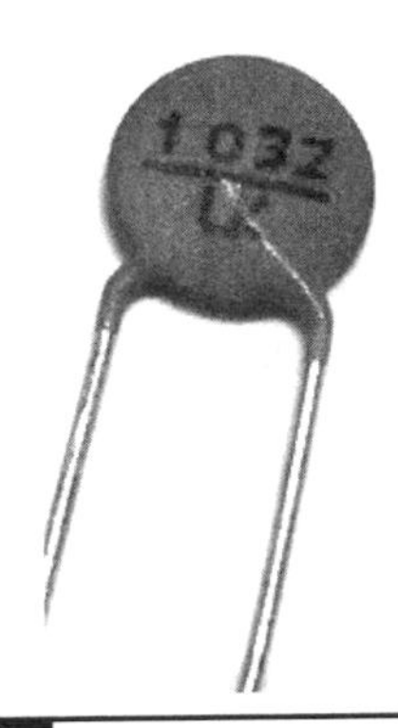

Figure L28-5

Or it might be marked with 0.01 or even u01. This refers to 0.01 μF. The label u01 uses the value marker as a decimal marker as well. Face it, there's not much space.

This capacitor is 10 times smaller than the one you have in the circuit right now. Capacitors this small do not have any polarity. There are no positive or negative legs. R4 is unchanged at 2.2 megohms.

Connect your power supply.

Notice the new setup shown in Figure L28-6.

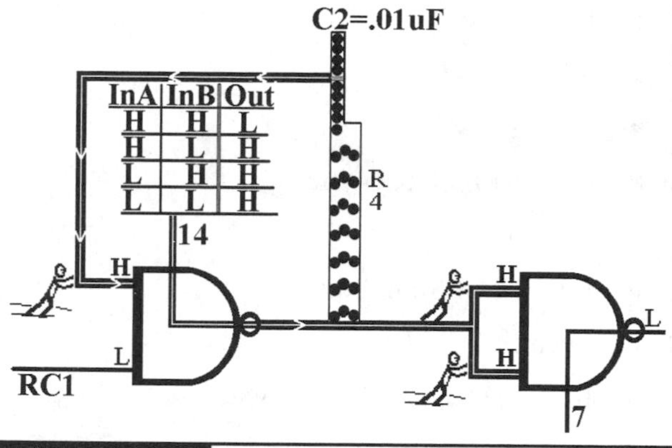

Figure L28-6

The system is at rest, but C2 is represented as a tenth the size as before. So what do you expect will happen?

The LED should flash very quickly for about 10 seconds, depending on your timing for RC1. Figure L28-7 shows the reaction by the NAND gate to the changing voltages on pin 1.

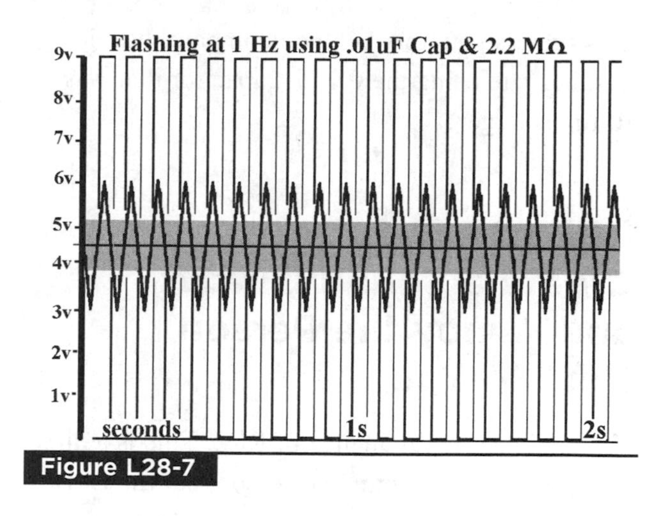

Figure L28-7

Ideally, it reacts exactly 10 times faster because the capacitor is 10 times smaller.

Exercise: Controlling the Flash Rate

Pull R2, the resistor, from RC1. That way you can count without worrying about the circuit timing out at the wrong time.

On your solderless breadboard, you will change components to affect the oscillation timing of RC2. Track your results in Table L28-1.

TABLE L28-1 Tracking Table

R4	C2	Comment	Timing Flashes in 10 s 1	2	3	Average
1 MΩ	0.1 μF					
2.2 MΩ	0.1 μF	Twice the resistance Expect half the rate				
4.7 MΩ	0.1 μF	Twice the resistance Expect half the rate				
10 MΩ	0.1 μF	Twice the resistance Expect half the rate				
10 MΩ	0.01 μF	Tenth the capacity Expect 10 times faster flash rate				
4.7 MΩ	0.01 μF	Half the resistance Expect twice as fast				
2.2 MΩ	0.01 μF	Half the resistance Expect twice as fast				
1 MΩ	0.01 μF	Half the resistance Expect twice as fast				

Is there a pattern when you compare it to the flashing rate using the 0.1 μF capacitor that was 10 times larger?

Lesson 29
Create a Sound Output and Annoy the Person Next to You

You adjusted the frequency of RC2. This is a direct continuation of the previous lesson, but the LED output has a flash rate too fast to see. Did you know that people see smooth motion if related pictures are presented at 24 frames per second? That is why movies are projected onto a screen at that rate. That is also why we have to move from the LED to a speaker. When the LED is 24 frames a second or faster, it might appear to dim a little, but you won't see it flash. Why does it dim? Because it is off half the time. Don't you realize you're sitting in complete darkness in the movies for half the time, too?

Modifying Your Circuit

Don't clear your breadboard. Figure L29-1 shows the schematic that you have been using (see the Parts Bin). Just replace the LED with the speaker and change the values of other components stated in the parts list.

PARTS BIN

- R1—20 MΩ
- R2—10 MΩ
- R3—470 kΩ
- R4—2.2 MΩ
- C1—1 μF electrolytic
- C2—0.1 μF
- D1—4148 Signal diode
- Speaker—8 Ω
- IC—4011 Quad NAND gate

Also, if you removed R1 for the exercise in Lesson 28, put it back in.

Disconnect your power to make these changes.

Going from left to right on the schematic.

1. You have a touch switch to activate the circuit.
2. The amount of time the circuit stays active is set by R2 and C1. R2 and C1 make the first resistor/capacitor circuit (RC1).
3. The rate of oscillation is determined by C2 and R4 (RC2).
4. Here, the voltage from pin 10 moves from V+ (high) to ground (low) at a frequency set by RC2.

Be cautious. Don't connect your speaker directly to a battery or the power supply. Small speakers are made with very fine wire.

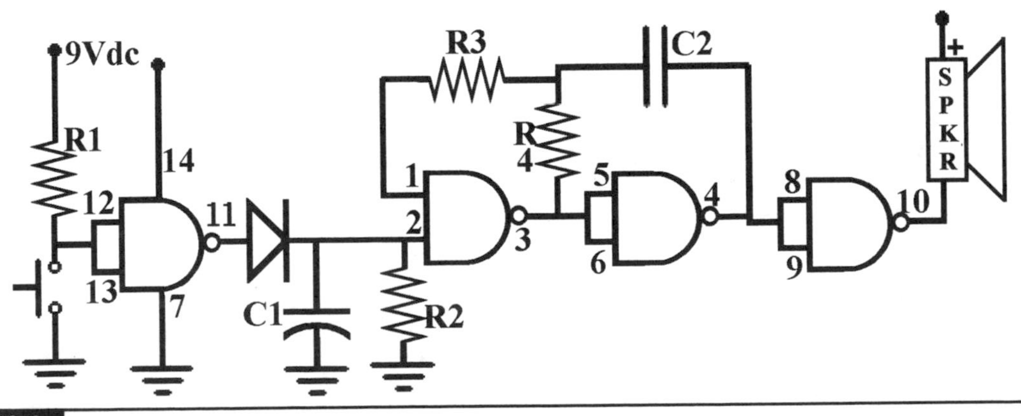

Figure L29-1

Too much current will heat the wire, possibly enough to melt it. Such a break would render the speaker useless.

5. Speakers respond only to voltage changes. Speakers do not produce sound just because V+ is applied to them. Buzzers have a circuit inside. They create their own noise. If you put your speaker to voltage, you hear a "crackle" as you connect and another as you disconnect. It is merely responding to changes in voltage. An excellent explanation about how a speaker works is posted at www.howstuffworks.com/speaker1.htm.

6. Replace the LED with the 8-ohm speaker. Note the polarity of the speaker.

   The speaker will click slowly and very quietly. You may have to use your fingers to feel the pulse. It should pulse about 15 to 20 times in five seconds.

   The speaker pulses each time the current is turned on, moving from low to high, and each time the current is turned off, moving from high to low. In the previous exercise you pulsed the LED at different speeds by changing the rate of oscillation in RC2.

Your exercise results should have shown this pattern (see Table L29-1).

Decreasing the resistance is like widening the drain. Figure L29-2 shows that by decreasing the resistor value we increase the oscillation speed because it takes less time to fill and drain the capacitor.

Exercise: Create an Annoying Sound Output

A quick definition. *Hertz* is a measurement of frequency, specifically defined as a measure of beats per second. For example, a system oscillating at 512 beats per second is more easily stated as 512 hertz.

As you do the following changes, note your observations in Table L29-2. Remember to detach power any time you make a change to your prototype on your breadboard.

TABLE L29-2 Observations

Resistor Value R4	Capacitor Value C2	Description
4.7 MΩ	0.01 μF	
2.2 MΩ	0.01 μF	
1.0 MΩ	0.01 μF	
470 kΩ	0.01 μF	
220 kΩ	0.01 μF	
100 kΩ	0.01 μF	
47 kΩ	0.01 μF	500 Hz
Now, one more change. Put your ear very close to the speaker. Listen for the quiet tone.		
22 kΩ	0.01 μF	1,000 Hz

It's annoying, but very quiet right now because the 4011 IC does not produce very much power at the output. So the volume is not much at all. But Figure L29-2 offers hope.

TABLE L29-1 Exercise Results

Resistor Values	Capacitor Values	Flashes in 10 s	Speed
10 MΩ	0.1 μF	1	Very slow
4.7 MΩ	0.1 μF	2	Double of previous
2.2 MΩ	0.1 μF	4	Doubled again
1 MΩ	0.1 μF	10	10 times faster than the 10 MΩ

You want volume??  **OK!**

But first, a good oscillator deserves an oscillosope.

The Lesson after that,
You Will Learn How To
Amplify That Output

Figure L29-2

Lesson 30
Introducing the Oscilloscope

This lesson:

- Introduces one of the most important tools in electronics
- Introduces you to the concept of what any oscilloscope can do
- Shows you how to build a probe to use with the Soundcard Scope
- Acts as an introduction to using the popular Soundcard Scope freeware

If you have an oscilloscope sitting on your desk at home, you are unique. If you have access to an oscilloscope, you are special.

Otherwise, I recommend the Soundcard Scope. You can download it from http://www.zeitnitz.de/Christian/scope_en. It was created Christian Zeitnitz, professor of physics at Wuppertal University.

Think of the Soundcard Scope as a special "skin" for your sound card. Different skins allow for different adjustable visual effects that show on your monitor as music plays. Soundcard Scope takes this idea forward a step. The Soundcard Scope skin reacts to the internal signals as well as "microphone" or "line" inputs to your sound card.

In the real world, we expect that as price increases, complexity and quality should increase. Soundcard Scope, however, is priceless. It still includes all of the major functions common to all

Disclaimer

Safety measures when using Soundcard Scope *and* the Soundcard Scope Probe:

The Soundcard Scope Probe is designed to be used only with your 9-volt systems and the proper probe. Make sure you follow testing procedures before you use it. Instructions for building the probe are given in this lesson.

Being software, Soundcard Scope itself cannot damage your hardware, but it is very easy to burn out at least your sound card when trying to investigate signals of unknown amplitude and DC offset.

So, you must always be extremely careful when establishing an electrical connection between your computer and external equipment. It is a good practice to use a conventional multimeter or real oscilloscope to find out whether signal levels are acceptable for your sound card.

Regarding connecting to things besides circuits produced in this book, it should be safe to connect to any audio/video equipment using standard line in jacks and cables. You may consider at least using the Scope Probe to decrease the voltage input. Make sure you have a stable signal source that allows you to control signal level manually, outside of the computer.

To avoid personal injury, always follow the usual safety rules when working with electric circuits.

SOUNDCARD SCOPE IS SUPPLIED TO YOU AS IS, AND IN NO CASE IS THE AUTHOR OF SOUNDCARD SCOPE OR THE ASSOCIATED PROBE RESPONSIBLE FOR PERSONAL INJURY, HARDWARE, AND/OR DATA DAMAGE, PROPERTY DAMAGE, OR PROFIT/LOSS ARISING FROM USE OR INABILITY TO USE THIS OSCILLOSCOPE SOFTWARE.

THE AUTHOR/DESIGNER DOES NOT GUARANTEE THE FITNESS OF SOUNDCARD SCOPE FOR ANY PARTICULAR PURPOSE. SOUNDCARD SCOPE IS NOT INTENDED FOR INDUSTRIAL OR COMMERCIAL USE.

IN GENERAL, USE SOUNDCARD SCOPE AND THE ASSOCIATED PROBE AT YOUR OWN RISK. ALWAYS CONSULT YOUR SOUND CARD MANUAL FOR DETAILS ON CONNECTING TO EXTERNAL DEVICES.

oscilloscopes, plus some added features. Its quality is adequate for our immediate needs. *One of its limitations is that it ignores stable DC voltage.* But it does respond to changing voltages in the audio frequencies 40 Hz to 15,000 Hz. Its biggest advantage is the price.

NOTE **For this particular oscilloscope software, the input is defined by Windows "Master Volume" sound mixer. The Soundcard Scope software does not communicate directly with the sound card. With this in mind, any Soundcard Scope problems should be dealt with through Windows' own sound card system. Check page 8 of the scope's manual.**

As the Digital Revolution became the digital standard, oscilloscopes changed too. The best-quality scopes still cost in excess of $20,000. An adequate beginner's scope would be a two-channel Velman handheld scope or external USB hardware, starting around $150. For classrooms, $2000 for a good quality scope is not unreasonable. Reconditioned analog equipment is an option frequently overlooked. Equipment that was "top of the line" 20 years ago can be purchased at reasonable cost. These older tools give quality outputs comparable to equipment that would cost over ten times as much new.

Your DMM measures voltage. It has no time component.

Oscilloscopes are used to give instant visualization of voltage compared to time. Its input clips to the system's output and displays a graph of the voltage represented on the Y axis (vertical) and time shown across the screen horizontally as the X axis. You are able to adjust the scales.

Oscilloscopes specifically create pictures of voltage changes over time. A good scope can show events that occur at a snail's pace, or it can freeze a frame to display microsecond variations.

Initial Walkabout

Enough of me blathering on. If you do not have alternative hardware, download, install, and start the Soundcard Scope now. At startup, the Soundcard Scope screen looks like Figure L30-1.

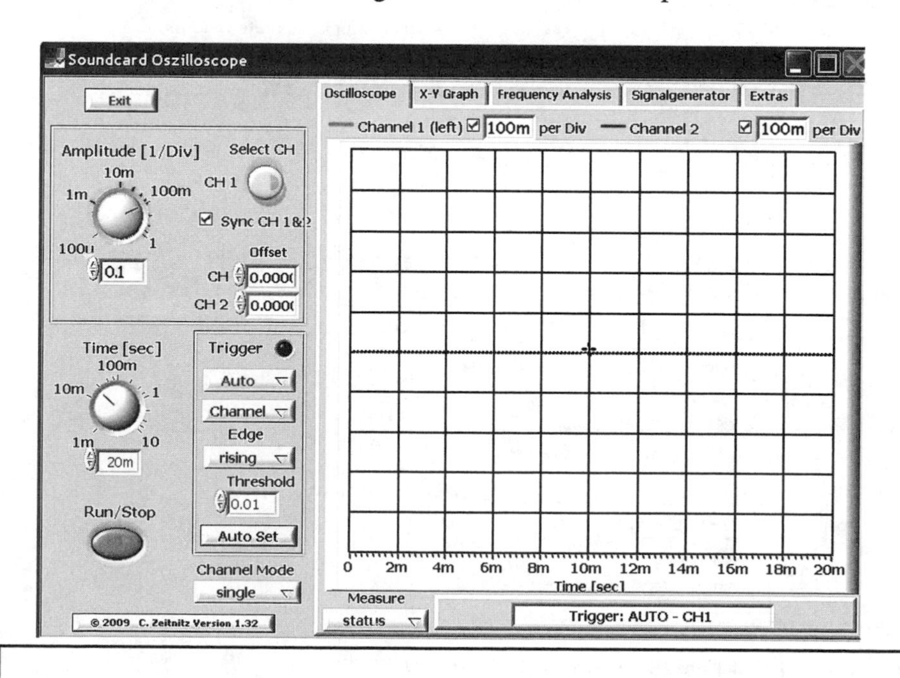

Figure L30-1

Be aware that this entire tool is not even a good doorstop unless you have a signal.

The Soundcard Scope has some valuable features that we are going to use immediately.

Open and adjust your master volume control to match Figure L30-2, and then close it.

Click the Signal Generator tab. Adjust the frequencies as shown in Figure L30-3.

Return to the oscilloscope screen. The left side of Figure L30-4 shows what to expect. It should be close, so don't worry if it's not exactly what you see. The signal might be too big to fit on the display screen. Don't change the signal size with the master volume control. Adjust the scale of the graph by inserting the number **.2** as indicated by #1 in the Amplitude control area. You can mouse over the knob on any of these to control them. Your display should now be similar to the right

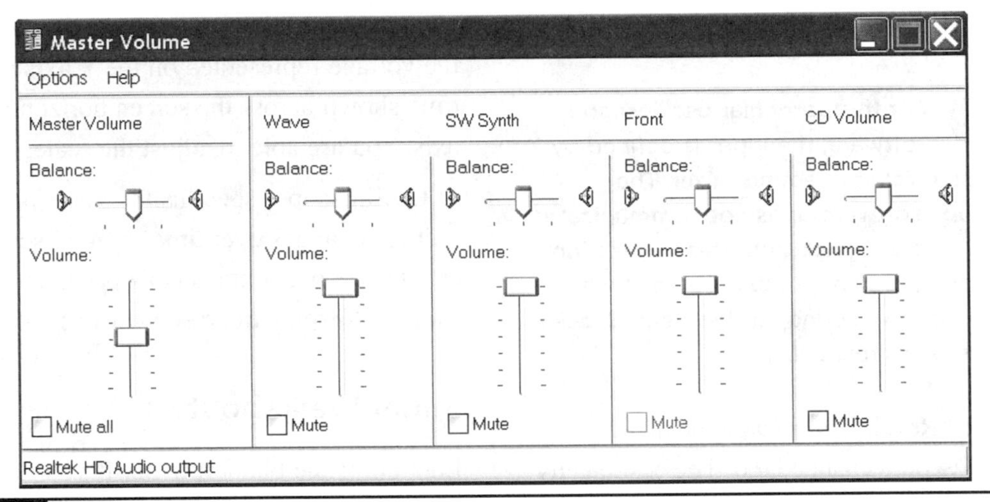

Figure L30-2

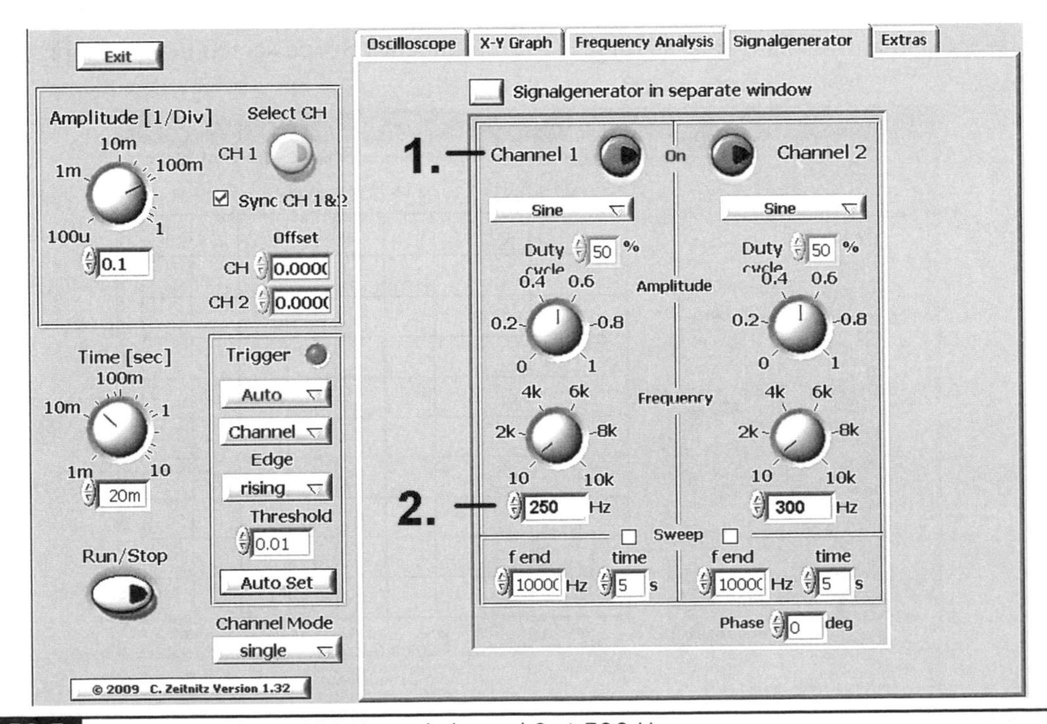

Figure L30-3 Set channel 1 as 250 Hz and channel 2 at 300 Hz.

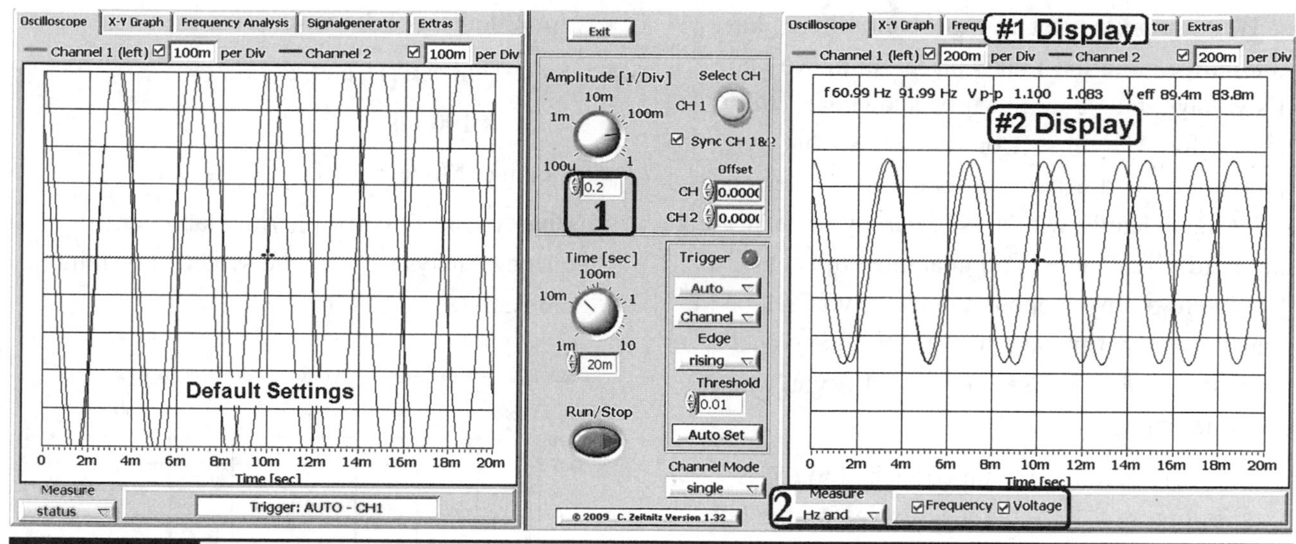

Figure L30-4

side of Figure L30-4. Just like on a piece of paper, the information in the graph doesn't change—just the scale. The notations in #1's display indicate that each division has doubled its value.

The more information you have, the better you can understand what is happening. Click the STATUS button under Measure, as indicated by #2 on Figure L30-4. Choose the Hz and Volts option. Check the Frequency and Voltage boxes that pop up next to it. An information bar appears at #2's display. The fact that the readout should be 250 and 300, respectively, doesn't matter. The first number is cut off because of limited space. The fact that the reading is off by .01 is irrelevant.

Wouldn't it be nice to play with one signal at a time? Again, it's not hard if you know how. In Figure L30-5, #1 shows the box next to Sync CH 1 & 2 has been deselected. Just above that, at #2, click the Select CH button, activating only CH 1. At #3, you can shift the amplitude of Channel 1 to further distinguish it from Channel 2's signal. Change its scale to .4 per division as noted by #3. Now you have the left screen of Figure L30-5.

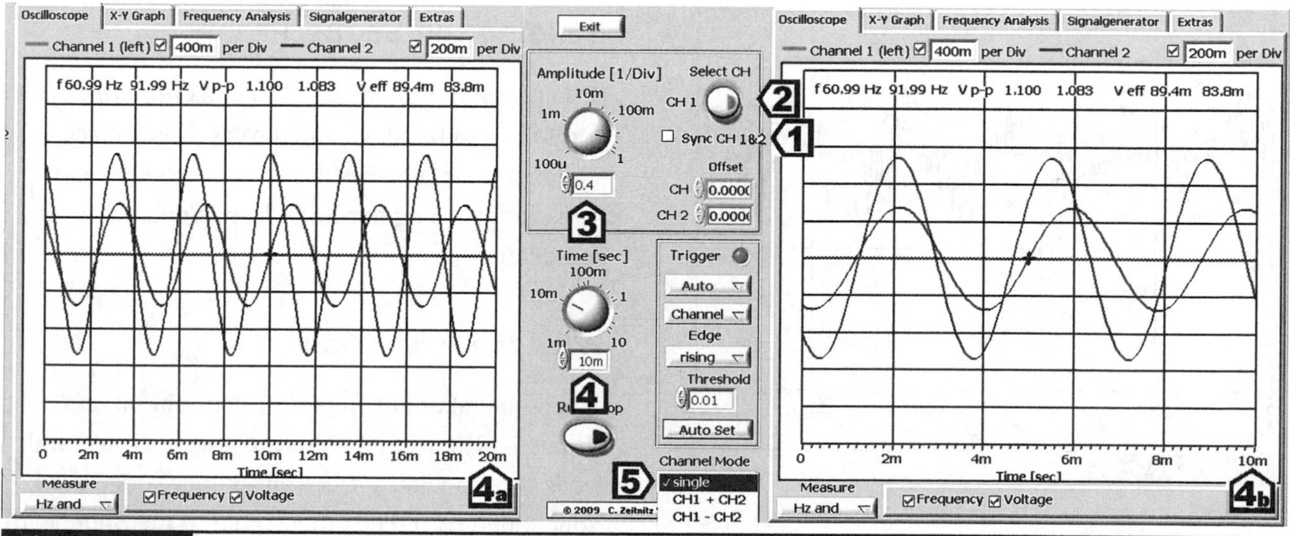

Figure L30-5

Wait a minute? Regarding the amplitude, have you noticed that the Y axis has no unit. We know it's voltage, but shouldn't it be showing .1 v/div or .4 v/div? Why just .1 or .4? Here's what's happening. As a generic program, it was designed to interact cleanly with a wide variety of software and hardware. There is no guarantee of accuracy. The intent of any scope is to picture voltage changes across time and to compare one signal's strength to another. Use your DMM if you want accurate voltage.

But there is a definite unit attached to the X axis. In Figure L30-5, the #4a display on the left screen shows all activity taking place in 20 ms. For those of you stuck on fractions, that's 1/50th of a second. If you want to get a closer look, change the value at #4 to 10 ms per division. The signal stretches as it is magnified across time. The screen now displays the voltage changes over 10 ms at #4b.

This scope is really capable. Here's one more thing to play with right now. At #5, click the Single button underneath *Channel Mode*.

Use the following settings, and you'll create the form in Figure L30-6:

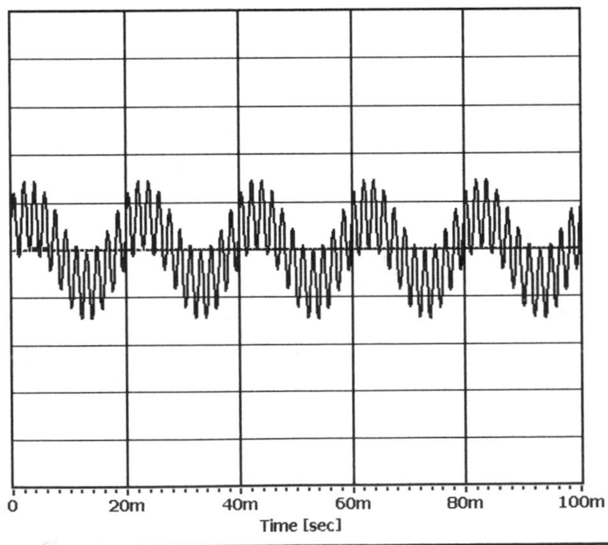

Figure L30-6

- Amplitude at .2/division
- Synchronize CH1 and CH2
- Time at 100 ms
- Channel Mode = CH1 × CH2

When you're finished playing, tab over to Frequency Analysis to see the screen shown in Figure L30-7.

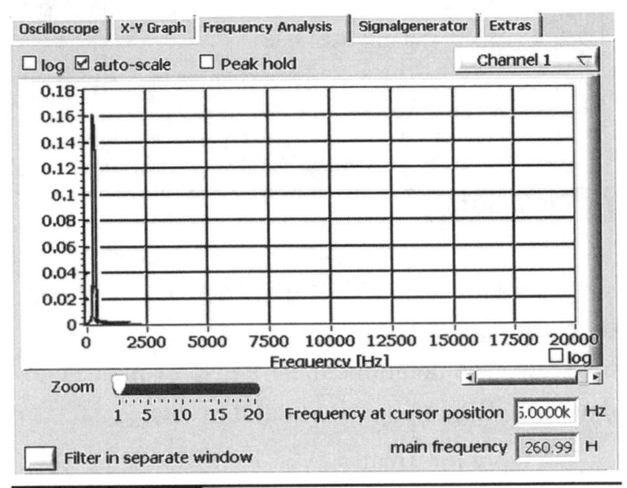

Figure L30-7

Sure enough, the Frequency Analysis feature has something to do with frequency analysis. Capable of only one channel at a time, this tool analyzes individual signals and displays component parts, showing relative strengths.

Building the Scope Probe

If you already have access to an oscilloscope, you don't need to build a scope probe. It is a piece of equipment that you need to be able to test your own circuits using the Soundcard Scope on your computer.

What the Probe Does

The probe takes any signal in your circuit, cuts out all but 1/11th using a simple voltage divider, and feeds that remaining fraction to your sound card. Your sound card feeds this signal to the Soundcard

Scope software, which interprets this signal and displays it on your screen.

Keep in mind that any sound card cannot accept more than 2 volts. Anything more than 2 volts will cause damage to your sound card. So if your output is 9 volts, only .8 volts is fed to your sound card.

There are three sections to the probe, as shown in Figure L30-8.

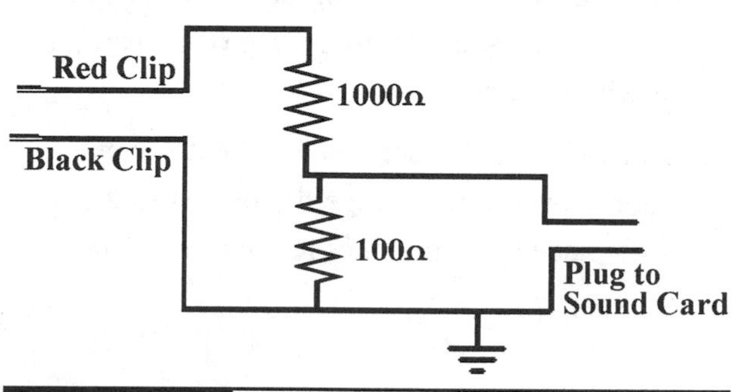

Figure 30-8

A more detailed set of photographs is available on the website www.mhprofessional.com/computingdownload.

The Connecting Clips

1. You need to have at least a three-foot length of speaker cord. Carefully remove one inch of insulation from both ends.
2. Mark one of the lines on both ends to identify it as the same line on both sides. Use this line for ground.
3. Disassemble the clips and slide the covers onto the wire.
4. Twist the end of each wire strand and push this strand through the hole at the base of the clip as shown in detail in Figure L30-9.

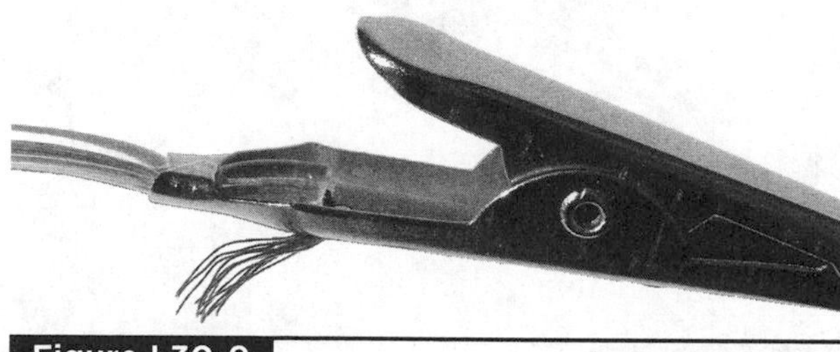

Figure L30-9

5. Lay the insulation into the saddle and use pliers to crimp the two sides of the saddle over the insulation. This physically holds the clip to the wire.
6. Now solder the wire strand at the bottom and clip the extra wire away. You should have something similar to Figure L30-10 now.

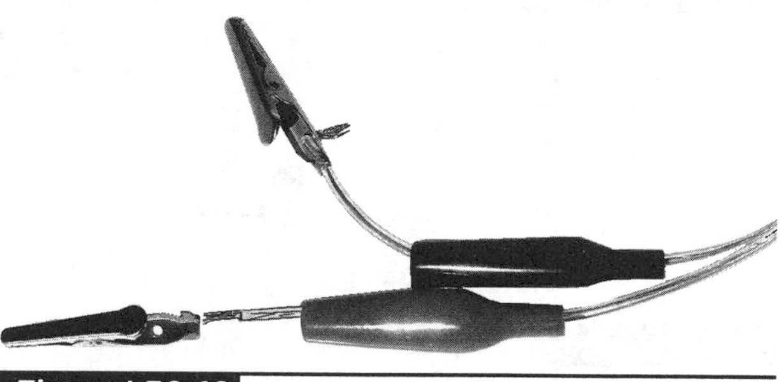

Figure L30-10

7. Slide the covers over the back of the clips. This is done easily if the jaws are clamped open onto something large.

That should finish the clips.

The Voltage Divider

The voltage divider is the heart of the probe. It is not a regular connector, but decreases the input voltage by a factor of 1:11.

Assembling the voltage divider for the scope probe requires the following:

1. Cut the dual cord six inches from the end opposite the clips.
2. Strip at least 1/4 inch (0.5 cm) of insulation off four ends.
3. Mark both sides of the ground line. This is the one connected to the black clip.
4. Figure L30-11 displays how to wrap the wire around the resistor leg before soldering. This is not critical, but it is very effective.

Figure L30-11

5. Slide skinny heat-shrink tubing onto each line before you solder. The heat-shrink tubing is easier to use than tape. The layout is shown here in Figure L30-12.

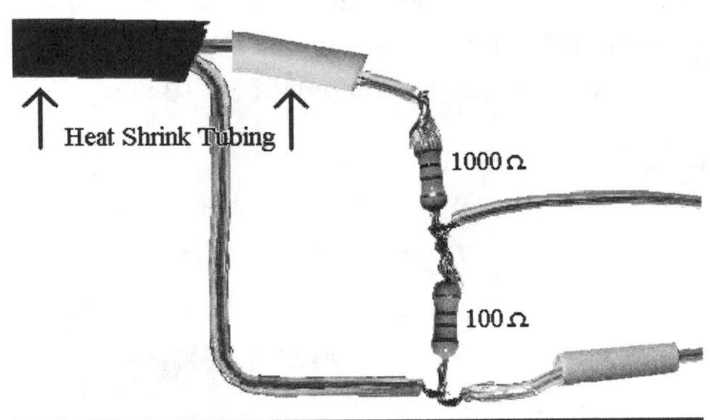

Figure L30-12

6. Move the heat-shrink tubing away from the heat of the soldering area until you are ready to shrink it into place. Test your parts placement immediately after you finish your soldering.

   Use the schematic diagram in Figure L30-13 as a guide to check that your scope probe is set up properly (see Table L30-1).

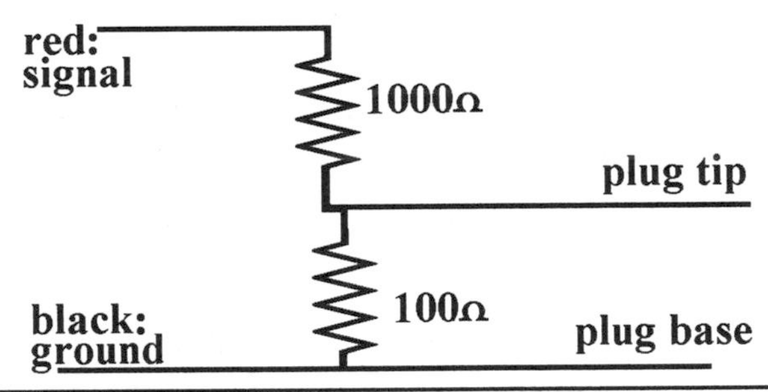

Figure L30-13

TABLE L30-1 Scope Probe Checklist

Probe Values	
Red clip to plug tip	1 kΩ
Plug tip to plug base	100 Ω
Ground line to plug tip	100 Ω
Signal to ground	1.1 kΩ
Ground to plug base	0 Ω

Then you can move the heat-shrink tubing into place as displayed in Figure L30-14.

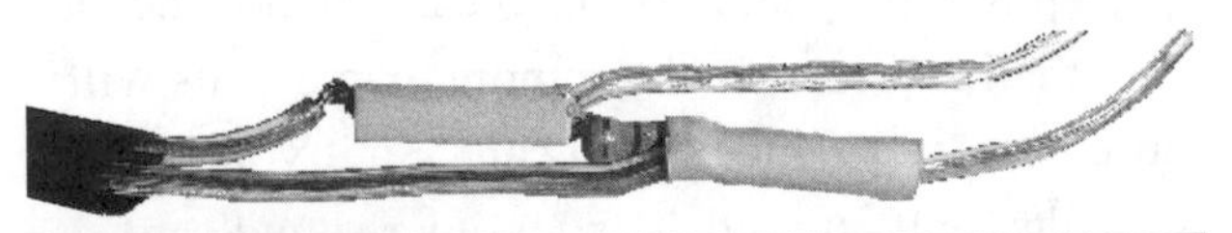

Figure L30-14

7. You can shrink the tubing either by using a hair dryer on its hottest setting or by caressing it with the hot solder pen.
8. Then slide the wider piece of heat-shrink tubing over your voltage divider. Shrink that into place over the other heat-shrink joints.

   DONE.

The Jack

Connecting the plug is the same as connecting the alligator clips.

Remember to slide the heat-shrink tubing onto the wire first.

1. Clamp and solder the ground wire to the long stem shown in Figure L30-15. The long stem is connected to the base, the lowest part of the plug. Now slide the heat-shrink tubing over the ground line connection. Make sure there are no stray wires.

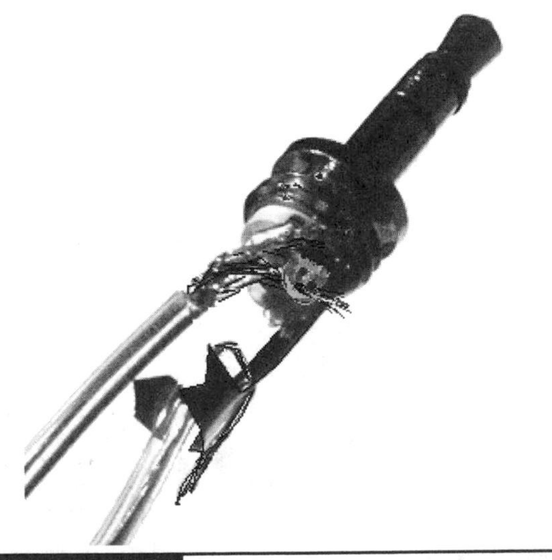

Figure L30-15

2. Lead the signal line between both of the other leads. We are going to have only one signal feed to the sound card, so both tabs will be connected. Solder that line to both tabs. There is no need to cover the last connection with heat shrink.
3. Slide the cover back down and screw it over the back.

Again, make sure your readings match Table L30-1 before you try it out. A wrong connection here can be disastrous.

You will use the scope probe in Lesson 31.

Lesson 31
Scoping Out the Circuit

NOTE **When you do testing and measurements, don't set your circuit near the computer. The "noise" generated by its power supply will corrupt your readings.**

We'll start off with a comment about what we can realistically expect from the Soundcard Scope. First, this is a "low power" circuit. Even though the circuit works at 9V, the total power output is around .02 watts. Even at that low power, 9V threatens any sound card. Second, the scope probe reduces the signal's voltage by 90 percent. That reduces the power to less than .002 watts. Third, your sound card might not respond to these really small power signals. So no matter how high you set the volume control, at best, you still will not see signals at Test Point 1 or 2. If you do get a signal, it will be badly distorted.

What I'm saying is that you might not be able to get a signal to display for Test Points 1 and 2 because of the following combined factors:

1. The power output of the circuit at these test points
2. Scope probe (adds to the problem)
3. Sound card
4. Windows' sound control
5. The Soundcard Scope

We'll start off with the basic circuit shown in Figure L31-1. All of the following work will refer to these test points. Necessary modifications to the circuit are listed here.

1. Setup is with R4 = 47kΩ and C2 = 10 nF.
2. Remove any connections to pin 10 so there is a clean output.
3. Connect pins 12/13 directly to ground so the circuit is continuously active.
4. Now open the Soundcard Scope. Don't adjust anything.
5. Connect the probe's plug into the microphone (line in) connection of your computer. If your computer asks you to define the input as "line in" or "microphone," choose microphone because it is much more sensitive.

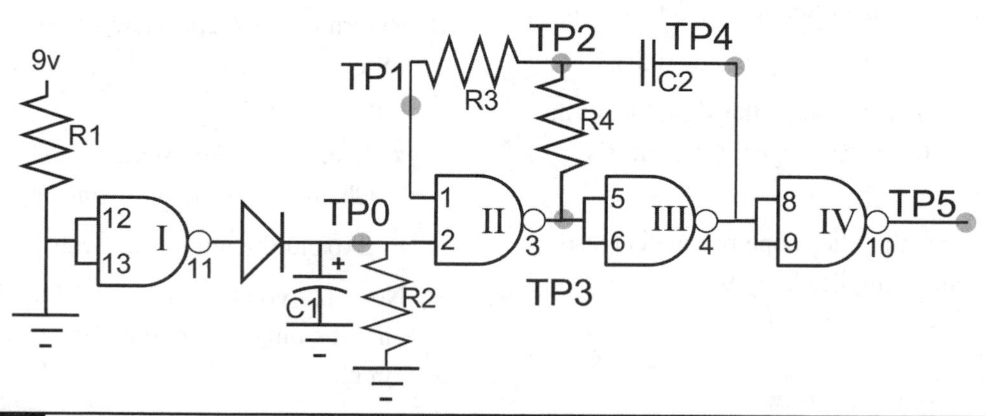

Figure L31-1

Connect your scope probe's clips as noted here:

- The black clip connects to ground and stays there.
- Place a piece of wire into the red alligator clip. Use that wire to touch the signal sources.

You will see something similar to Figure L31-2.

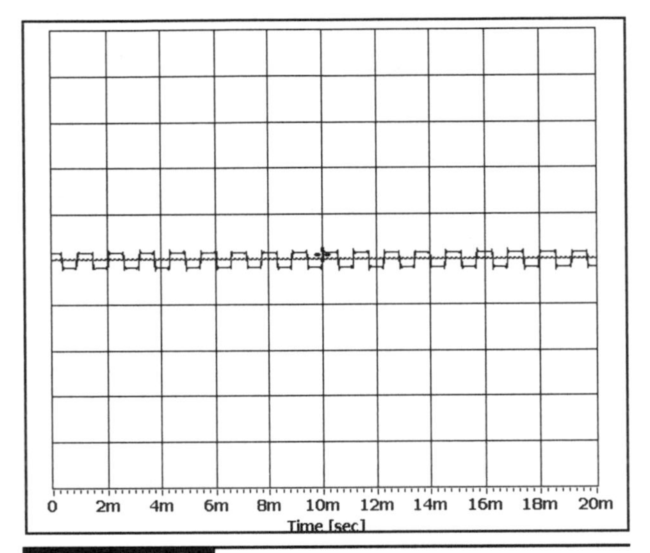

Figure L31-2

6. Now go directly to the Signal Generator. Make the adjustments shown in Figure L31-3
   - Set the channel you are using to 1000 Hz.
   - Adjust the amplitude knob to .1.
7. Shift over to the Frequency Analysis screen.
   - Increase the magnification by 5 to 10 times.
   - Scroll to the lower frequencies (left).
   - Figure L31-4 compares the signal from the circuit to the 1k Hz Signal Generator's signal.
   - Remember that the scope probe decreases the circuit's amplitude by 90.

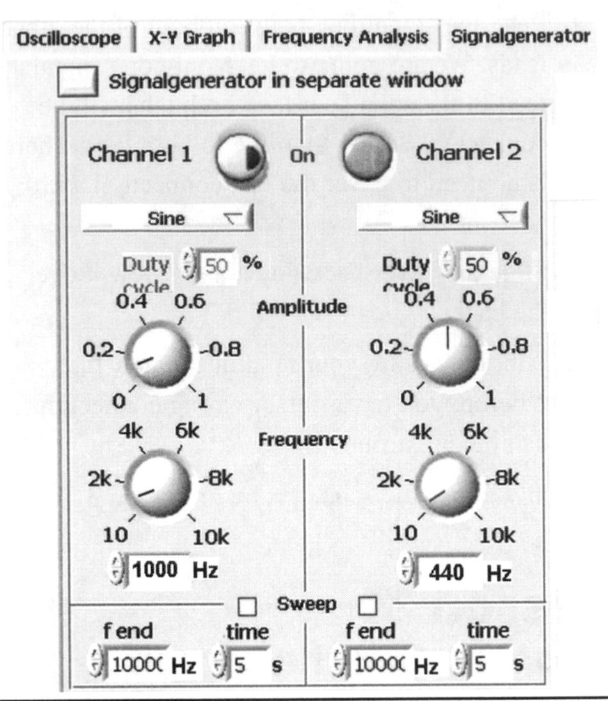

Figure 31-3

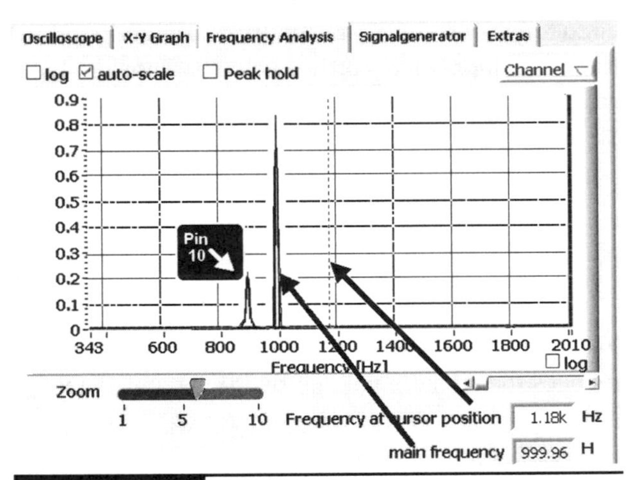

Figure 31-4

8. Return to the scope view and make the following adjustments to get the display on Figure L31-5:
   - Expand the time scale (X axis) by 10 times. Change the time/div from 20 ms to 2 ms.
   - Expand the amplitude scale by 10 times.
9. Now that you have a reasonable view of the circuit's output, scope out the rest of the circuit.

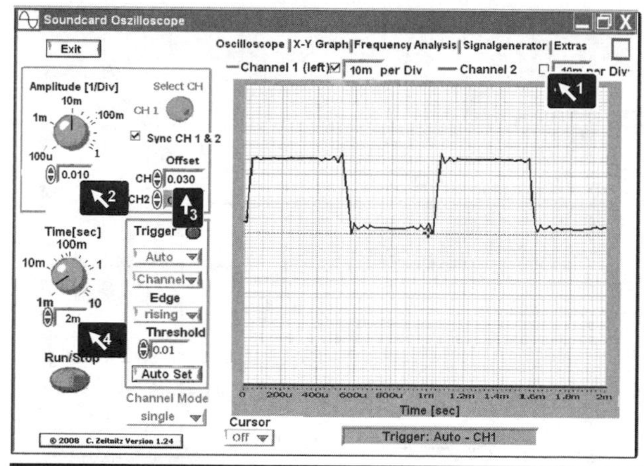

Figure 31-5

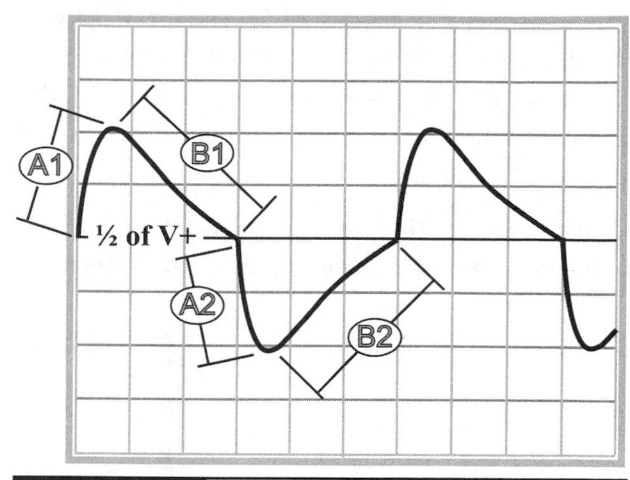

Figure 31-6

Test Point 0

It is important to understand the scope reading at Test Point 0. First, check out the voltage using your DMM. It should be steady between 8 and 9 volts. When you first touch the lead of your scope probe to TP0, the signal jumps, but then quickly settles back to the center and sits there.

Why? It's easy to understand what's happening. The Soundcard Scope responds only to electronic signals in the sound frequency range.

Is the signal at TP0 changing? No frequency = 0.0 Hz. Then why the sudden jump just as you connect? Obviously, the disconnected probe's input has no real voltage. The input voltage jumps as you connect. The scope is reacting to that voltage change. But with no more activity, there is no more signal.

Test Point 1

We know that the oscillating input to pin 1 actively changes the gate's output at pin 3. But remember: Pin 1 is an input, a gauge that compares its voltage to the voltage supplied to the IC. What we see in Figure L31-6 is the changing signal to a NAND gate's input.

The rising slope indicated by A1 shows the increase in voltage as the cap charges from the half V+ mark. The downward slide, B1, displays the capacitor's discharge as the voltage returns to the half V+ mark. As the signal passes the half V+ mark, Pin 1 interprets this as a Lo input. Pin 3 changes the phase from Lo to Hi. C3 starts charging, moving upwards until the signal passes the half V+ mark. The cycle then repeats itself.

Test Point 2

Test Point 2 is the junction of R3, R4, and C2. Even though the action in Figure L31-7 looks dramatic, it is actually showing the same action as Test Point 1, even though it has more amplitude. Each signal unit lasts the same amount of time, but the charging cycle is much faster, and the

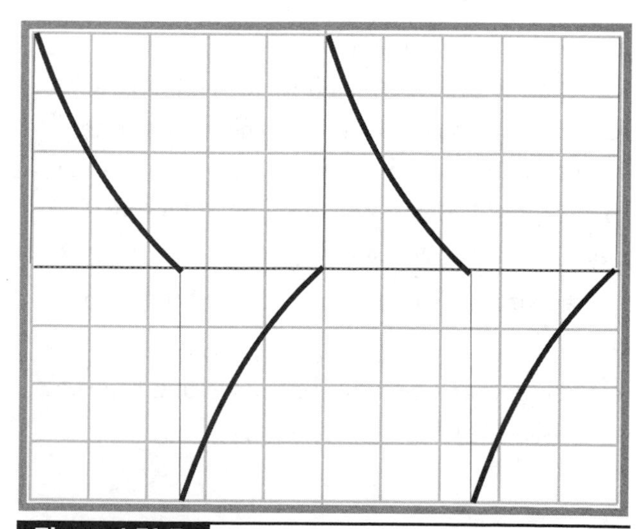

Figure L31-7

discharge takes up the rest of the time. The actual events take exactly the same amount of time. Because R3 is 10 times more resistant than R4, the charging action at that point is 10 times slower and the signal is smaller.

We'll overlay these test points in Figure L31-8 to examine how the signals relate to each other.

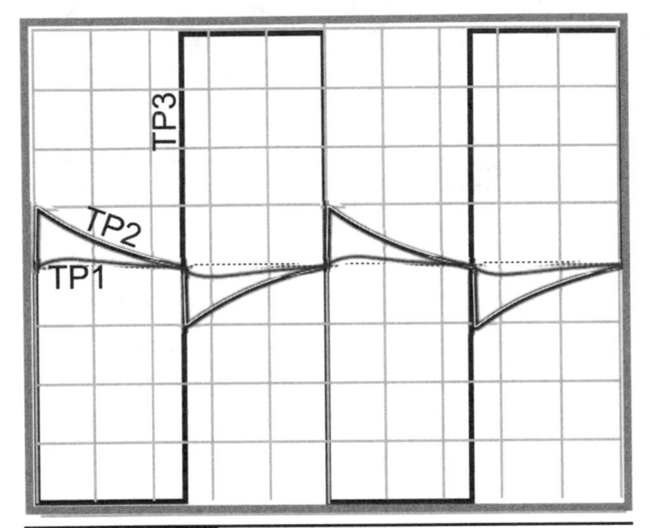

Figure 31-8

Remember what happens when pin 1's input reaches half V+? As a voltage comparator, pin 1 interprets that as Lo and kicks the NAND gate's output (pin 3) to Hi.

Test Point 4

The signal at Test Point 4 looks a bit smaller and wobblier. It's been affected by the action at RC2. Keep in mind that TP4 is both the *output* of the third NAND gate and *input* to the fourth NAND gate. Figure L31-9 compares Test Point 4 to Test Point 5, the last gate's output.

Two significant changes happened to the input signal.

- The output signal has been inverted, flipped upside down.

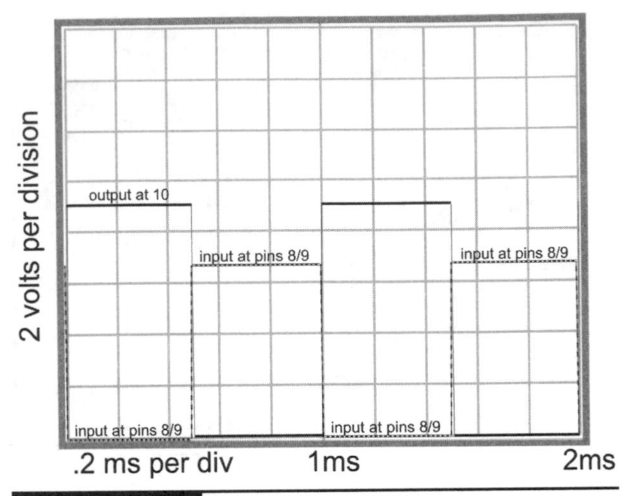

Figure L31-9

- The output signal has been refreshed. That means that the weak input signal has been restored to full strength.

Now, let's return to making annoying sounds.

Lesson 32
Using a Transistor to Amplify the Output

Transistors are a natural choice as amplifiers. Their action and application are explained. Either transistor would work, but the PNP is used because it offers certain advantages. You get a great response with dogs howling at 1,000 Hz. It "hertz" their ears.

Right now you have a very quiet alarm. The output from the 4011 NAND gate provides a small amount of power. It is enough to turn on an LED. You have already found it is definitely not enough power to give even a small speaker any volume. But it provides plenty of power to turn on a transistor.

Modifying the Circuit

Make the modification shown in Figure L32-1. Be sure to insert the PNP 3906 transistor the right way.

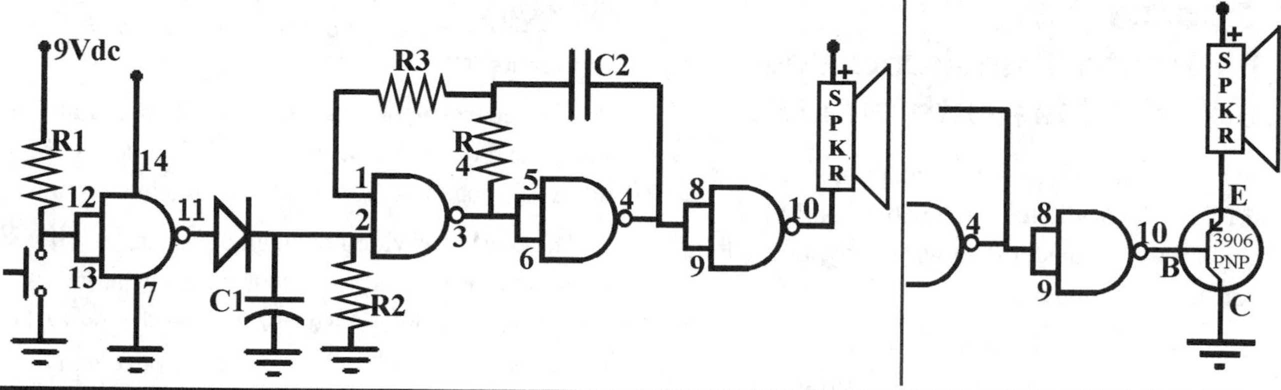

Figure L32-1

Why use the PNP 3906 transistor? Think . . . when the system is at rest, pin 10 is high. The two opposing voltages stopped any movement of current, and that's why the LED stayed off. Figure L32-2 shows how this action is used to our advantage. A high output keeps the voltage from moving through the transistor. The 3906 is turned off. This way, you don't drain the battery.

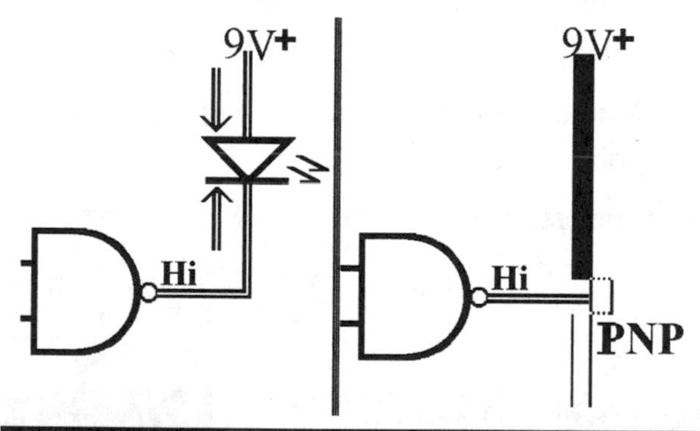

Figure L32-2

What to Expect

This circuit should be quiet when you connect the power. At most, it might squeak for a fraction of a second. If it continues, disconnect the power and check the circuit.

When you activate the circuit, the sound should be annoyingly louder. Don't worry if, as it returns to rest, the speaker "burps."

How It Works

When pin 10 goes low, the transistor turns on and the following happens:

1. This allows much more voltage and current to pass through.
2. That results in more power passing through the speaker coil.
3. The greater power produces more electromagnetic force in the coil.
4. That produces more movement of the coil and cone, producing a louder output.

If you chose to use the NPN 3904 transistor, here's what happens. While at rest, the high output from pin 10 would keep the current flowing from voltage directly through the speaker coil to ground. This is shown in Figure L32-3.

This would quickly drain the battery and annoy you.

How much better not to drain the battery quickly and go annoy some other people now.

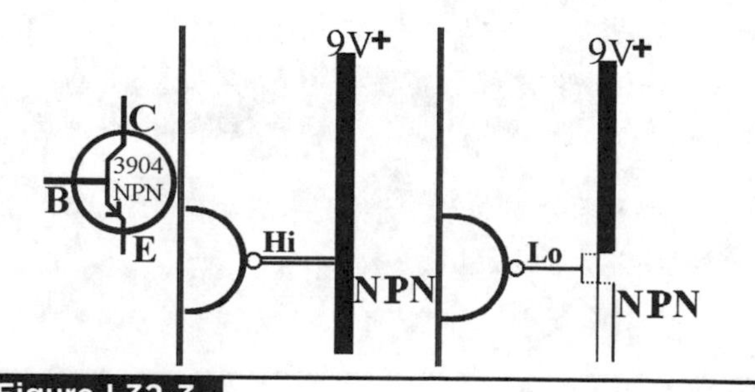

Figure L32-3

Lesson 33
The Photo Transistor: You Can't Do This with an LDR

Now is a good time to revisit the phototransistor and further examine its unique abilities. Many digital systems transfer data wirelessly, through the air or through optical cable.

Surprisingly simple substitutions are made in Figure L33-1 to change over to the infrared LED. Use the same values for RC2 as you did in Lesson 31, producing a frequency in the 1kHz range, and keep pins 12 and 13 connected directly to ground for an uninterrupted signal.

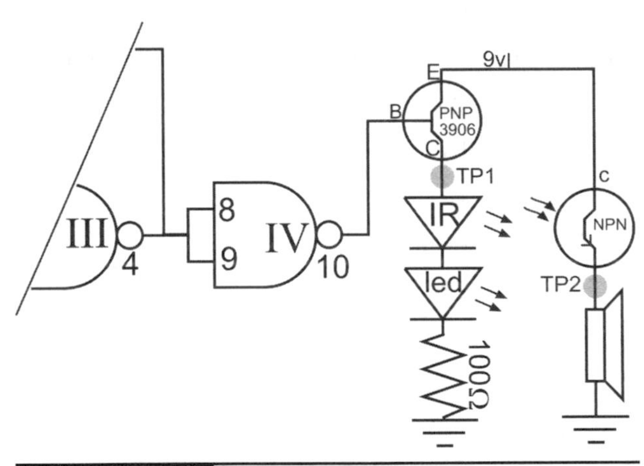

Figure L33-1

Remember that the short leg/flat edge of the phototransistor represents "C."

The speaker/phototransistor circuit is separate, at the far end of the SBB. Figure L33-2 indicates the actual distance I've given between the two circuits.

The power provided through a transistor will burn out our infrared LED, but we do want it as "bright" as possible. Also, we use the second LED to tell us that something we can't see is actually working. We could add three more regular or IR LEDs in series to cut the voltage. Each LED uses about 1.7 volts. That would work. But a single 100-ohm resistor will eat up the remaining voltage.

If you really want to see an IR LED working, look at a TV remote control's output through a digital camera, and then push a button. The IR LED might appear bright or hardly on at all.

1. Make sure the IR LED and phototransistor are lined up directly with each other. The best results come from having both components looking at each other through a straw.
2. Open the Soundcard Scope and plug in your scope probe.
3. Connect the scope probe to TP1, shown in Figure L33-1.

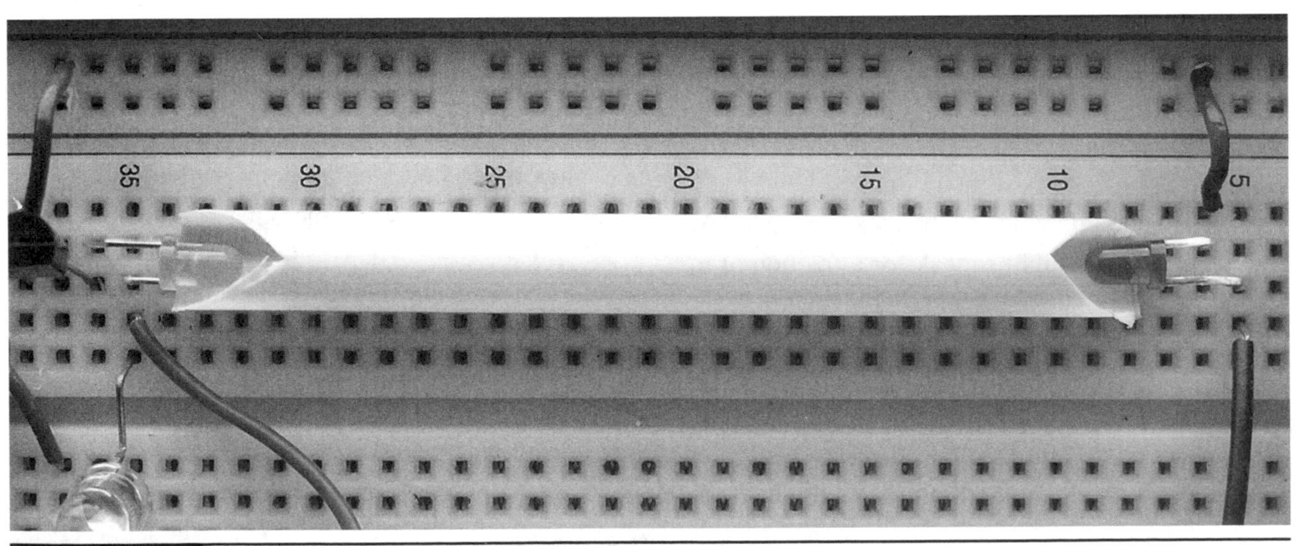

Figure L33-2

4. When you apply power to the circuit, you should get both sound and activity on the scope screen.

Shift over to the Soundcard Scope's Frequency Analysis tool, and you'll see something similar to Figure L33-3.

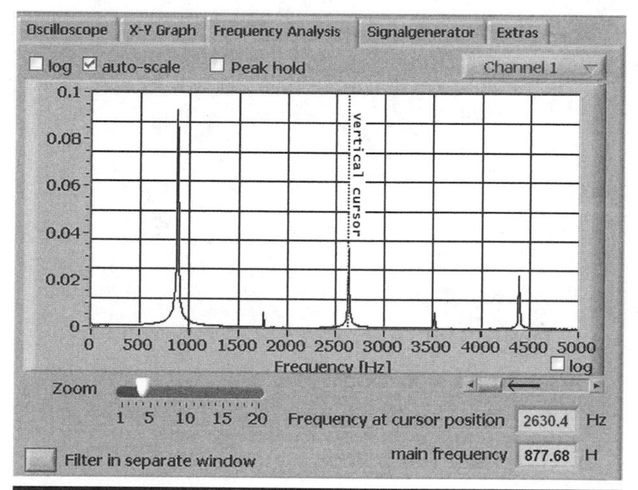

Figure L33-3

Two pieces of data give you valuable information:

1. The amplitude, marked on the left side.
2. The frequency. My circuit produces 878 Hz, which is within acceptable tolerance.

Shift the probe over to TP2, between the phototransistor and speaker. The sound continues, but you lose the signal. Real scopes are more sensitive and you would see the signal. To make the signal more available to us, match the partial schematic in Figure L33-4 by replacing the speaker with a 10 kΩ resistor.

Look at those two readouts again.

1. The amplitude has dropped significantly. You could tell that just from the volume of the speaker. That's to be expected.
2. But look at that frequency—it should precisely match the 4011's output.

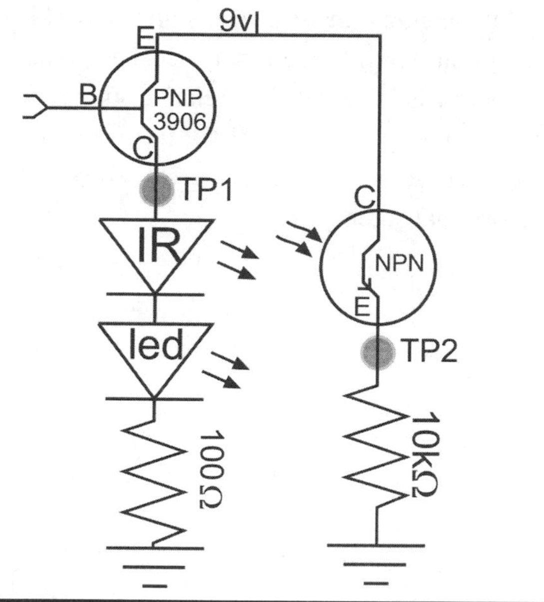

Figure L33-4

Put your finger between the IR LED and the phototransistor. The amplitude diminishes but the frequency remains unchanged.

Have you noticed that more than one frequency registers on the scale? What about those other frequencies? I've highlighted the first one in back in Figure L33-3. These are harmonics created by the main frequency. They resonate strongly enough to be picked up by the scope.

Use your mouse to move the dashed vertical line to directly cover the first harmonic.

The vertical cursor's readouts correspond to the vertical cursor's placement. It indicates a frequency of 2630 Hz. And you're saying "Yeah? So what?"

Multiply my base frequency by three.

$878 \times 3 = 2634$

After you check that out for your system, do the same for the second large harmonic. What number works? Is there a pattern?

The initial frequency and the harmonics are created at first by the 4011 circuit. That signal is transmitted by the IR LED. The phototransistor receives and reproduces that signal precisely.

Any analog transistor reacts precisely (I keep using that word) to the input at its base. The frequency is identical. The relative strength within the signal stays constant. If the received signal was amplified, it would be an accurate reproduction of the original signal.

My suggestion is that you don't use this component as part of the 4011 project. Save it as a design component for Part Four, where we play with amplifiers. It can be used to transmit voice or data signals and other neat things.

SECTION 10

Digital Logic Project

WOULDN'T IT BE EASIER for projects just to give you a circuit board, tell you where you have to put the parts, and then solder them in? In fact, that's a great plan if your career goal is to be a solder jockey in a Third World country. However, that career opportunity is vanishing as such workers are being replaced by robots.

Lesson 34
Design—Systems and Samples

Design is the most interesting portion of the entire field of electronics. Different components build into specific unit parts. Like Lego, every part fits together. Have fun. Use your imagination.

1. The reader develops different applications by defining different inputs and outputs.
2. Common modifications to the processor are explored.
3. Examples of reasonable enclosures are provided and discussed.

Inputs

There are four main areas (see Table L34-1) to explore in designing your own project.

Contact Switches

Any of the contact mechanisms can be substituted for the push button shown in the schematic of Figure L34-1. The value of R1 should be 100 kΩ.

You do have the regular push button available shown in Figure L34-2. Aside from being boring, however, it is hard to rig these buttons to turn on with anything other than a push of the finger.

TABLE L34-1 Four Main Areas

	Processor		
Input	**RC1**	**RC2**	**Output**
Contact Switches	Turn On/Timed Off	No Oscillation	Low Power LEDs Music Chip
Light Detector	Time On Delay	Oscillation Rate	
Dark Detector	Touch Off		Amplified Power
Touch/Moisture	Turn Off/Time On		Buzzer Speaker Motors Relays

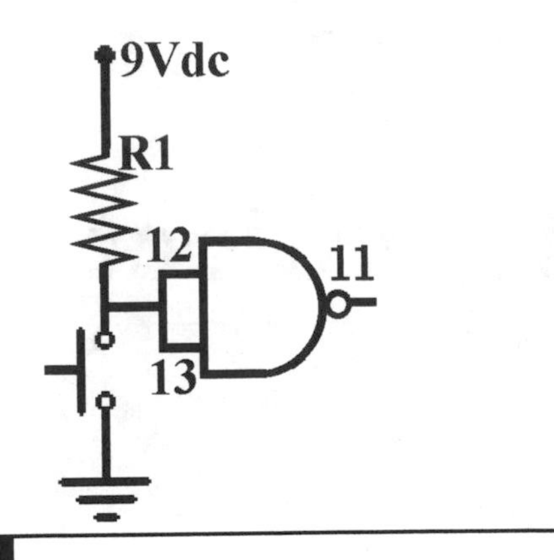

Figure L34-1

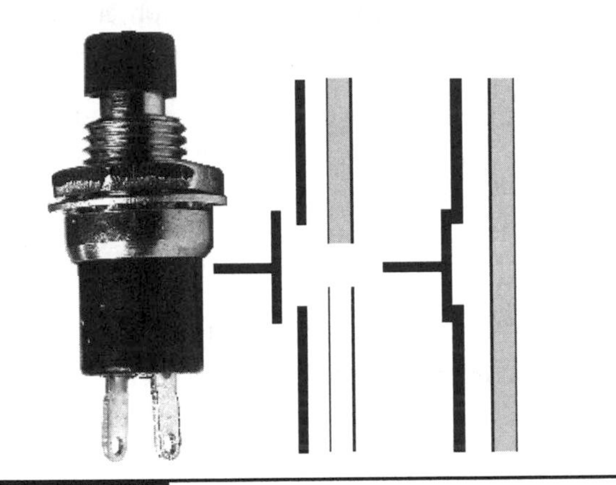

Figure L34-2

Then there is the motion detector, displayed in Figure L34-3. These can be made by balancing a weight on the end of a spring. Almost any metal weight will do, but a tapered screw is most easily attached. The best springs for this purpose are inside retractable pens. But you can't solder to spring steel. For the weight, turn the screw into the spring until it catches. For the bottom, wrap a piece of copper wire around the base. Solder the wire to the PCB. This style of switch can be made to be surprisingly sensitive.

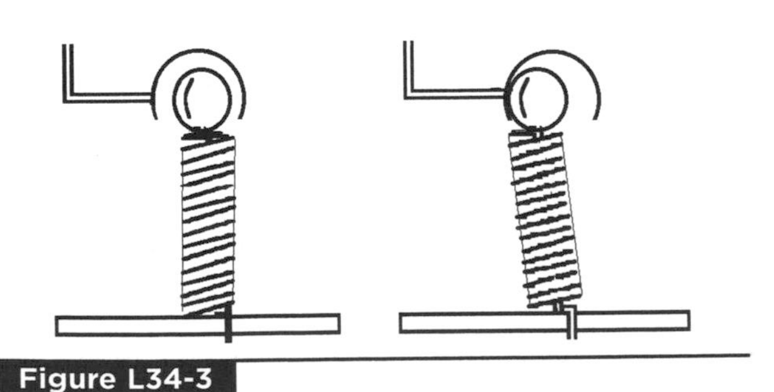

Figure L34-3

Turn the spring mechanism upside down. This pendulum setup shown in Figure L34-4 is not nearly as sensitive as the upright spring, but it uses the same concept.

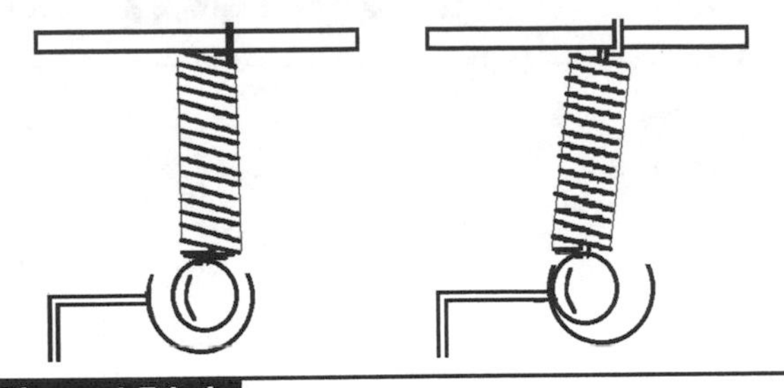

Figure L34-4

Two springs can be attached to a nonmetal support. They can pass over a metal contact bar, as demonstrated in Figure L34-5.

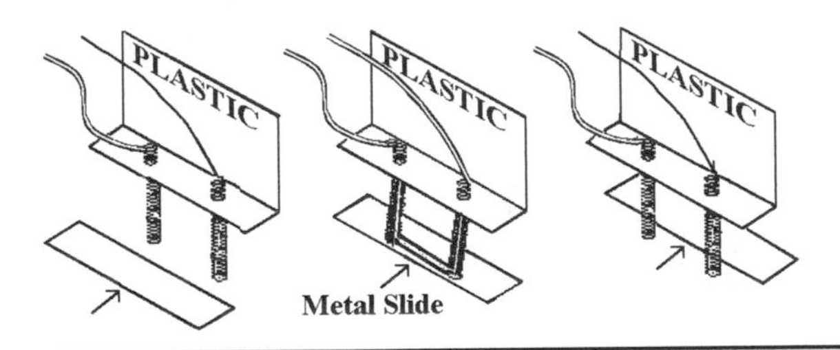

Figure L34-5

Either the spring support or the metal bar can slide. This is the perfect setup for doors or drawers.

Microswitches are exactly that. They are very small. They are readily available for free. Any broken mouse provides two of them. If you go out and try to purchase these new, they will cost upward of $4 each. Each switch has three contacts. Look closely at the photo in Figure L34-6. One is called the "common" because it is shared between the two other contacts. Like common property. Depending on your choice, this switch can act as a normally open or normally closed push button.

A coin-activated switch is a bit trickier to make. Vending machines often use a lever on a microswitch. The coin pushes the lever down, which in turn pushes the contact switch down. A simpler device is shown in Figure L34-7.

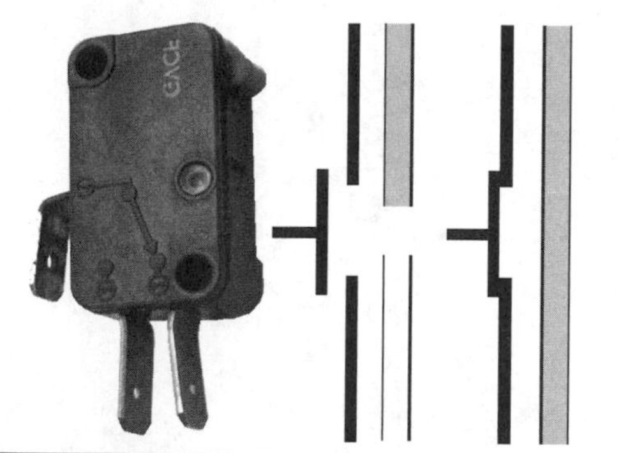

Figure L34-6

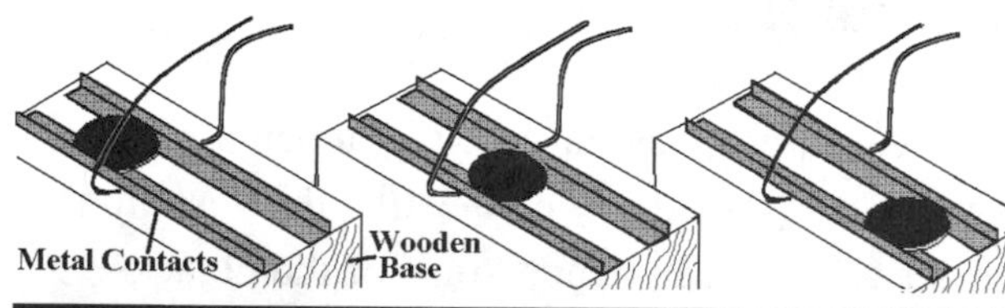

Figure L34-7

It works and is simple to build. Mount the two metal strips onto a simple wooden or plastic base.

Light-Dependent Resistor

The LDR can easily be set up to get triggered by lights turning on and off.

Light Detector

The LDR is the base of this light-sensitive switch. The circuit shown in Figure L34-8 will become active when it is exposed to light.

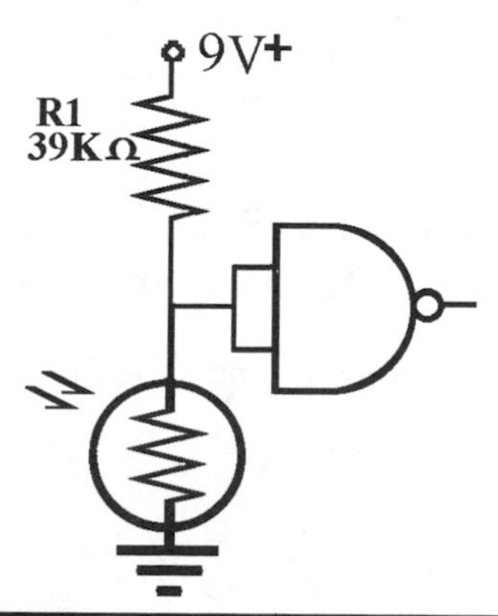

Figure L34-8

Unwanted light may turn it on rather than the event you intended. If the LDR is to be used in a generally well-lit area, it is best to use a cowling, as shown in Figure L34-9.

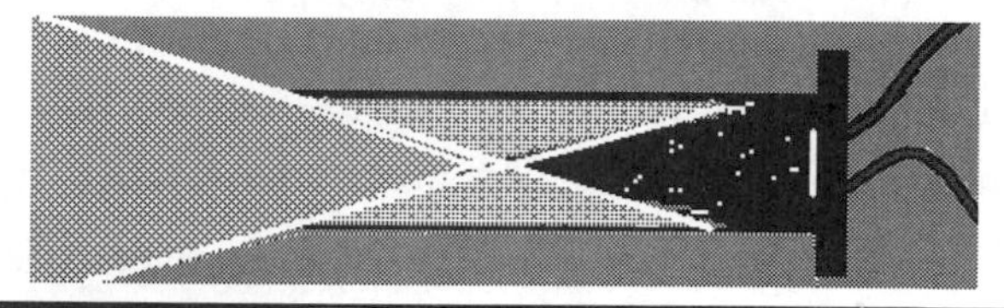

Figure L34-9

Depending on the light source, it might be necessary to use a lens to concentrate the light source onto the LDR. This is demonstrated in Figure L34-10.

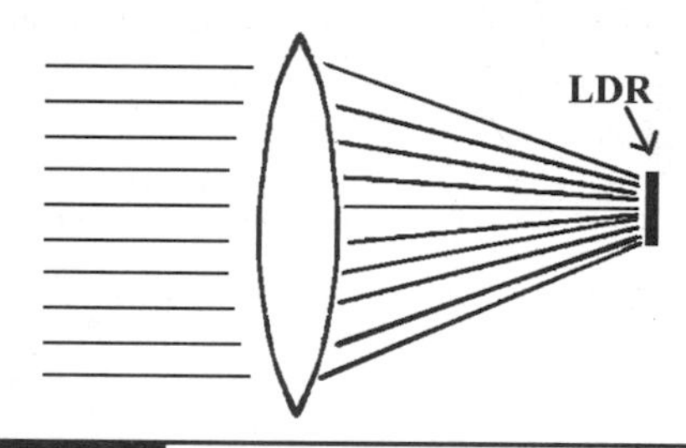

Figure L34-10

Just remember that a lens works only with a preset light source and won't successfully focus generalized light.

Dark Detector

The circuit shown in Figure L34-11 is identical to L34-8, except that the two components have traded places.

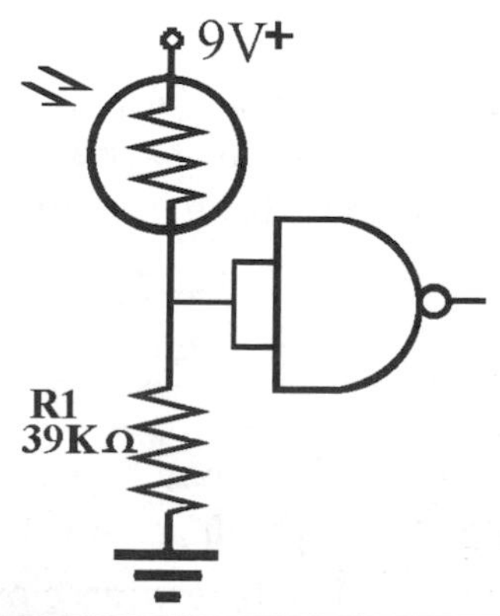

Figure L34-11

With the reversing of the LDR and 39-kilo-ohm resistor, the voltage divider is reversed as well. A cowling, shown in Figure L34-9, is even more important here. The circuit stays at rest as long as a steady light falls on the LDR. If you want to use a steady light source such as a laser pointer, the cowling guarantees the circuit will react to the breaking of that one light source.

The best source for light over a long distance is the laser pen. Using mirrors, the beam can even travel around corners. The system is shown in Figure L34-12. A laser pen can be powered with a wall adapter matched to the same power rating as the batteries that normally power it. Every cell inside a laser pen has 1.5 volts. For example, if it has three cells, you need to find a wall adapter that provides 4.5 volts.

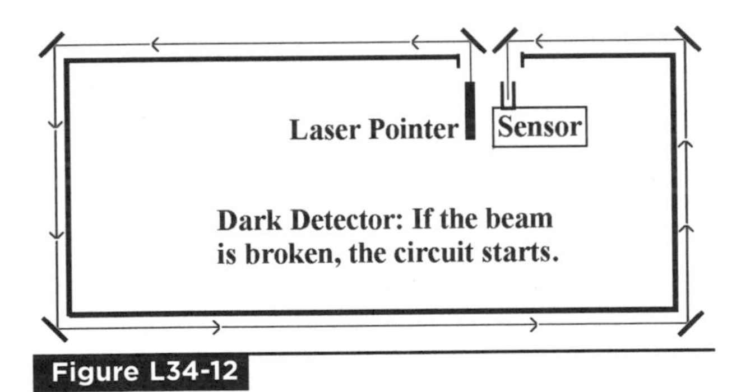

Figure L34-12

The beam here is shone from inside a window and travels around the outside of the house. A speaker in both next-door neighbors' homes was set off for 10 seconds at 1,000 Hz. The system was able to be keyed off outside by the owner.

CAUTION Be careful. Many laser pens claim to meet safety specifications, but really can damage your eyes if you are exposed over time.

Touch Switch

The schematic shown in Figure L34-13 works as a touch switch. This setup will also work with water spills. All it needs is two bare wires close together, but not touching.

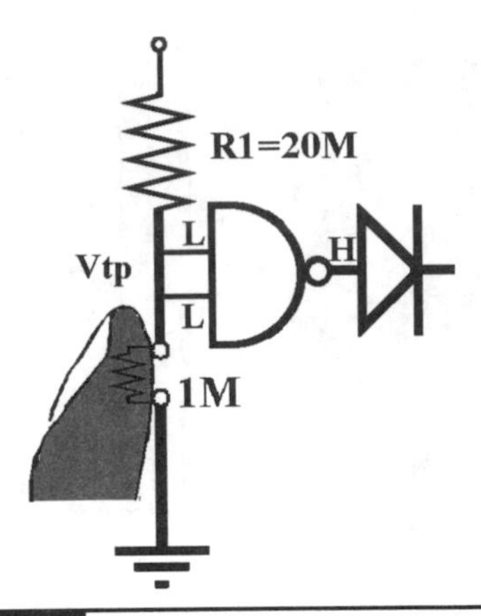

Figure L34-13

A clean, professional-looking, touch-sensitive switch can be made by connecting the wires to the underside of broad-headed pins or thumbtacks pushed through black plastic. These are displayed in Figure L34-14.

Figure L34-14

Processors

You can choose from several physical modifications and a wide variety of RC timing options or no RC timing at all.

Possibilities for the First Resistor/Capacitor Circuit—RC1

Figure L34-15 shows the basic RC1 setup.

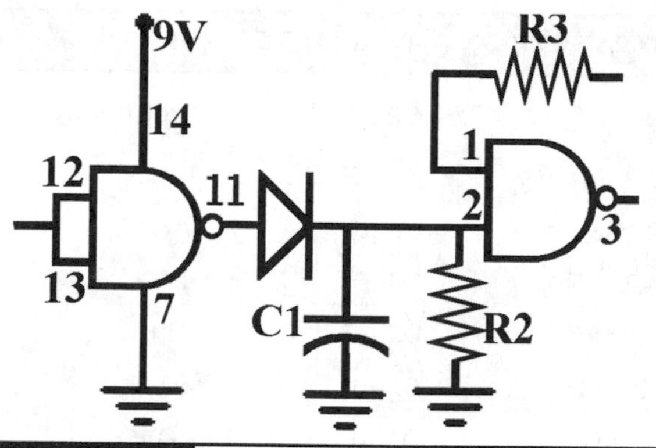

Figure L34-15

Table L34-2 provides a rough guide for timing RC1. Remember that this is only a rough guide. It is not a precise time.

TABLE L34-2 RC1 Timing

R2	C1	Time
20 MΩ	10 μF	120
10 MΩ	10 μF	60
4.7 MΩ	10 μF	30
20 MΩ	1 μF	12
10 MΩ	1 μF	6
4.7 MΩ	1 μF	3

The schematic in Figure L34-16 shows how to manually speed up the timed off. Use pinheads for the touch switch. Your finger acts like a 1-megohm resistor. If R2 is 10 megohms, it will drain C1 10 times faster. Or you could use a PBNO to drain C1 instantly.

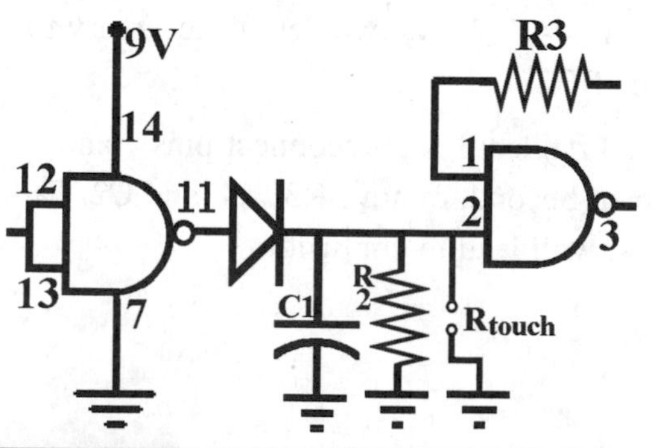

Figure L34-16

The next circuit is similar. As you can see in Figure L34-17, the simple modification causes several changes. This alarm stays active until you turn it off. C1 keeps the inputs of the second gate high until you touch the points to drain C1. Your finger is the only drain. A hidden touch point of two pinheads or a push button (normally open) is all that you need. Use a small capacitor at C1 (0.1 μF), and when you touch the pinheads, the alarm will appear to turn off instantly (about a half-second).

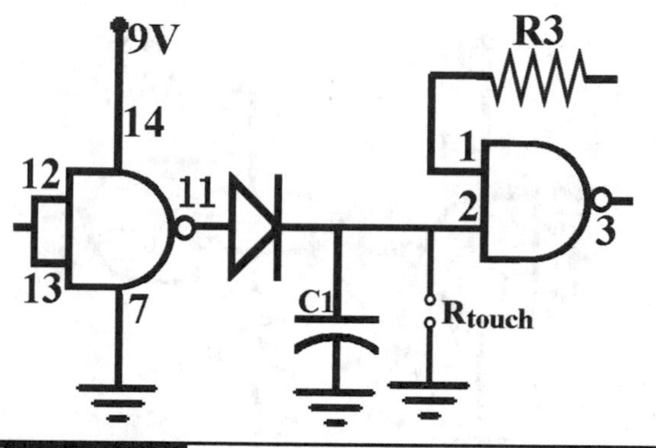

Figure L34-17

The schematic in Figure L34-18 is also impressive. It is a delayed time on. It can be used effectively in the light-sensitive switches to slow down the triggering speed.

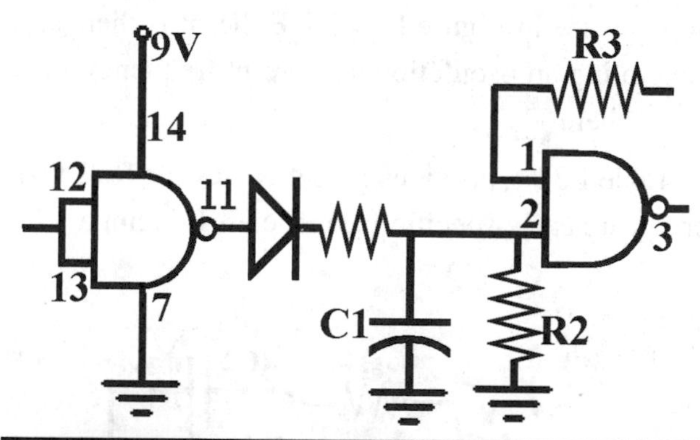

Figure L34-18

This modification can be used to delay the activation of the circuit. It can be used to give you time to set the circuit in a car alarm, for example, and give you time to close the door.

The value of the extra R must be, *at most*, a fifth the value of R2 to work because the extra R and R2 become a voltage divider. The input at pin 2 must rise clearly above the half-voltage mark. To give more time, use a larger capacitor.

If you choose, you can remove C1 as shown in Figure L34-19. This effectively destroys RC1. When the inputs are low at pins 12 and 13, the circuit is active. The circuit immediately turns off when the inputs go high.

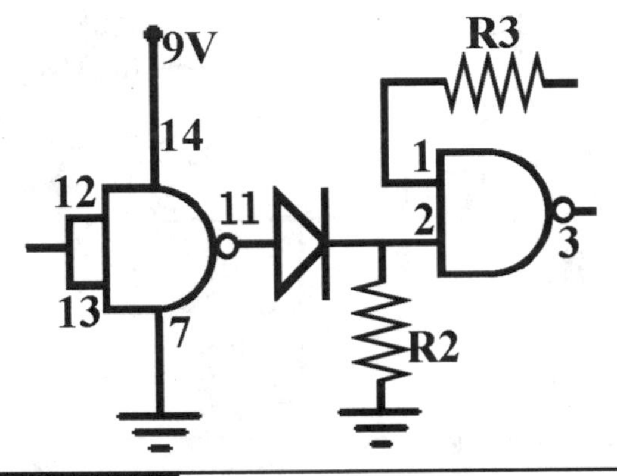

Figure L34-19

Timing and Modifications for the Second RC Circuit

There is limited potential for modifying RC2, as you can see in Figure L34-20. Either it is there and generating an oscillation at a preset frequency, or it is not there.

Table L34-3 provides preset values for RC2 that produce nearly specific frequencies. It is not a precise time. You won't be able to use it as a reliable pitch pipe for tuning.

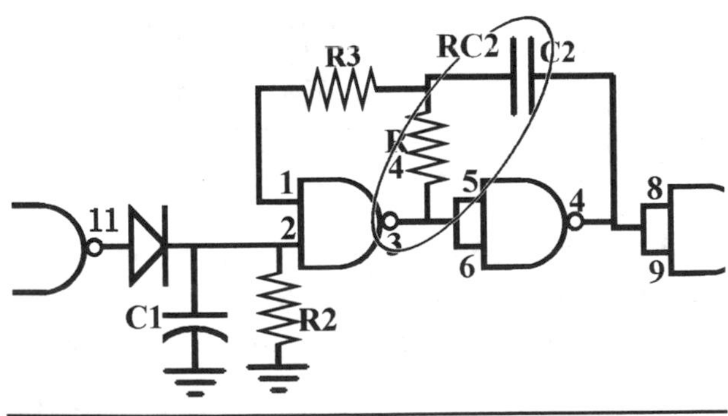

Figure L34-20

TABLE L34-3 Values for RC2

R4	C2	Frequency
2.2 MΩ	0.1 μF	1 Hz
2 MΩ	0.1 μF	2 Hz
470 kΩ	0.1 μF	4 Hz
220 kΩ	0.1 μF	10 Hz
100 kΩ	0.1 μF	20 Hz*
47 kΩ	0.1 μF	40 Hz
22 kΩ	0.1 μF	100 Hz
1 MΩ	0.01 μF	20 Hz
470 kΩ	0.01 μF	40 Hz
220 kΩ	0.01 μF	100 Hz
100 kΩ	0.01 μF	200 Hz
47 kΩ	0.01 μF	400 Hz
22 kΩ	0.01 μF	1,000 Hz
10 kΩ	0.01 μF	2,000 Hz
4.7 kΩ	0.01 μF	4,000 Hz
2.2 kΩ	0.01 μF	10,000 Hz

* The eye can't distinguish flashing from continuous motion for anything faster than 24 frames per second.

For certain applications, it is obvious that you want to remove RC2. No oscillating, please. For example, you don't want to listen to the first phrase of "happy birthday t'" (wait two seconds) "happy birthday t' . . . " as your circuit works through a two-second on, two-second off cycle. Figure L34-21 shows two details necessary to disable RC2.

The first detail is to reconnect pins 1 and 2 together. Second, remove R3, R4, and C2. Failure to do so will lead to confusion.

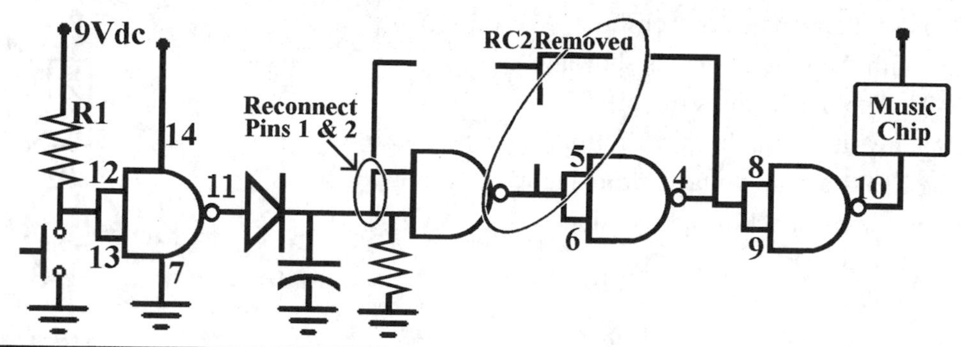

Figure L34-21

Outputs

Table L34-4 describes the outputs by comparing oscillation needs against power requirements.

TABLE L34-4 Comparing Oscillation Needs Against Power Requirements

	Low Power Output	High Power Output Needs a Transistor
Oscillating	Slow Flashing	Speaker for Alarm (1,000 Hz)
	LEDs	Buzzer (slow pulse @ 1 Hz)
		Relay (slow pulse @ 1 Hz)
Not Oscillating	Music Chip	Car Alarm
		Relay (no pulse)
		Low-Power DC Motor

Low Power

A low-powered output is good for only low-powered applications.

LEDs

The output of a 4011 chip can power more than 10 LEDs, but not many more. Even so, there are two ways to wire these up: the right way and the wrong way. Figure L34-22 shows the right way to connect more than four LEDs.

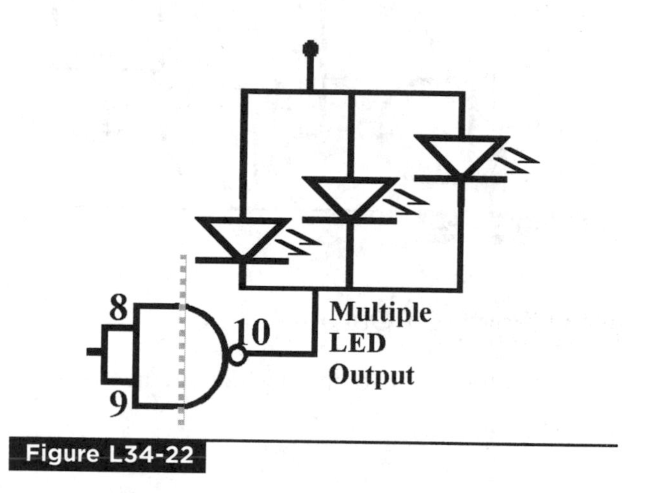

Figure L34-22

Music Chip

Carefully remove a music chip from a greeting card. Don't break any wires. Examine the music chip.

- Tape the wires to the speaker in place. Don't bend them.
- Note the circuit's connection to the "+" side of the battery.
- Remove the battery from the music chip. Remove the small stainless steel battery holder, crimped in place.
- Solder two wires, respectively, to the circuit's + and − battery connectors.
- Connect the ground side to pin 10.
- Connect the V+ side to voltage of your circuit.

- If the sound is scratchy, place two or three LEDs in line with the music chip, as shown in Figure L34-23. The music chips run off 1.5 volts. Too much voltage can keep them from working. The LEDs use up voltage, dropping it down to where the music chip can function properly.

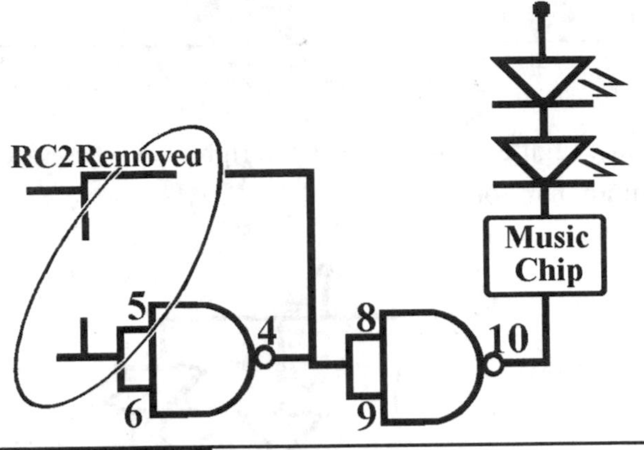

Figure L34-23

High-Powered Outputs

Adding a transistor to amplify the circuit's output is pretty easy too.

Buzzer

A buzzer has different needs than a speaker for output. All a buzzer needs is voltage. It produces its own signal. If you want the buzzer to turn on and off, use a slow oscillation of 1 Hz. It will "beep" once a second, on and off. A signal faster than 10 Hz will only confuse the buzzer and give muddled results at best. Figure L34-24 shows the setup.

An amplified output to a buzzer should oscillate at 1 Hz. That can be created with RC2 values of R4 = 2.2 megohms and C2 = 0.1 μF (see Table L34-3).

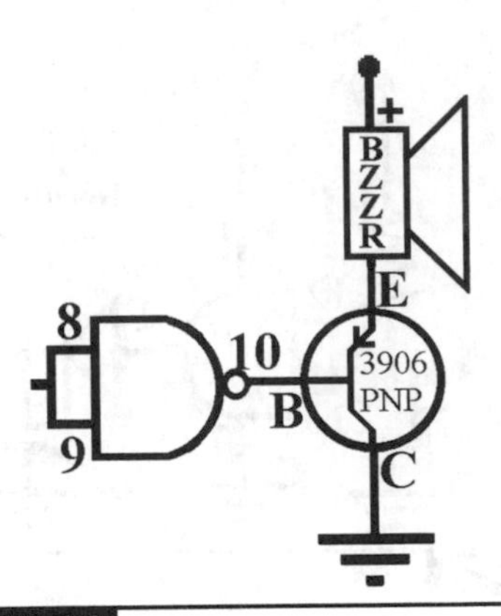

Figure L34-24

Speakers

Speakers need a signal to be heard. If you put only voltage to a speaker, you will hear a crackle as the voltage is turned on. Nothing more. The speaker needs a signal generated by RC2. A 1,000-Hz signal generated by using a 0.01-microfarad capacitor and 22-kilo-ohm resistor is a very noticeable sound. The PNP transistor shown in Figure L34-25 amplifies the strength of the signal.

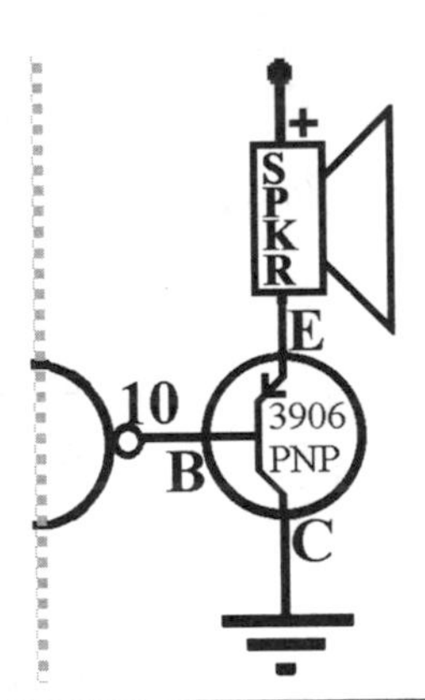

Figure L34-25

Relays

Relays allow us to use the 9-volt system to control power for another system. The on/off to the second system is connected through the relay.

Relays can be used in a variety of places, but they are best used in the following:

- Car alarms. A slowly pulsed relay connected to a squawker creates a sound unique from all the other car alarms we've come to ignore.

- Control the power to 120-volt circuit. This can be used for Christmas lights. The sun goes down; the lights automatically come on.
- Nonoscillating toy motor circuits, instead of direct connection to pin 10. The best results happen here when the motor uses a separate power supply and won't work off 9 volts.

This option presents easy rewiring. The on/off power to the motor is routed through the relay.

Here is a quick explanation of how a relay works. As the current flows toward ground, a magnetic field expands, creating an electromagnet that closes a switch. The *diode* shown in Figure L34-26 is *vital* because there is a close relationship between electric current and magnetism. When the electricity is turned off, the collapsing magnetic field actually pushes the current backward. The reversed diode across the relay helps to control the backward surge of voltage pressure and current created by the collapsing electromagnetic field. If it is not in place, the transistor will quickly burn out.

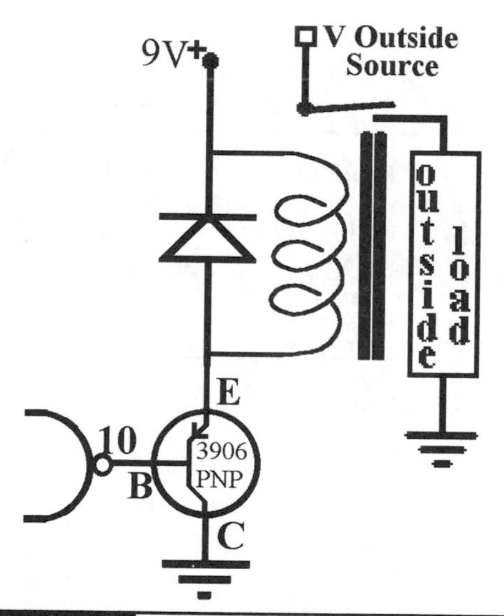

Figure L34-26

Motor

Depending on your needs, a small motor might work directly connected to a transistor, as shown in Figure L34-27. The best motors for this purpose

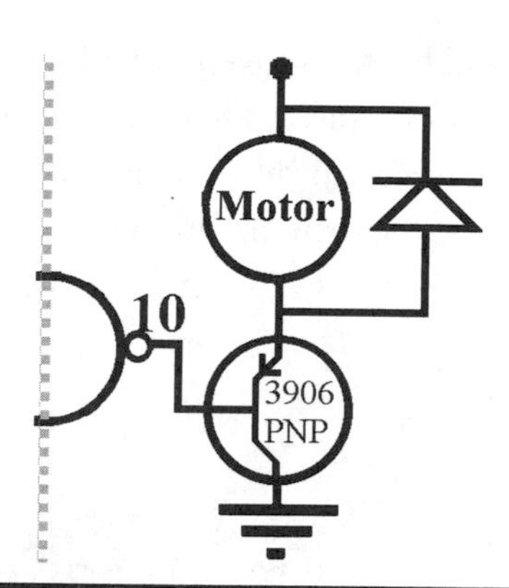

Figure L34-27

are the miniature vibrator motors made for cell phones. These can be purchased through electronics surplus suppliers found on the Internet.

For most small motors, a relay would provide much better results.

Electric motors use electricity and also generate electricity.

Try this with the motor outside of the circuit, if you are interested.

Connect the DMM across the motor leads and spin the shaft first one way and then the other. The motor also uses current and magnetic fields to create movement.

The reversed diode across the small motor helps to control the backward pressure of the extra voltage created by the motor. If it is not in place, the transistor would quickly burn out.

Examples

Each of the systems shown was conceived and designed by people just learning electronics.

A Pop Can Motion Detector

The weight on a spring input was "tuned" precisely so it would start as someone walked by the table that it was on. By the time they stopped and turned around to look where the noise was coming from, it would stop. All they would see was normal-looking junk on the table.

The pop had been removed via a hole in the bottom. The rim on the top had been sanded down so the lid was removed intact. A picture is shown in Figure L34-28. The related schematic is displayed in Figure L34-29 (see also Table L34-5).

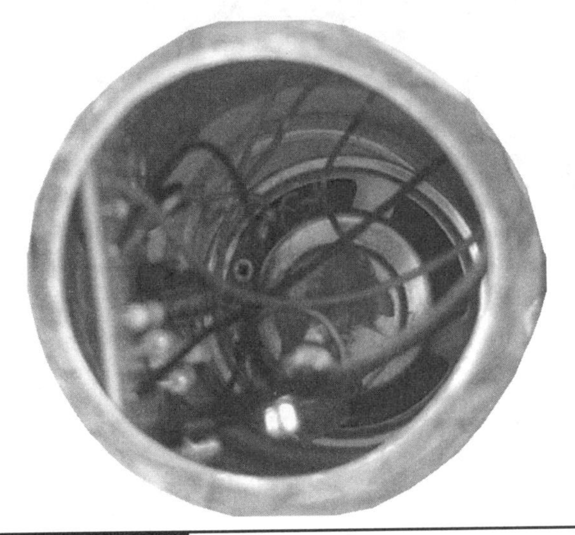

Figure L34-28

TABLE L34-5 Related to Schematic Displayed in Figure L34-29		
Input	**Processor**	**Output**
Motion detector	RC1 = 5 s	Speaker (amplified)
	RC2 = 1,000 Hz	

The Gassy Cow

This is definitely a young man's idea of fun. An MPG file on the website www.mhprofessional.com/computingdownload shows the real action of this fun toy. Words simply don't do it justice. A picture is shown in Figure L34-30. The related schematic is displayed in Figure L34-31 (also see Table L34-6).

Figure L34-30

TABLE L34-6 Related to Schematic Displayed in Figure L34-31		
Input	**Processor**	**Output**
Light Detector	RC1 = Instant On/Off	Speaker (amplified)
	RC2 = 80 Hz	

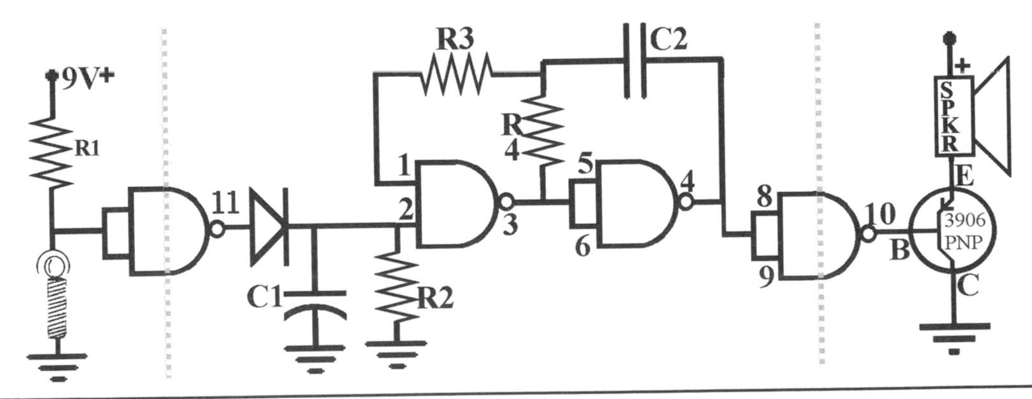

Figure L34-29

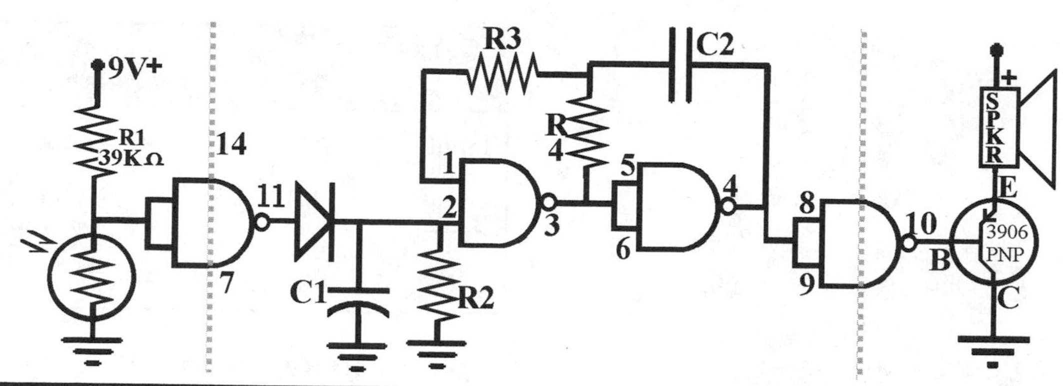

Figure L34-31

Shadow Racer

An MPG file on the website www.mhprofessional.com/computingdownload shows the action of this race car. Wave your hand over the top, and away it goes. It has an on/off switch; otherwise, it would want to go all night. A picture is shown in Figure L34-32. The related schematic is displayed in Figure L34-33 (also see Table L34-7).

TABLE L34-7 Related to Schematic Displayed in Figure L34-33

Input	Processor	Output
Dark Detector	RC1 = 10 s	Small Motor
	RC2 = Disabled	

Figure L34-32

Jiggle Me Teddy

This one proves that other familiar toys are no great works of genius, just great works of marketing. A picture is shown in Figure L34-34. The special motor setup for creating the jiggle is shown in Figure L34-35. The related schematic is displayed in Figure L34-36 (also see Table L34-8).

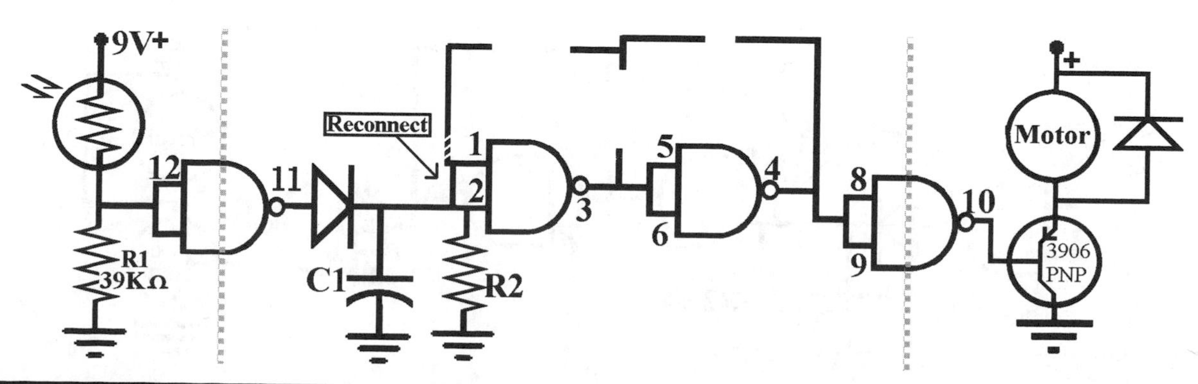

Figure L34-33

Figure L34-34

Figure L34-35

There are two film canisters inside the teddy bear. The motion detector is a spring in a loop. The motor has a weight soldered onto its shaft. Both are sealed inside film canisters that keep them from getting caught up in the stuffing.

TABLE L34-8 Related to Schematic Displayed in Figure L34-36		
Input	**Processor**	**Output**
Motion detector	RC1 = 10 s	Motor with Eccentric
	RC2 = Disabled	

Supercheap Keyboard

This is a particularly challenging application. An MPG demonstration is given on the website www.mhprofessional.com/computingdownload.

The initial input is a touch switch, just to make the circuit active.

The second input changes the value of R4 in RC2. That changes the output frequency. RC2 is given stability by having a 20-megohm resistor connected. A picture is shown in Figure L34-37. The related schematic is displayed in Figure L34-38 (also see Table L34-9).

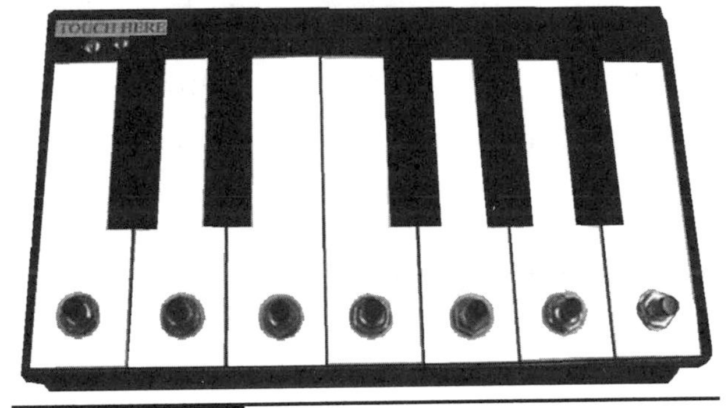

Figure L34-37

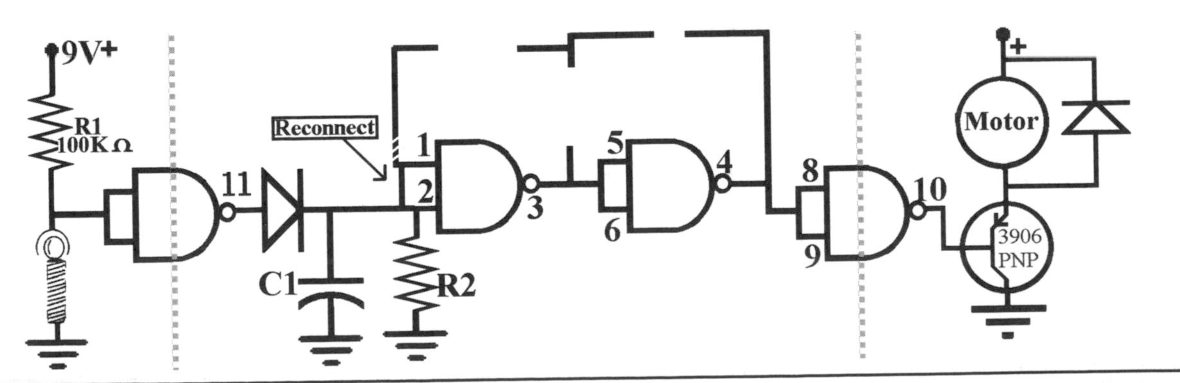

Figure L34-36

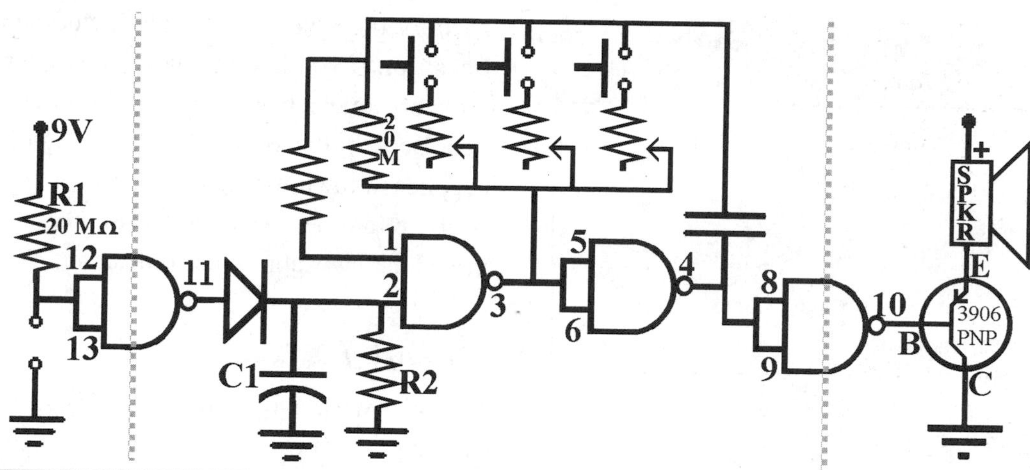

Figure L34-38

TABLE L34-9 Related to Schematic Displayed in Figure L34-38		
Input	**Processor**	**Output**
Touch Switch and push buttons (RC2)	RC1 = Instant On/Off	Speaker amplified
	RC2 = Various	
	C2 = 0.01 QF	
	R4 = Various	

Heartthrob Teddy

Even though this is a basic application, it is worth mentioning that as a child's toy, it is still a favorite. Kiss the bear on his nose and his heart throbs. A picture is shown in Figure L34-39. The related schematic is displayed in Figure L34-40 (also see Table L34-10).

Figure L34-39

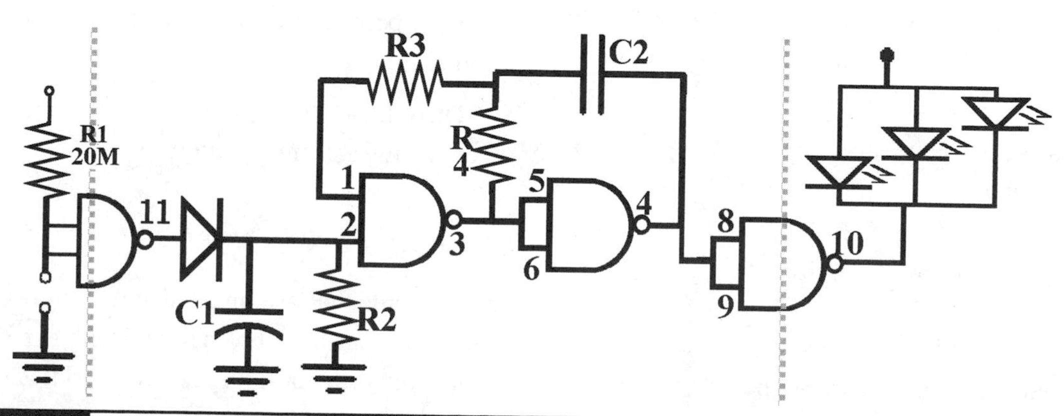

Figure L34-40

TABLE L34-10 Related to Schematic Displayed in Figure L34-40		
Input	**Processor**	**Output**
Touch Switch (pinheads on nose)	RC1 = 10 s	LEDs
	RC2 = 2 flashes/ second	

Lesson 35
Consider What Is Realistic

Completing this project should take no more than a few hours of applied time.

Designing the Enclosure

Use what is available around you. Consider the projects shown as examples. None of them required the students to "build" the enclosure. Each of these applications was designed with the idea of putting the circuit into something premade.

KISS

Remember the KISS principle: Keep It Simple, Students! Is your initial design realistic?

Can this circuit do what you imagine it can? It is not a clock! It is *not* a radio! If you want to have two different outputs, this system is severely limited. You could have a buzzer pulsing and an LED flashing easily, but it becomes too complex to have the circuit control a music chip and a motor at the same time.

Keep It Simple, Students! This is an application of your learning.

Parts, Parts, Parts

What parts are available to you?

If you purchased the kit through www.abra-electronics.com, you have what you need for the basic application, including the transistor. If you live in a larger city, there is probably an electronics components supplier near you. Look in the Yellow Pages. If you don't have a supplier in town, find and order your components over the Internet. An excellent source, reasonably priced, is www.abra-electronics.com.

The Level of Difficulty

Consider what is realistic when designing and building your project.

A simple idea applied with imagination will better impress people than a complex idea that is never finished.

What is a "counterproductive" design?

A motion detector in a toy car. It starts when you jiggle it and keeps going because it jiggles itself. It is like a screen door on a submarine. It helps keep the fish out.

If you find the material fairly easy, create a simple project now so you can keep moving.

If you find the material difficult, create a simple project now so you won't get bogged down here and can keep moving through the course.

Time

Real limitations have to be balanced against the available time.

There are two general things I have found in life.

Most people are always in such a rush. Do this! I gotta do that!

There is never enough time to do it right, but there is always time to do it over.

Note Regarding the LDR

Many items that use an LDR need an on/off switch. Think of your little brother or sister. Time for bed. Lights out—the doll's eyes just keep flashing.

Or you can just keep the lights on at night.

Safety

What if you want to use a relay to switch on a 120-volt AC circuit? My answers are:

1. Who do you know that is comfortable working with 120 volts? Get their help! Safety first!
2. If you don't have direct, knowledgeable help, modify your expectations. Use the relay to switch a smaller voltage! Safety last!
3. What you don't know can hurt you! Safety always!
   a. You need to have a proper relay and enclosure for the 120 volts.
   b. Mount the PCB properly into the enclosure. At 9 volts, you can be sloppy.
   c. Enclosure considerations need to meet certain standards for higher voltages.
   d. Remember soldering considerations for 120 volts. You can have messy soldering for a 9-volt system and have it work, or not, and still be safe.
   e. A messy 120-volt system could spark up an otherwise uneventful day.

Practice Designing Systems

Describe the two systems in Tables L35-1 and 35-2.

Assume that these sample systems use only a few parts you don't already have.

1. Given a musical doorbell, explain the system.
2. Your system can be used to create a car alarm. Support each description with a schematic.
3. Many car alarms turn off when they are tilted more than 20 degrees, allowing for quiet towing. Imagine some type of automatic input that could cut the power in this situation. Describe how this would work.
4. No matter how you apply this system, the electronics are similar. The application is defined by the inputs, outputs, and the enclosure.

Develop five applications for this project. Look at all of them carefully. Scribble a quick system diagram for each of them. Then toss three. Now commit to one of the two remaining. Move forward.

TABLE 35-1 The Music Chip Is Its Own System, Powered by the 4011's Hi Output

Input	Processor 1 (RC1)	Processor 2 (RC2) There is no oscillator (RC2) in this system.	Output Connect pin 10 to chip's V+
Push Button (N.O.)	Timed to the length of the music chip's tune.		Amplified to regular speaker.
	Schematic	Schematic	Schematic

TABLE 35-2 The Amplified Output Has Power to Turn on a Squawker Designed for Alarm Systems

Input	Processor 1 RC1	Processor 2 RC2	Output
Description	Description	Description	Description
Schematic	Schematic	Schematic	Schematic

Lesson 36
Building Your Project

Here you get to solder your parts onto the printed circuit board. Yes, the same PCB is being used for all of the many different applications. The processor is essentially the same. You have developed an application by defining and designing different inputs and outputs. You also have to find a reasonable enclosure to use for your defined application.

This lesson deals with parts placement. Close attention is paid to the different variations of inputs, processors, and outputs. Figure L36-1 shows the bottom view of the printed circuit board. Figure L36-2 shows the same PCB from the top view. Note that the voltage and ground lines have been displayed with different textures.

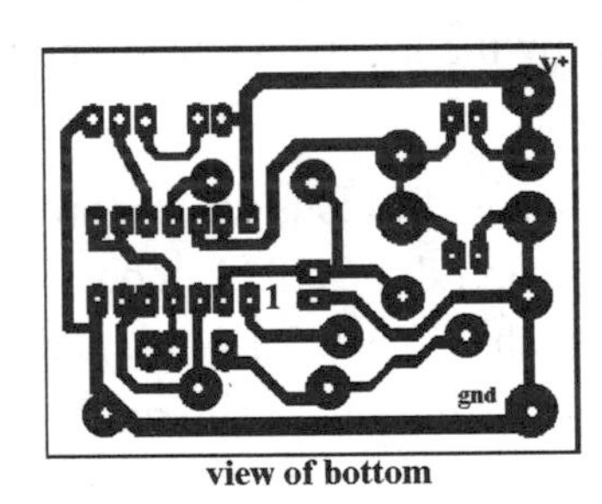

Figure L36-1

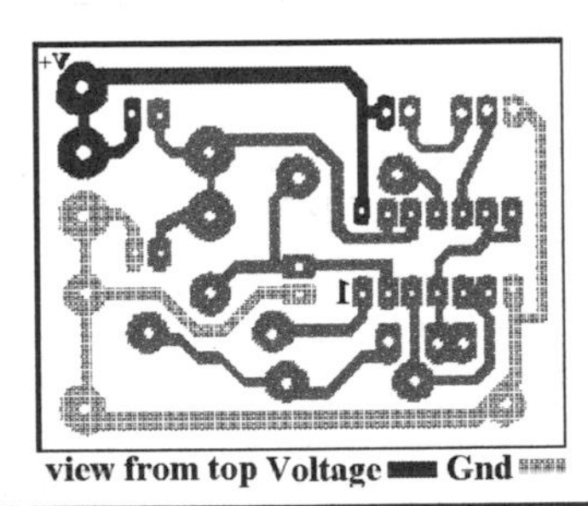

Figure L36-2

The parts placement shown in Figure L36-3 is for a standard application with low-power output. Note that the chip seat is soldered into the PCB. The 4011 can be inserted and removed as needed.

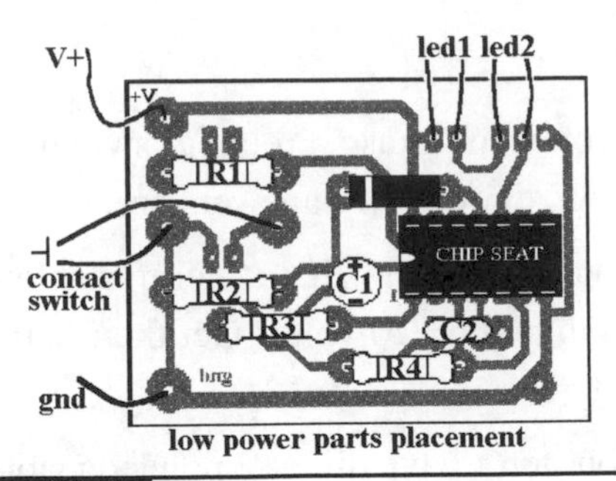

Figure L36-3

Parts are shown in place for high-powered output in Figure L36-4. Again, this is a standard application for the high-power output.

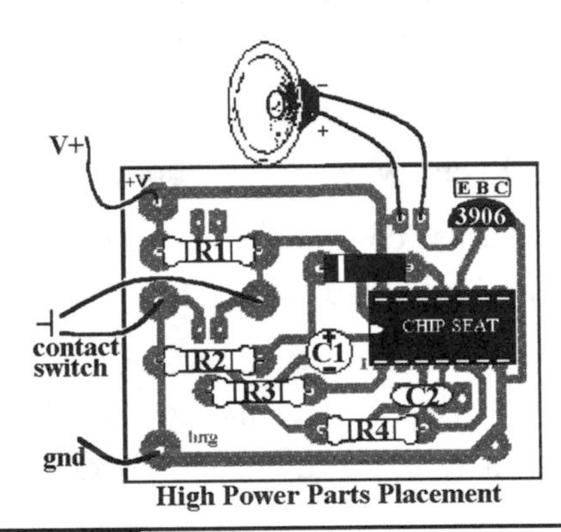

Figure L36-4

Inputs: Variations and Parts Placement

A close look at Figure L36-5 reveals the small difference in parts placement between the light detector and the dark detector. Use the same pads for hooking up the touch switch as shown in Figure L36-6.

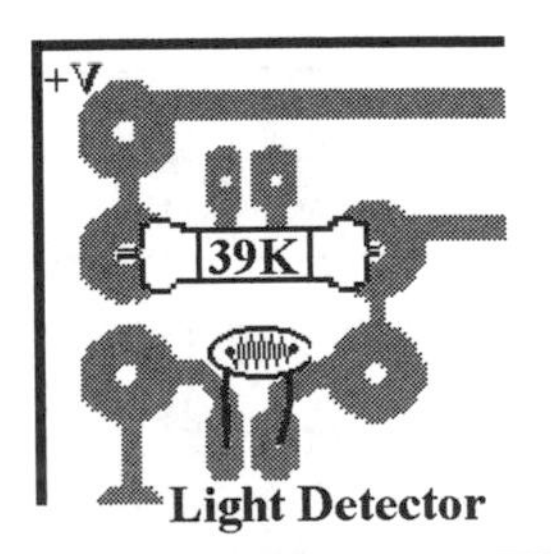

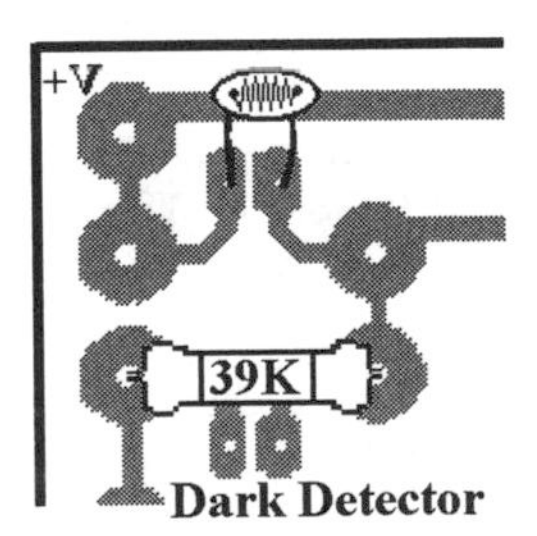

Figure L36-5

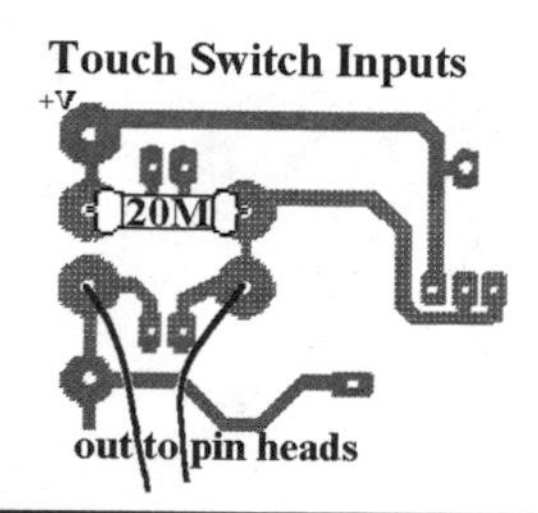

Figure L36-6

RC1: Variations, Timing, and Parts Placement

Remember that this is only a rough guide. It is not a precise timer (see Table L36-1).

TABLE L36-1 Rough Guide

Resistor	Capacitor	Time Output
20 MΩ	10 μF	120 s
10 MΩ	10 μF	60 s
4.7 MΩ	10 μF	30 s
20 MΩ	1 μF	12 s
10 MΩ	1 μF	6 s
4.7 MΩ	1 μF	3 s

If you want to add the ability to adjust the timed-off setting manually, Figure L36-7 shows where to connect the wires that would lead to the contact points that you would touch.

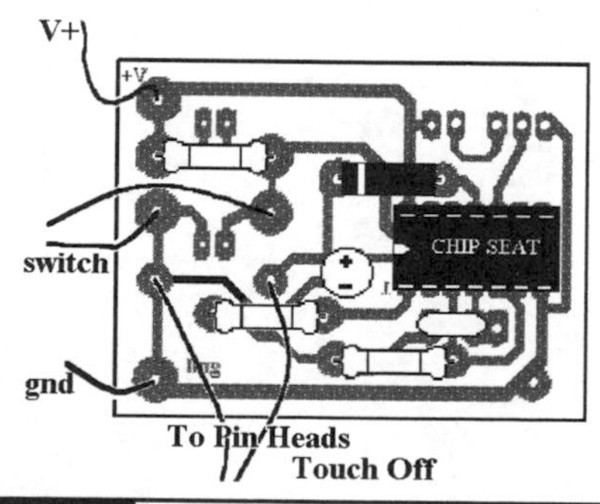

Figure L36-7

RC2: Variation and Timing

Remember that this is only a rough guide (see Table L36-2). The 4011 oscillator is not a precise timer.

TABLE L36-2 Rough Guide

Resistor	Capacitor	Time Output
2.2 MΩ	.1 μF	1 Hz
1 MΩ	.1 μF	2 Hz
470 kΩ	.1 μF	4 Hz
220 kΩ	.1 μF	10 Hz
100 kΩ	.1 μF	20 Hz*
47 kΩ	0.1 μF	40 Hz
22 kΩ	0.1 μF	100 Hz
1 MΩ	0.01 μF	20 Hz*
470 kΩ	0.01 μF	40 Hz
220 kΩ	0.01 μF	100 Hz
100 kΩ	0.01 μF	200 Hz
47 kΩ	0.01 μF	400 Hz
22 kΩ	0.01 μF	1,000 Hz
10 kΩ	0.01 μF	2,000 Hz
4.7 kΩ	0.01 μF	4,000 Hz
2.2 kΩ	0.01 μF	10,000 Hz

* The eye sees continuous action at anything faster than 24 Hz.

The most common option that people want, however, is to disable RC2. The most effective way to reconnect pin 1 and pin 2 is displayed in Figure L34-8.

If at first your circuit does not work, then it is time to do some troubleshooting. Refer back to Lesson 26.

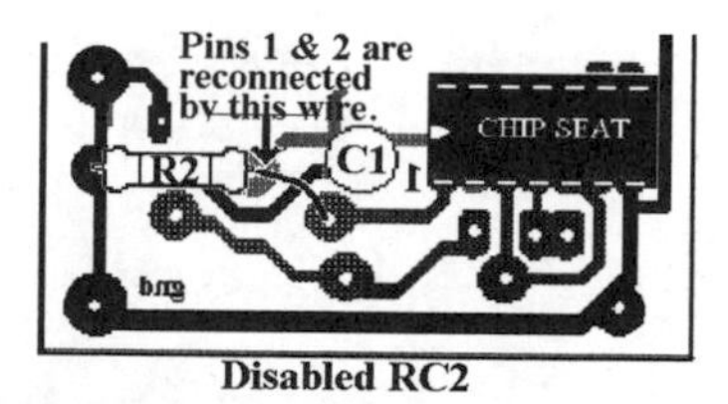

Figure L36-8

PART THREE

Counting Systems in Electronics

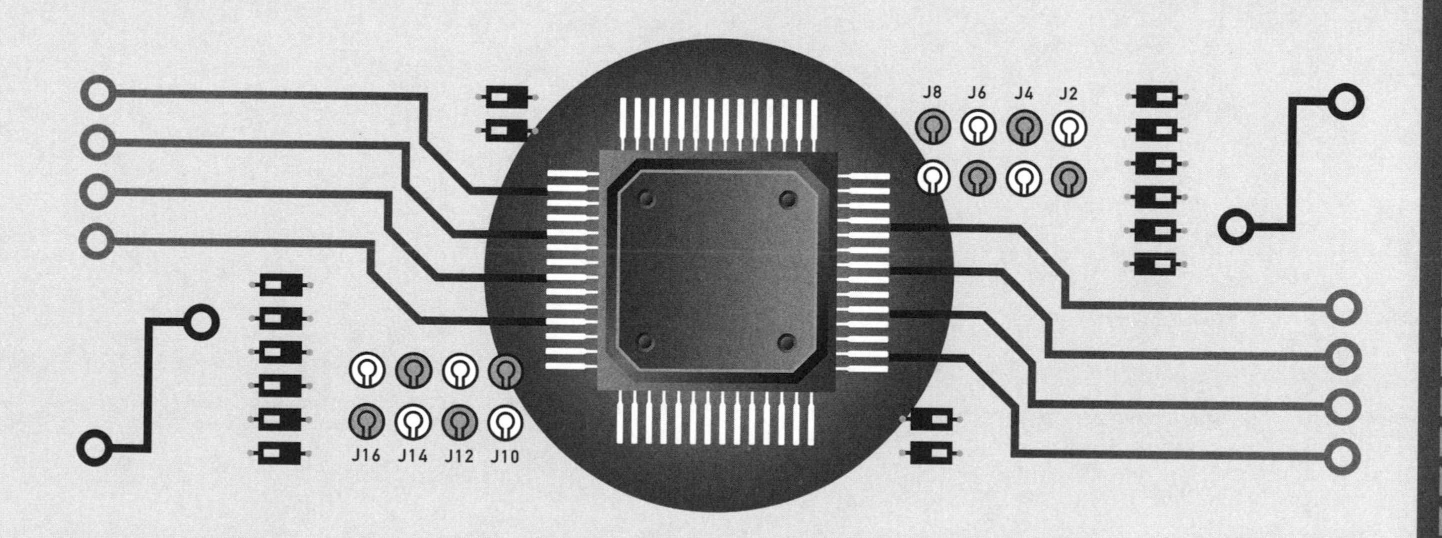

Systems within Systems

Remember that the intention of this course is to deepen your understanding of electronics. The best way to understand electronics is by thinking of it as "systems."

Systems can be simple, but they are usually complex, composed of multiple layers and subsystems, as shown in the figure below.

1. Visible subsystems include:
   a. Cars and trucks, and highways
   b. Trains and tracks
2. And the not so visible:
   a. Manufacturing
   b. Maintenance and repair
   c. Producing and transporting fuel
   d. Marketing

The Parts Bin on the next page has the complete inventory for Part Three. Make every part count.

This small portion of the transportation system provides an appropriate analogy.

PARTS BIN FOR PART THREE

Description	Type	Quantity
5.1 z 1N4133	Zener diode	1
1N4005	Power diode	5
2N-3904 NPN transistor	TO-92 case	1
LEDs	5 mm	15
7 SEG CC .56"	7-segment display	1
100 Ω	Resistor	1
470 Ω	Resistor	15
1,000 Ω	Resistor	1
22,000 Ω	Resistor	1
47,000 Ω	Resistor	1
100,000 Ω	Resistor	15
220,000 Ω	Resistor	1
470,000 Ω	Resistor	1
1,000,000 Ω	Resistor	1
2,200,000 Ω	Resistor	1
10,000,000 Ω	Resistor	5
20,000,000 Ω	Resistor	4
.1 μF disk or film	Cap	2
1 μF radial 15 v	Cap	3
22 μF radial 15 v	Cap	1
4017 Walking ring	IC	1
4046 VCO (A to D)	IC	1
4511 7-segment control	IC	1
4516 (D to A)	IC	1
4011 Quad NAND gate	IC	1
Oscillator input	PCB	1
Dual input	PCB	1
4046 + Timed Off	PCB	1
7-segment system	PCB	1
4017 Walking ring	PCB	1
PBNO momentary	Hardware	1
Battery clip	Hardware	1
LED collars set	Hardware	10
Socket (14 pin DIP)	Hardware	1
Socket (16 pin DIP)	Hardware	4

SECTION 11

Introducing an Analog-to-Digital Converter

HERE IS A PREVIEW OF SOME of the fun possibilities that can be designed with the knowledge you will learn in Part Three. You also get a reminder regarding the care and feeding of your CMOS ICs.

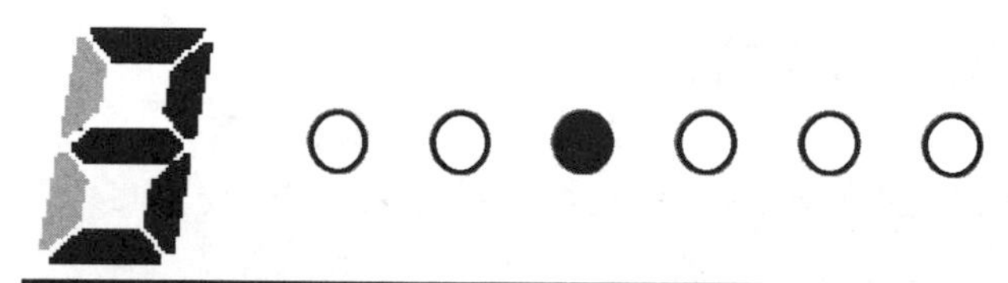

Figure L37-1

Lesson 37

Introducing Possibilities—Electronics That Count

The DigiDice shown in Figure L37-1 is the basic system that you will build as a prototype before you begin to design your own application.

Here, a varying analog voltage input is changed into a random number generator (see Table L37-1).

Thousands of applications and toys can be developed from these components. You've certainly seen some of these at the mall or casino.

There are simple fortune tellers, lottery number generators, light chasers, animated signs, slot machines, and many more. You might have even spent some money on them. Ideas are explored in further depth in Lesson 47.

Your focus should be understanding how the components can relate to each other, much as smaller pieces shown in Figure L37-2 relate to larger structures and models they are used to build.

Each part is a piece of the larger unit, and each unit can be a piece of the larger system.

TABLE L37-1 Random Number Generator

Input	Processor	Output
Push button	**1.** Roll down (4046 IC) controlling 2 and 3	**1.** Fast cycling through 6 LEDs.
	2. Walking ring 6 LEDs (4017 IC)	**2.** Fast cycling through numbers 0 through 9 shown on the number display.
	3. Decimal-counting binary (4510 IC)	**3.** Cycling slows steadily to a completely unpredictable stop.
	4. Binary-counting decimal (4511 IC); seven-segment display	**4.** Displays of both fade about 20 s after the cycling stops, and the system waits to be triggered again.

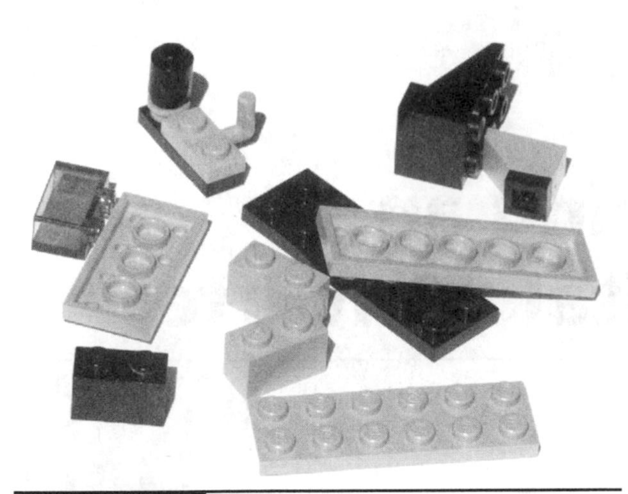

Figure L37-2

Safety First, Last, Always

From the *CMOS Cookbook* by Don Lancaster, page 50:

"New CMOS ICs from a reliable source are almost always good and, with a little common-sense handling, practically indestructible. Possibly you will get two bad circuits per hundred from a quality distributor, maybe a few more from secondary sources unless you are buying obvious garbage. In general, the ICs are the most reliable part of your circuit and the hardest part to damage.

"If your CMOS circuit doesn't work, chances are it is your fault and not the IC's. Typical problems include:

1. Forgetting to tie down inputs
2. Forgetting to debounce and sharpen input clocking signals
3. Getting the supplies connected wrong, totally unbypassed, or backward
4. Putting the ICs in upside down
5. Doing a PCB layout topside and reversing all the connections
6. Missing or loosening a pin on a socket or bending a pin over
7. Misreading a resistor (have you ever noticed the color-code similarity between a 15-ohm resistor and a 1-megohm resistor?)

"And of course, causing solder splashes and hairline opens and shorts on a printed circuit board.

"The key rule is this: Always BLAME YOURSELF FIRST and the ICs LAST. Always assume that there is something incredibly wrong with your circuits when you first power them up. You'll be right almost every time. In fact, if things seem to work perfectly on the first try, this may mean that the real surprises are hiding, waiting to get you later or when it is more expensive to correct them. Anything that 'has' to be correct is usually the mistake. And what seems like 'impossible behavior' is really the poor IC trying its best to do a good job. With a little help and the right attitude you can help the ICs along."

Remember: What seem like impossible outputs really are impossible outputs.

Lesson 38
RC1—Creating the Switch

Here you are introduced to the zener diode, used inside an RC circuit. It modifies the input to the 4046 IC.

The switch for this circuit is an RC that provides voltage from 4 volts to 0 volts as the resistor/capacitor circuit drains. But doesn't my power supply provide 9 volts? How do I get 4 volts? For this task, you will use a zener diode. Three common types of diodes are displayed in Figure L38-1 (also see Table L38-1).

A zener diode allows for two-way traffic, but only if there is enough voltage pressing backward. Here's what that means. When voltage is applied to

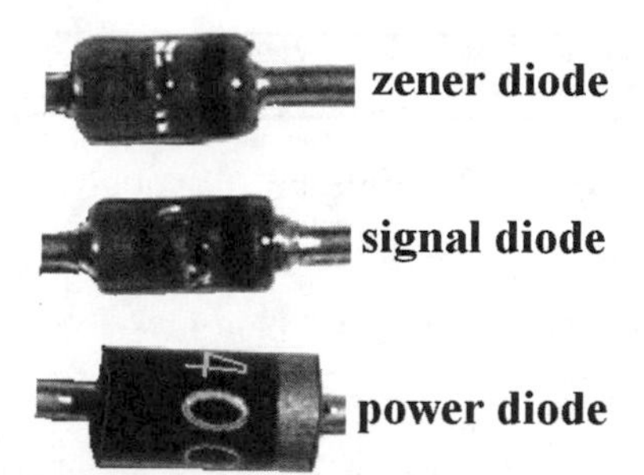

Figure L38-1

TABLE L38-1	Common Types of Diodes
Zener diode	One-way valve used as counterflow lane during rush hour. It is for light traffic.
Signal diode	One-way valve to handle light traffic (smooth, one-lane, unpaved road).
Power diode	One-way valve able to handle large traffic demands (four-lane highway). Marked as 1N400#.

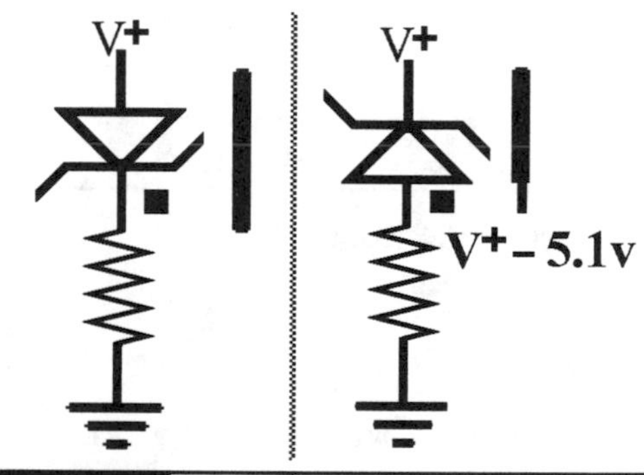

Figure L38-2

the positive side (anode) of the zener diode, all the voltage passes through. This is shown in the left side of Figure L38-2.

But on the right side of Figure L38-2 the voltage is applied to the negative side (cathode) of the zener. In that situation only a predetermined amount is blocked. That is called the zener diode's *breakdown* voltage. Simply put, that is when the diode's properties break down. The zener diode we are using has a breakdown voltage rated at 5.1 volts.

It is important to note that zener diodes are labeled the same way as other diodes, but they must be put in "backward." So even though the black line still refers to the diode's cathode, as shown in Figure L38-3, a zener cathode is pointed toward the positive. Note the extra squiggle on the cathode bar that identifies the diode as a zener.

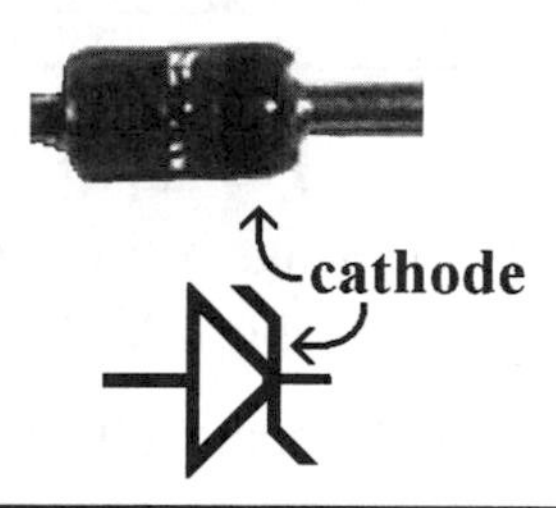

Figure L38-3

Zener diodes are available in ranges of 2 to 20 volts.

It is very important to note that your power supply must be 7 volts or more. If you use a power diode as a "protection" device, it will drop voltage by nearly 1 volt. And 7 volts minus 1 volt leaves 6 useful volts. Now the zener blocks 5 volts. Thus, 6 minus 5 volts leaves 1 volt; 1 volt left to play with. You can't do much with 1 volt. This is particularly a concern if you are using a 9-volt battery as your power source.

Exercise: RC1—Creating the Switch

1. You are going to measure the breakdown voltage of your zener diode. It is rated at 5.1 volts.

   Figure L38-4 has two different setups for you.

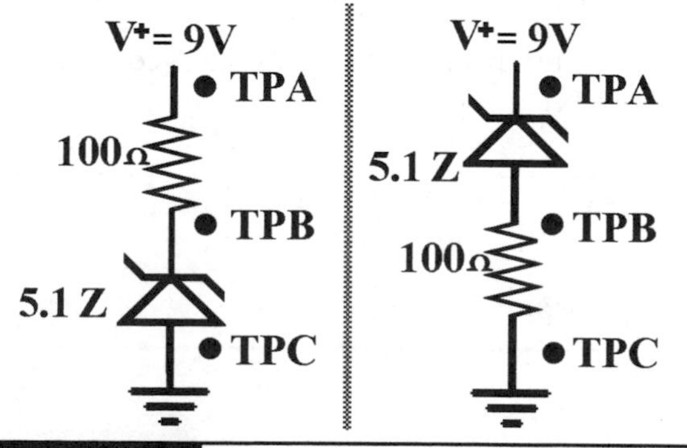

Figure L38-4

TABLE L38-2 Two Different Setups

For the setup on the left	For the setup on the right
Total V from TPA to TPC = _____	Total V from TPA to TPC = _____
V used from TPA to TPB = _____	V used from TPA to TPB = _____
Volts used across zener = _____	Volts used across zener = _____
How accurate is the rating? _____	How accurate is the rating? _____

The voltage used by any component in a circuit does not change because of its position in the circuit (see Table L38-2).

2. Now check out the actual effects of the zener diode (see Table L38-3). Figure L38-5 shows two circuits. Breadboard each separately and do the measurements. Note the LED in each.
3. Define "breakdown" voltage.

   ____

4. If a zener diode has a breakdown voltage of 7.9 volts, how much voltage remains after the diode when V+ is 12 volts? __________

Breadboard This Circuit

Build the circuit shown in Figure L38-6 on your breadboard (see the Parts Bin).

PARTS BIN

- PB—Normally open push button
- D1, D2—Power diodes
- R1—10 MΩ
- DZ—5.1 Zener diode
- C1—1 μF electrolytic

TABLE L38-3 Two Different Setups

For the setup on the left	For the setup on the right
TPA to TPD (voltage to ground)	TPA to TPD
TPA to TPB	TPA to TPB
TPB to TPC	How much voltage is available to the LED at this point?
How much voltage is available to the LED at this point?	TPB to TPC
TPC to TPD	TPC to TPD
Describe the brightness of the LED compared to the next setup. ______________________________	Describe the brightness of the LED. ______________________________
	Why is the LED a different brightness from the previous setup? ______________________________

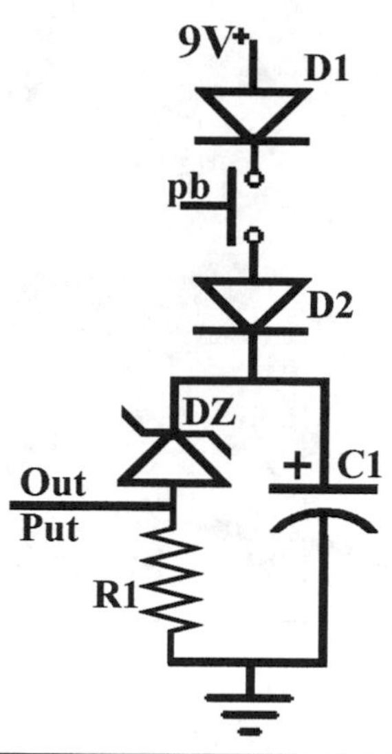

Figure L38-6

This is the switch that you will use for the larger system. But first, do some initial voltage testing at the output. Check for a few critical items.

1. What is the peak voltage?
2. How long does it take to drain to 0 volts?
3. How sensitive is it?
   a. Is any voltage fed through if you touch across the bottom of the push button?
   b. Substitute an LDR into the spot for the push button. Anything?
4. Change the values for R1 and C1. Does the RC act the way you would expect?
5. Take the zener diode out. Replace it with a wire. Do the measurements again. What is the biggest effect that the zener has on this switch?

Lesson 39
Introducing the 4046 Voltage-Controlled Oscillator

The 4046 chip takes the analog voltage from the RC switch and changes it directly into digital "clock" signals (Table L39-1). It is the "analog-to-digital" converter. How does it count? Digitally!
0 – 1 – 0 – 1 – 0 – 1 – 0

The voltage input at pin 9 determines the frequency output at pin 4. Breadboard this circuit (see the following Parts Bin).

Pay attention to the schematic in Figure L39-1. Only three pins are connected on the top side of the IC. Don't accidentally build on top of the 4046.

TABLE L39-1 System Diagram of Our Application of the 4046 IC

Input Pin 9 from RC1	Processor VCO	Output Rolldown Clock Signal
The rate of voltage drain is controlled by the value of R1 and C1 (RC1).	The VCO in the 4046 compares the input voltage at pin 9 to the supply voltage at pin 16.	The length of the rolldown is controlled by RC1.
The voltage from RC1 is moving from about 4 volts to ground.	There is a direct relation between the input voltage and the speed of oscillation at pin 4, the output.	The output starts quickly but "rolls down" to a complete stop.
	The higher the input voltage, the faster the oscillation output. The maximum oscillation is controlled by R2 and C2 (RC2).	**Definition:** A clock signal is a rise from 0 V to V+ in less than 5 μS (1 μ = 1 micro = 0.000001).

PARTS BIN

- R1—10 MΩ
- R2—220 kΩ
- C1—1 μF electrolytic cap
- C2—0.1 μF film or disk cap
- D1—1N400# power diode
- D2—1N400# power diode
- DZ—1N1N4733 5.1 Zener diode
- PB—Normally open PB
- LED—5 mm round
- IC1—4046 CMOS

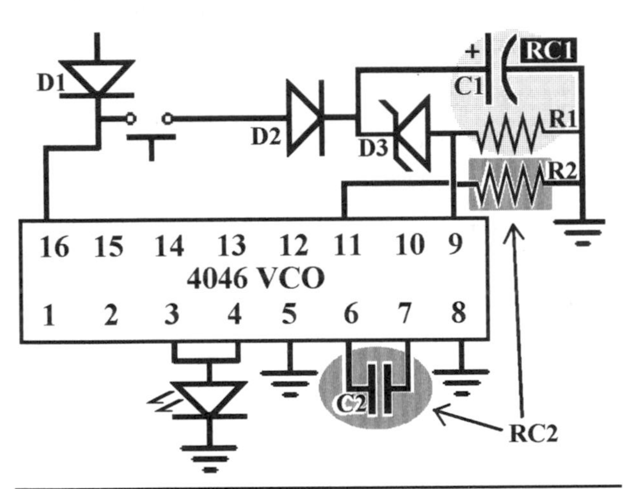

Figure L39-1

Also, note the space restriction that is being imposed. You should squish this initial circuit into the first 15 rows of the breadboard as noted in Figure L39-2.

What to Expect

The circuit should work like this. You push and release the plunger. The LED turns on. As you wait, you notice that the LED appears to be flashing, but it's so fast, you're not sure. As you patiently watch, the flashing becomes very obvious as it slows down. It should slow down to a crawl before it stops. It could stop with the LED on or off. But it will stop. To see an animation of what to

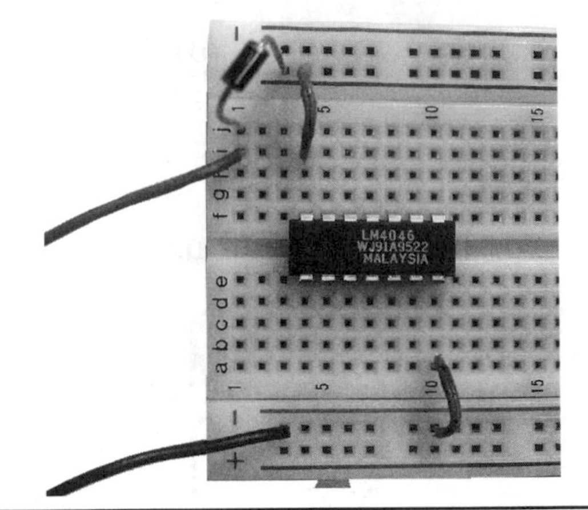

Figure L39-2

expect when the circuit is working, go to www.mhprofessional.com/computingdownload.

1. We are using only the voltage-controlled oscillator portion of the 4046.
2. Breadboard this circuit and briefly test that it is working.
3. Be sure to scrimp on the space as shown in Figure L39-2. This circuit should occupy only 15 rows of the SBB.
4. All components should be laid out north-south or east-west.
5. Wiring should not cross over the top of major components.
6. Wire length should be kept short enough so that your little finger can't fit underneath.

In this schematic, the wire connecting R2 to pin 11 jumps across the wire connected to pin 9.

4046 Data Sheet

Now is the time to look at the 4046 data sheet. Even if your circuit is not immediately successful, *work through the data sheet first*.

The diagram in Figure L39-3 is relevant to all ICs. Like any manufactured product, each chip is marked with all the important information. Who would have thought?

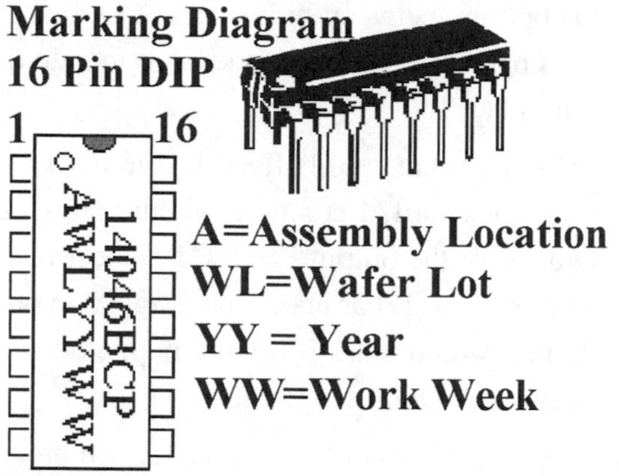

Marking Diagram
16 Pin DIP
1 16
14046BCP
AWLYYWW
A=Assembly Location
WL=Wafer Lot
YY = Year
WW=Work Week

Figure L39-3

The system diagram and pin assignment are both shown in Figure L39-4. As with any data sheet, these are specific to the 4046 voltage-controlled oscillator and phase comparator IC.

1. The 4046 IC is a dual-purpose chip. It has two major processors that can be used independently of each other. For our purposes, we will use the 4046's VCO ability to convert analog voltage input to digital frequency output.
2. As with all ICs, unused inputs must always be tied to an appropriate logic state (either V+ or ground). Unused outputs must be left open.
3. The inhibit input at pin 5 must be set low. When the inhibit is high, it disables the VCO to minimize power consumption in a standby mode.
4. Analog input to the VCO is at pin 9.
5. Digital output of a square wave is at pin 4.
6. The frequency is determined by three factors.
   - The voltage value at pin 9 (as compared to V+ and ground).
   - Cx at pins 6 and 7 and Rmax make the RC that sets the maximum frequency.
   - Cx and Rmin make another RC that sets the minimum frequency. The higher the resistance, the slower the minimum flash rate. In this circuit Rmin is empty; and the resistance is infinite. The minimum flash rate is 0 (full stop) (see Table L39-2).

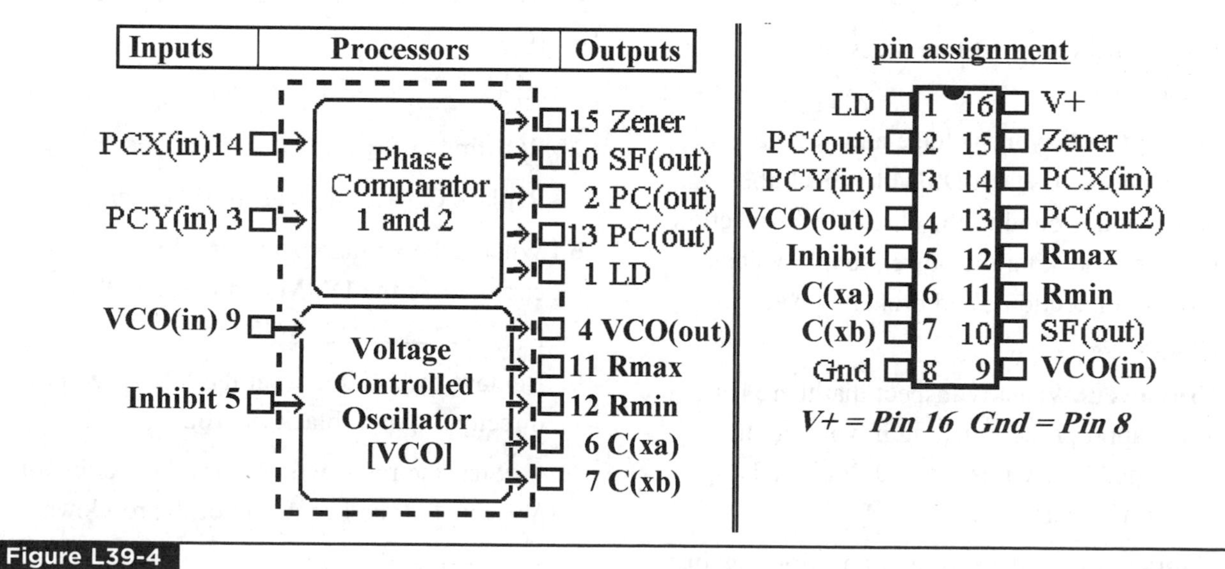

Figure L39-4

TABLE L39-2 Minimum Flash Rate

Input	Processor	Output
Voltage fed to pin 9	Input voltage at pin 9	The VCO produces a square wave at pin 4 [VCO_{out}]
	[VCO_{in}] is compared to V+ and ground.	
	Maximum frequency output is determined by Cx and Rmx.	
	Minimum frequency output is determined by Cx and Rmn.	

An example of the expected output on an oscilloscope is provided online at www.mhprofessional.com/computingdownload. Examples are available for both real scopes and Soundcard Scope.

Troubleshooting PCCP

These are always the first four steps of troubleshooting:

1. **Power.** Check your voltage and ground connections. Use the DMM to check the power and ground to the chip. This inspection includes checking the power supply voltage, proper connections, and broken power connectors.
2. **Crossovers.** Visually inspect that there isn't any inappropriate touching in your circuit. Don't get your wires crossed. We're talking electronics here.
3. **Connections.** Check your connections against the schematic. For example, the schematic shows there are *only* three connections to the chip from pins 9 to 16. There are only five on the bottom, between pins 1 and 8. Did you miss one or get an extra one? Are they in the right place?
4. **Polarity.** What could affect the circuit if it were backward? Let's think about that. Hmm? Diodes, or the output LED. C1 is electrolytic, so it is polar. What about the chip? Do you think it would work if it were popped in backward? Don't try it out to see.

Ninety-five percent of all problems will be found if you work through the PCCP method. Predict what the output should be. Use your DMM to double-check the results that you expect.

Exercise: Introducing the 4046 VCO

1. Here's a question to think about. Look carefully at the system diagram for the 4046. Why is pin 3 connected to pin 4? __________
2. In digital logic, there are two states. They are not Hawaii or Alaska. What are the two logic states? ___________ and ___________
3. Why must unused inputs be conditioned high or low? ______________
4. RC1 determines the length of the rolldown.

   R1 = ___

   C1 = ___

   RC1 timing ____

   Replace C1 with a value 10 times larger.
5. What is the voltage being provided to your circuit? Use the DMM and measure this at pin 16. __________ volts.
6. Remove the LED and set the DMM red probe directly to pin 4, black to ground.

   Start the rolldown. Record the reading of V+ pin 4 at the beginning of the rolldown. __________ V

   As the rolldown begins, the voltage reading at pin 4 should be half of the voltage at pin 16.

7. An oscilloscope (Soundcard Scope) is the best tool to help understand this. With the LED removed, attach the scope probe's red clip to the output at pins 3 and 4. The black clip goes to ground. The plug goes into the "Line In" connection of the sound card. The settings are shown in Figure L39-5.

   Figure L39-6 shows why the DMM reads half of the V+. The output is V+ half the time and 0 volts for the other half. The signal is switching quickly enough that the DMM averages the voltage and reads it as half of V+.

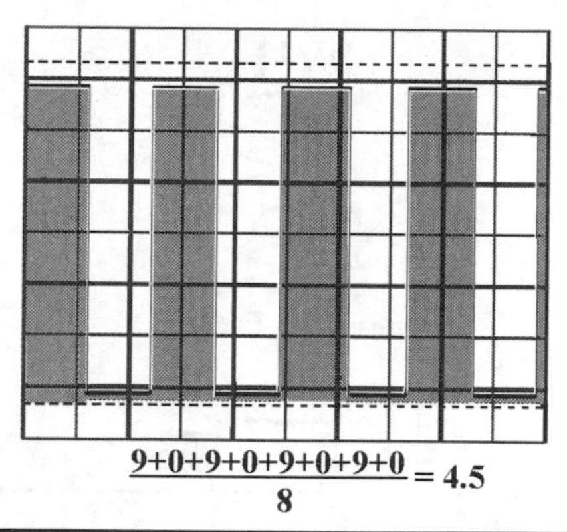

Figure L39-6

8. From the glossary, define "clock signal."

   ____

9. A square wave is made by a clock signal occurring at a steady frequency. Draw a representation of a square wave in the oscilloscope face presented in Figure L39-7.

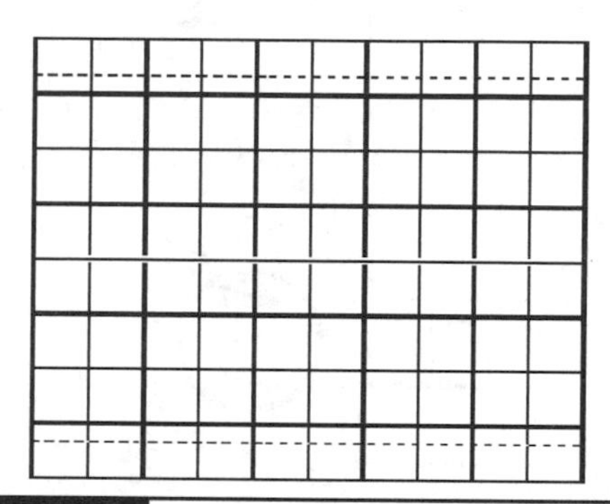

Figure L39-7

10. Modify your breadboard slightly to represent the changes shown in Figure L39-8. This allows you to adjust the voltage using a voltage divider. If the voltage is kept stable at the input to pin 9, what happens to the output at pin 4? ____________________

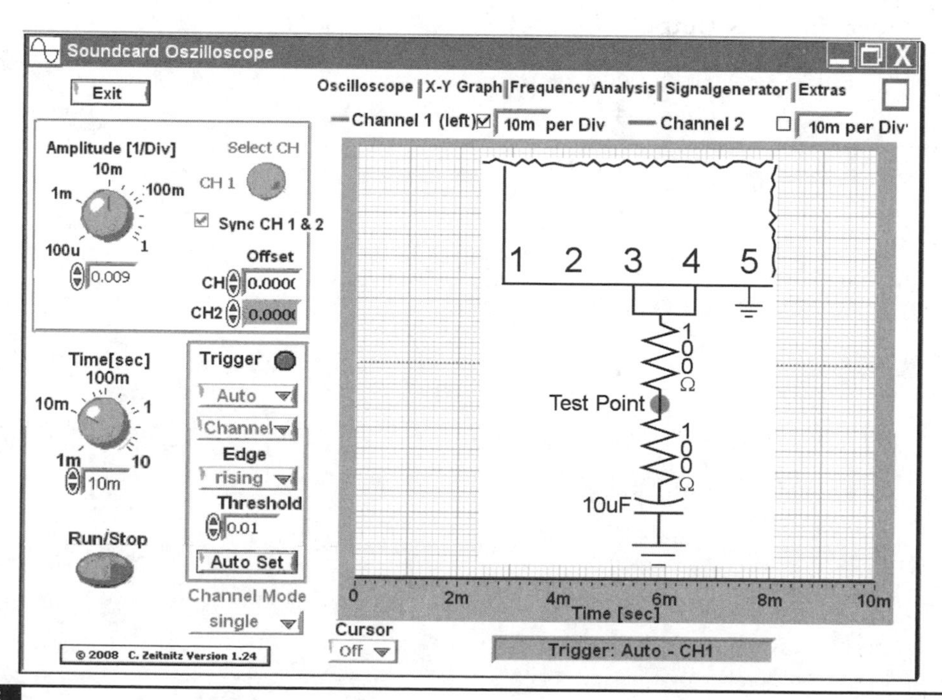

Figure L39-5

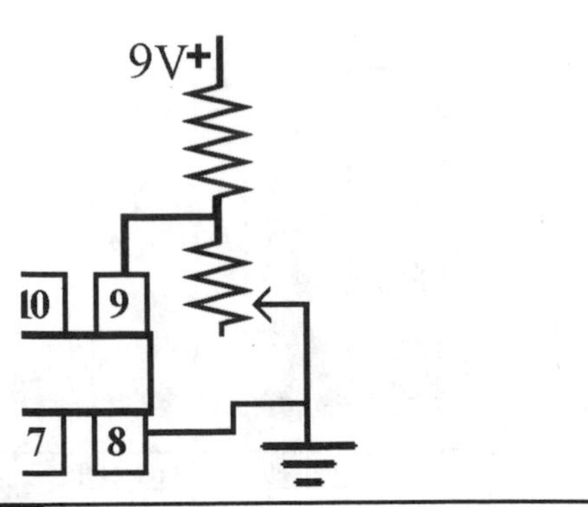

Figure L39-8

11. Consider the schematic presented in Figure L39-9.

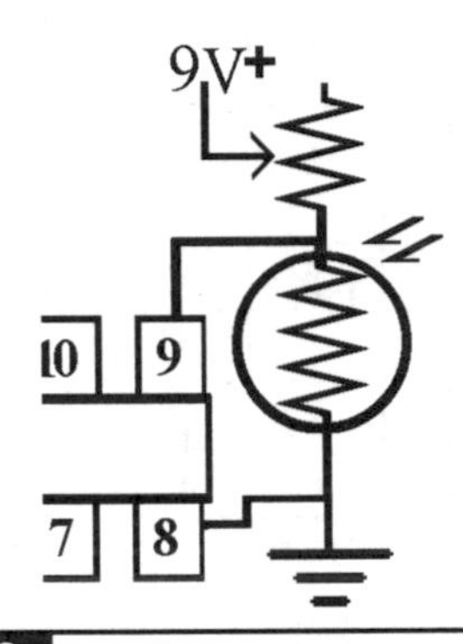

Figure L39-9

Input	Processor	Output
Amount of light	VCO	Frequency changes

12. RC2 controls the maximum frequency. It is made of Rmx and Cx, noted in the schematic as R2 and C2. Predict what would happen if the resistance were increased by 10 times.

    Frequency would be

    a. 10 × C faster

    b. 10 × slower

    Try it out. Change R2 from 220 kilo-ohms to 2.2 megohms. Did it make the frequency faster or slower? Did you predict correctly? Think about it. Did the increased resistance speed up or slow down the drain of C2? Reset R2 back to 220 kilo-ohms.

13. On the 4046 data sheet, pin 12 controls the minimum frequency, Rmn. Right now pin 12 is empty. No resistor means "infinite" resistance. Air is a pretty good resistor. Place a 10-M ohm resistor from pin 12 to ground. Start the rolldown. What is the result?

    If there is less resistance, predict whether the minimum frequency is going to increase (faster) or decrease (closer, to no flashing) when it stops.

S E C T I O N 1 2

The 4017 Walking Ring Counter

SO WE HAVE A CLOCK signal output. It rolls down to a complete stop. Now to expand from 1 LED to 11 LEDs. They don't all go on at once. It actually counts: 1, 2, 3, 4, 5, 6, 7, 8, 9, 10, 1, 2, 3, . . . and higher if you want to use the carryout capabilities.

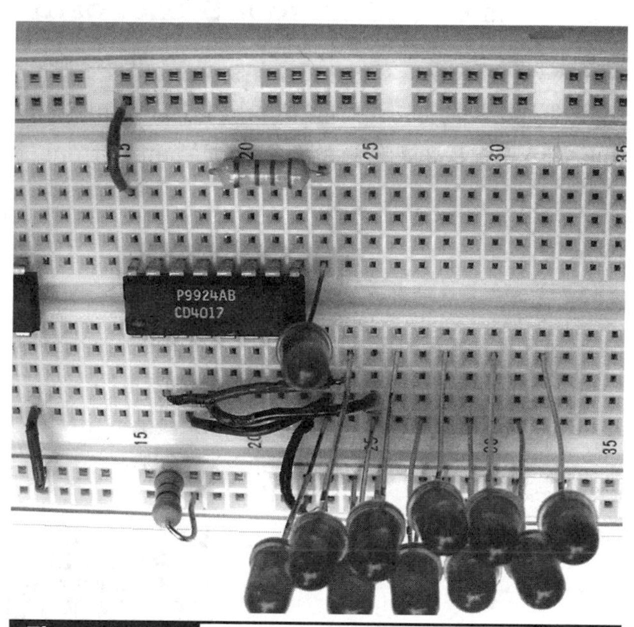

Figure L40-1

Lesson 40

Introducing the Walking Ring 4017 Decade Counter

Here is the system diagram for the prototype application of the 4017 IC (see Table L40-1).

Be very careful not to use too much space. There are still two major circuits to fit onto your board. All of the cathode sides of the LEDs can be ganged onto the unused line on the bottom as shown in Figure L40-1 (see the Parts Bin). The common resistor can connect the LED line to ground.

TABLE L40-1 Prototype Application

Input	Processor	Output
Roll down clock signal LEDs from the 4046 output	1. Each clock advances the high output sequentially, walking through outputs 0 through 9.	1. Fast cycling through 10.
	2. When the count reaches output 9, it starts counting from output 0 again.	2. Cycling slows steadily to a complete stop, randomly at any of the 10 outputs.

PARTS BIN

- IC2—4017 Walking ring decade counter
- R3—47 kΩ
- R4—10 kΩ
- R5—470 Ω
- R6—470 Ω
- LEDs—11

Add This Circuit to Your Breadboard

The 4017 is a very simple IC. As you can see by the schematic presented in Figure L40-2, it has one major input. That is the clock signal output from the 4046 VCO.

What to Expect

When you trigger the 4046 VCO by pushing and releasing the button, you will see the 10 LEDs lighting in sequence, appearing to have a light zipping down the series of LEDs. That is, in fact, exactly what is happening. As the VCO output slows, the zip slows as well. It will stop randomly, and one LED will be left on.

An animated schematic is available for viewing at www.mhprofessional.com/computingdownload. It shows the expected sequence of lights with a regular clock signal input. Even if your circuit does not immediately work to your expectations, now is the time to look at the data sheet.

4017 Data Sheet

The system diagram and pin assignments are shown in Figure L40-3.

For normal operation, the enable and the reset should be at ground.

For every clock signal input, the output increases from Out(0) to Out(9) by single steps. *The counter advances one count on the positive edge (ground to voltage) of a clock signal.* Only a

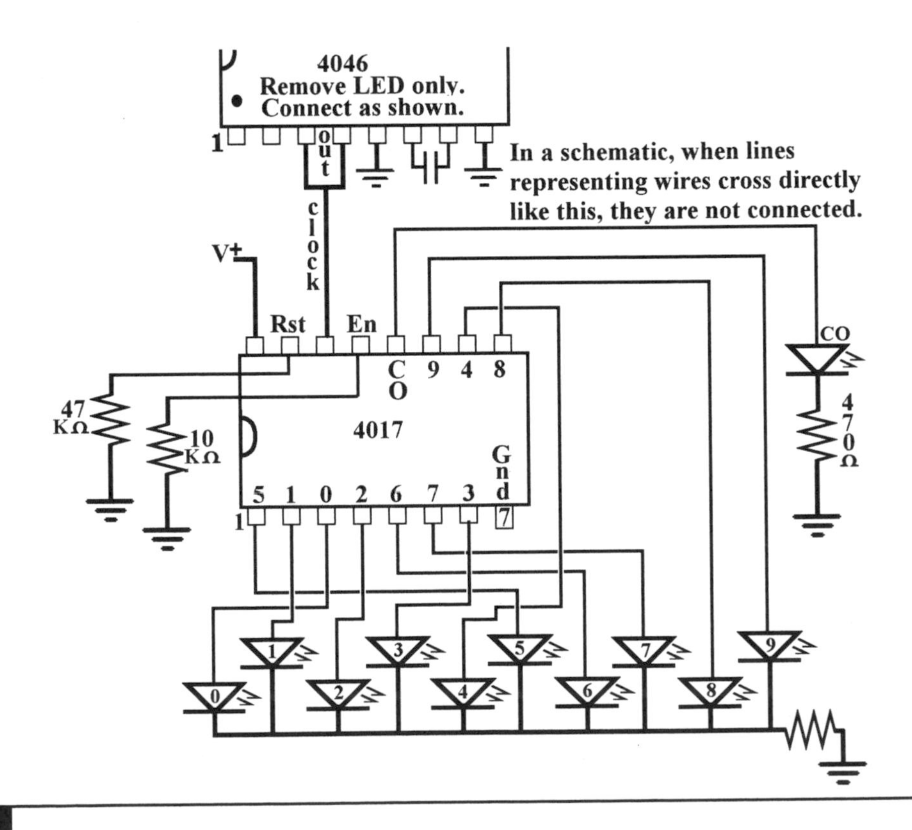

Figure L40-2

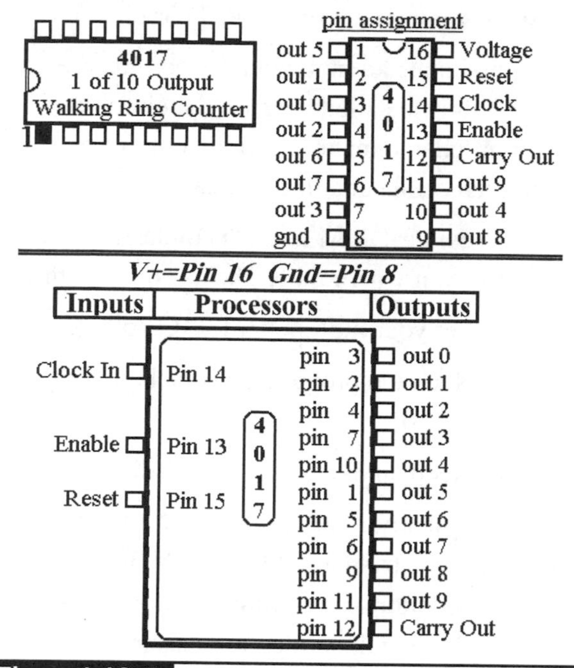

Figure L40-3

single output pin from Out(0) to Out(9) is high at any moment. By definition, if the output is not high, it will be low. When the count is at Out(9), the next clock input recycles output to Out(0), giving us the term "walking ring" because it is walking in circles.

Carryout

When Out(9) is reached, pin 12 carryout sets high. The carryout terminal is high for counts 0 through 4 and low for counts 5 through 9.

Because the carryout clocks once for every 10 clock inputs, the 4017 is often referred to as a "decade" or "divide by 10 counter." For example, it can be used to convert a 10,000-Hz signal input at pin 14 to a 1,000-Hz signal output at pin 12.

Making the reset high for a moment returns the counter to Out(0), setting pin 3 output to high. All other outputs drop to ground. The reset must be returned to ground to allow counting to continue. To stop the advance of the count, place a high on enable, pin 13. Returning enable to low will allow the count to continue.

As with any digital circuit, the inputs must be conditioned. That simply means they must be connected to either high or low at all times.

Of course, every input has its preferred state.

- Pin 13 enable should normally be set low; otherwise, the 4017 count will be frozen.
- Pin 15 reset should normally be set to low; otherwise, the 4017 will ignore the count. It will be stuck at 0.

The clock must be bounceless and have only one ground-to-positive movement for each desired count.

The 4017, like all CMOS chips, uses very little power. When it is at rest, it uses 0.002 watt of power at 5 volts.

Troubleshooting

What if the circuit is not working properly? Before you jump in, there are some bigger items to consider. Now is the time to start thinking systematically. Yes, you need to work through the circuit using the PCCP troubleshooting method, but the circuit just doubled in size. There are some necessary questions to ask first.

1. Do you have paper and pencil to record information down as you proceed?
2. What is working? Is there still a clock signal?
   - How can you tell? If there is no clock signal, then the problem is in the VCO.
   - If there is a clock signal and the LEDs are not working, then the problem is in the walking ring.
3. Once you have narrowed down the error to a single system area, start the PCCP process. And even though it may sound redundant, check the power anyway.

Here are some specific comments regarding the 4017 circuit:

- Make sure that pin 14 has the clock signal input from the 4046 (pins 3 and 4).
- Make sure that all of the LED cathodes (negative side) are connected together.
- Be certain that all LEDs are connected to ground through the single 470-ohm resistor.

The most common errors are as follows:

1. Reversing a single LED in the 4017 setup.
2. Not connecting pin 15 (reset) to ground. The schematic shows that it is tied to ground through a 47-kilo-ohm resistor. If pin 15 is connected to V+, output 0 remains high and nothing else moves. If pin 15 is not connected high or low, the input will react to static electricity in the air and give unstable results. An unconnected input creates "ghosts" that are difficult to explain.
3. Not connecting pin 13 (enable) to ground. This pin is also conditioned to ground through a 10-kilo-ohm resistor. If pin 13 is connected to V+, there is no output at all and the LEDs are all off. If pin 13 is not connected high or low, "ghosts" will be created from this as well.

Exercise: The 4017 Walking Ring Counter

The following drawings in Figure L40-4 show the most common styles used in schematics.

1. On the data sheet area, mention is made of the need to "condition" inputs. Explain what a conditioned input would look like.

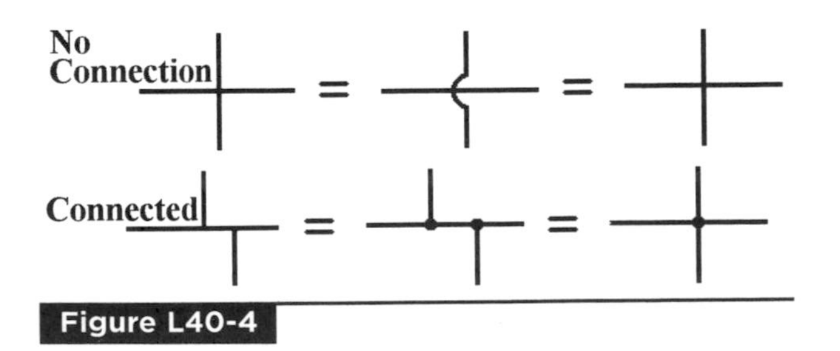

Figure L40-4

2. What controls any integrated circuit?
3. If the inputs are not conditioned, this creates "ghosts," so called because the circuit acts in an unreliable manner and the outputs are unpredictable and unexplainable.
   - Use your DMM as an "ectoplasm-meter" to find out where these ghosts come from.
     a. Set your multimeter to AC voltage.
     b. Keep the probe wires attached to the DMM.
     c. Keep an eye on the DMM display.
     d. Quickly move your DMM toward and away from an old style TV or computer monitor several times. Flat screens just don't leak enough to measure. Flourescent lights give good readings too.
     e. Place the DMM right up against the monitor and hold it still.
   - What was the highest AC voltage displayed? ____________AC mV
   - Do this again with DC voltage. ________DC mV

An unconnected input is not high or low. That input reacts with the signals and static in the air around us.

Some of the signals are from:

- Radio and television signals
- Cell phones
- The electrical wiring in the walls

4. Provide an explanation—what causes these ghosts at unconditioned inputs?

   ____

5. What is the result of having unconditioned inputs?____________________

6a. There are three inputs to the 4017. What are they? (Here is some information from the 4017 data sheet.)

a. _________ at pin _____conditioned by its connection to the 4046 output.

b. _________ at pin _____ conditioned ____ in normal operation.

c. __________at pin _____ conditioned ____ in normal operation.

6b. State the purpose of each input. (Here is some information from the 4017 data sheet.)

a. __________/ purpose________________

b. __________/ purpose________________

c. __________/ purpose________________

7a. What is the strict definition of the clock signal?

____

7b. Which shape in Figure L40-5 best shows a clock signal?

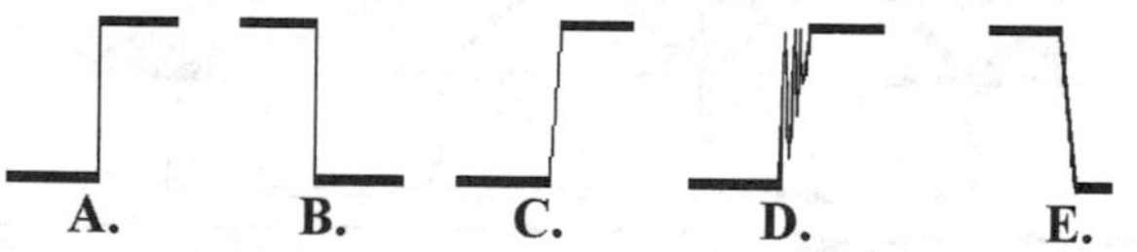

A. B. C. D. E.

Figure L40-5

8. Set the DMM to VDC. With the red probe to 4017's pin 1, trigger the circuit. Are the results the same as the output of pins 3 and 4 of the 4046? _______

9. The output from the 4046 VCO on the oscilloscope should have been similar to the diagram shown in Figure L40-6.

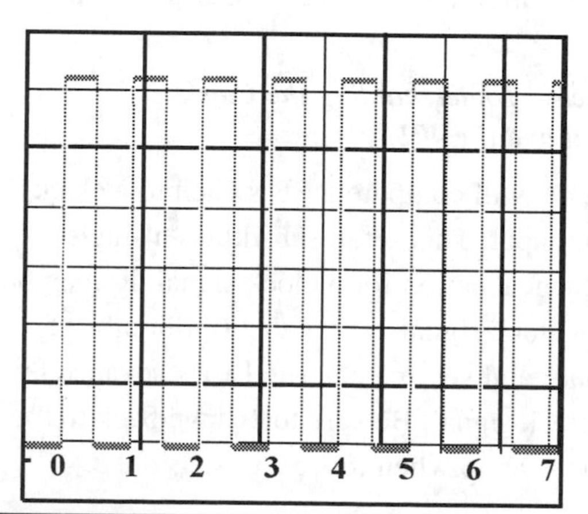

Figure L40-6

If you set up the oscilloscope just to measure at pin 1 of the 4017, what would you expect to see?

a. The same clock signal output frequency

b. Only half of the output frequency

c. High for the first five, low for the second five

d. Only one-tenth of the output

10. Consider the carryout (pin 12).

a. It is high for half the cycle and low for the other half.

When does the carryout of the 4017 change from low to high?__________

When does the carryout of the 4017 change from high to low?__________

b. Explain why the output of pin 12 (carryout) is so different from the other output pins 0 through 9. (Refer to the 4017 data sheet.)

____

____

Lesson 41
Understanding the Clock Signal Used by the 4017

Hickory dickory dock. The system responds to a clock.

A clock signal is widely used in electronics to trigger events throughout larger systems.

Definition: A *clock signal* is a clean digital signal that raises from zero to full voltage instantly. It must take fewer than 5 microseconds. That is 0.000005 second (1 microsecond = 10^{-6} second).

A clock signal is all of the following:

- Is a very clean signal
- Does not bounce or echo
- Triggers one event

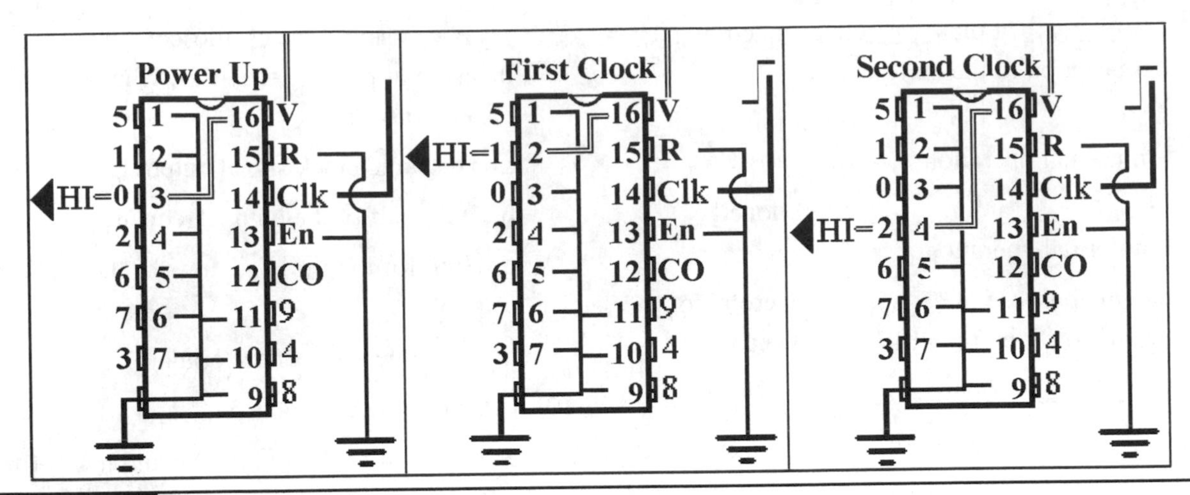

Figure L41-1

The clock signal is generated by the 4046 VCO circuit and is used to trigger the 4017 circuit. Each clock signal advances the "high" output by one as shown in Figure L41-1.

Each of the 4017's output provides a clean clock signal as it changes from low to high. When the output changes from 9 to 0, the carryout at pin 12 "clocks" and remains high until the count advances from Out(4) to Out(5). Any of the outputs can be used as clock signal inputs to trigger other circuits.

When reset is momentarily set to high, the count resets to 0.

When the enable is set to high, the count stops until it is returned to low.

Remember, all inputs need to be conditioned to either high or low.

Refer back to the data sheet provided in Lesson 40 as you work through this lesson.

To understand the circuit better, take one of the LEDs out and put it in backward.

- Trigger the circuit.
- How many LEDs are lit at one time? _____
- There is a rolldown . . . but *now* there are two lights on all the time: one of the other nine LEDs and the reversed LED.

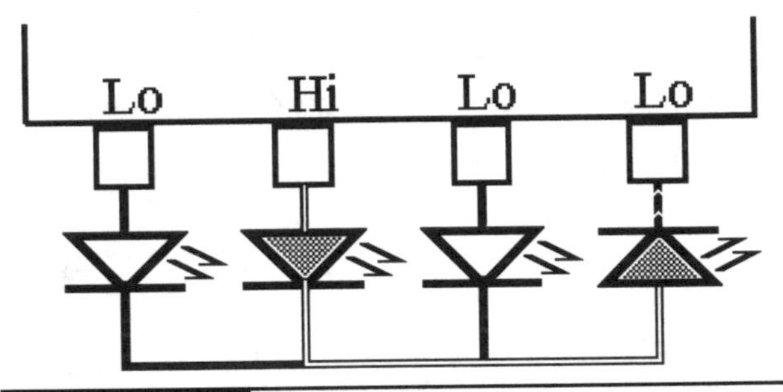

Figure L41-2

- The drawing in Figure L41-2 best explains what is happening.

Remember: Only one output is high at any time; the rest are low.

The reversed LED acts as ground. So the pin powering the LED in the count is draining through the backward LED connected to an output that is low. Return the LED to its proper position.

Exercise: Understanding the Clock Signal and the 4017

Here we will compare a PB switch to a clock signal input. This setup substitutes a single mechanical action for a clock signal. Remember that a clock signal is very fast (millionths of a second) and very clean. The LEDs advance by one per clock signal. Be sure to change back to the original setup when this exercise is done.

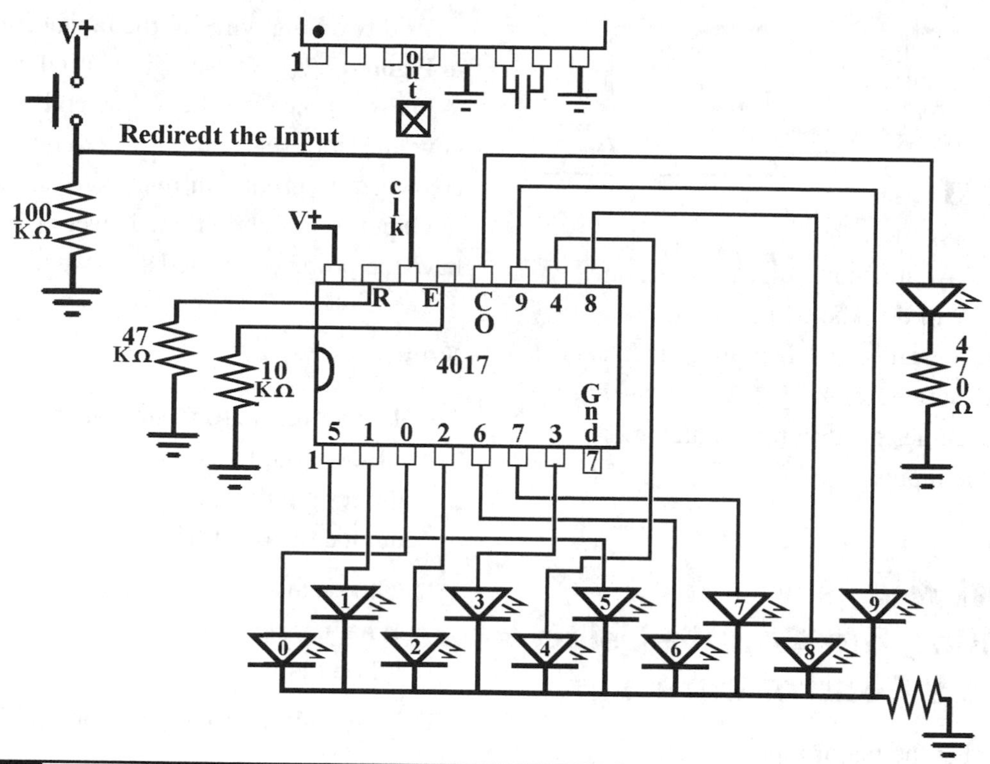

Figure L41-3

1. Attach a 100-kilo-ohm resistor and the PBNO to the clock input of the 4017 as shown in Figure L41-3.
2. Remove the wire that carries the clock signal from the 4046 (pins 3 and 4) to the 4017 (pin 14). When everything is set up, attach the power.
3. What LED is lit now? LED # ________
   - Press the plunger down in a definite movement.
   - Don't release it.
   - What LED is lit now? LED # ________
   - Release the plunger.
   - Did the count advance when you released the plunger? That was a voltage movement from V+ to ground.
4. You expect a push button to be a fast and clean movement from ground to V+. Is it a clock signal?____
   - You cannot use any mechanical device to provide a clock signal.
   - A clock signal must be generated electronically.
5. Push and release the plunger a few more times.
   - You can expect that the count will advance by a single step . . . sometimes by two or three steps.
   - The contacts of a physical switch like a push button do not provide a clean enough signal when a clock signal is required.
   - *An unclean input has bounce* as shown in Figure L41-4. Such a bounce is too fast to be noticed on inexpensive oscilloscopes. It would be impossible to see on Soundcard Scope. But you see the results of the bounce as the 4017 counts more than one step in each push of the plunger.

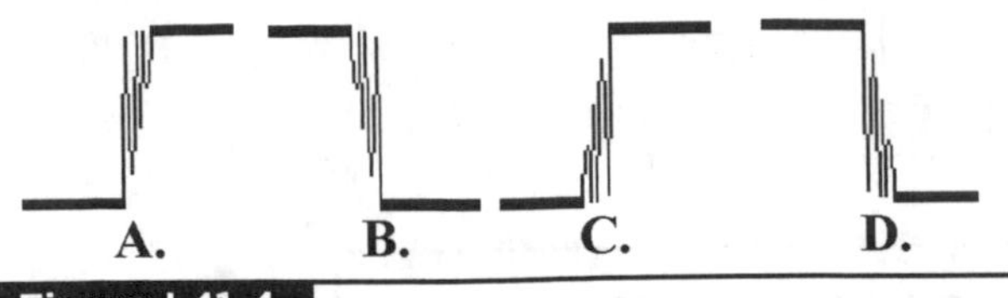

Figure L41-4

If an input says it needs a clock signal, only a clock signal will do. The CMOS 4093 is a specialized IC designed to "debounce" mechanical switches. It is nothing more than a fancy 4011 chip. It has been designed to react much more precisely to the inputs.

Lesson 42
Controlling the Count—Using the Chip's Control Inputs

You've seen that the major input is a clock signal. There are two other inputs. Let's visit them and see what they do.

Add two long wires to the breadboard as shown in Figure L42-1. These will be used as connections to V+ when needed. Leave the ends loose. Notice how pin 15 (reset) and pin 13 (enable) are each connected to ground through a resistor. That way the input is still conditioned, even if you do not have the probe connected somewhere.

Reset

1. Trigger the 4046 to start the rolldown. All 10 LEDs should flash in sequence. Now connect the reset probe to the anode (+) side of LED number 7. What happens?
2. Choose another LED and do the same thing. What happens? What does the reset do when it is set high for a moment?
3. Wait until the rolldown stops. Touch it to V+.
4. What happens if you keep the reset high and start the rolldown?

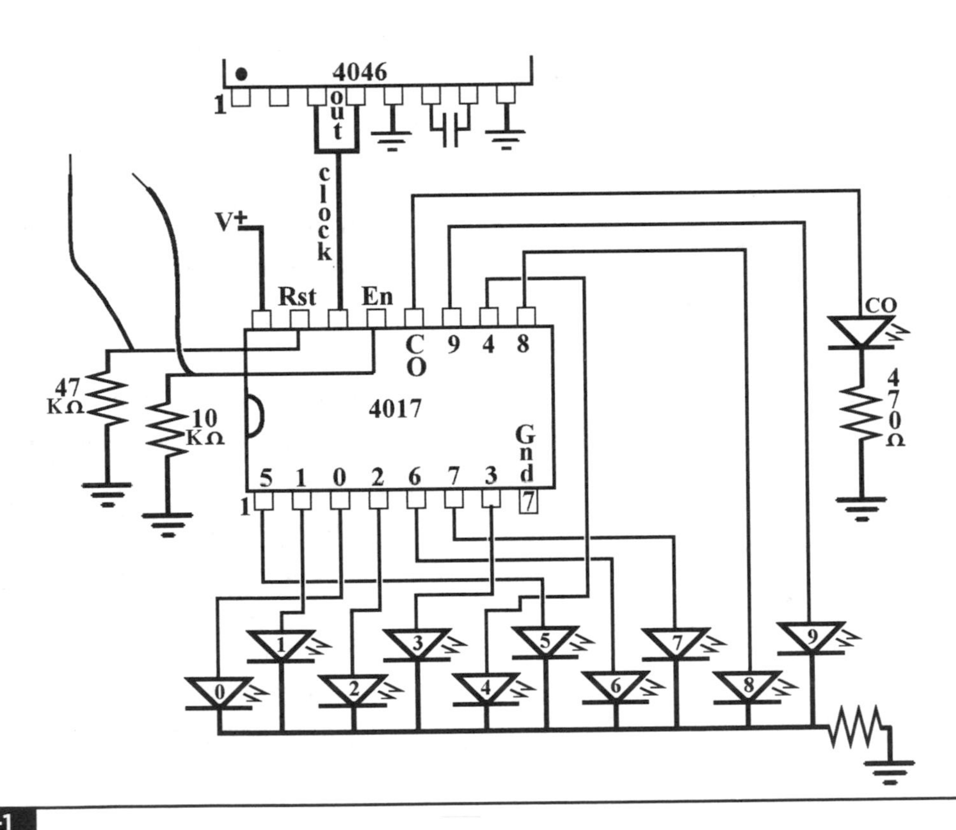

Figure L42-1

Take a look again at how the data sheet defines the purpose of the reset. It does exactly what it says it does. And the same is true for the enable function.

Enable

1. Pull the reset wire out while you play with the enable. Start the rolldown. Now connect the enable probe to the anode (positive) side of any LED. What happens?
2. Choose another LED. What happens? What does the enable do when it is set high for a moment?
3. Wait until the rolldown stops. Touch it to V+.
4. What happens if you keep the enable high and start the rolldown?

   So what good are these? That is up to you to decide. Maybe you don't want to count zero to nine all the time. Maybe you only want it to count to six, like rolling a die. Maybe you have a hexapod robot. It has a preset cycle for its six legs. You can use the 4017 to trigger this system. Figure L42-2 shows a really cheap, simple, yet effective idea for applying the 4017 in a video security surveillance system.

   The system needs to have the "high" produced by any source to reset the system. The reset does not need a clean clock signal. Any momentary high will do. The "high" from pin 10 is used here. Why don't we see the LED connected to pin 10 light up? Because of the speed, the circuit reacts in microseconds. That is too fast for us to notice any response in the LED.

5. Predict what happens when the reset wire is connected to pin 1. Pin 1 is related to what output?
   a. Out 0
   b. Out 2
   c. Out 4
   d. Out 5
   e. Out 10
6. Underline which LEDs light up. 0, 1, 2, 3, 4, 5, 6, 7, 8, 9
7. What is the state of the carryout for this setup?
   a. On all the time
   b. On for the first two, off for the second two
   c. Doesn't light up

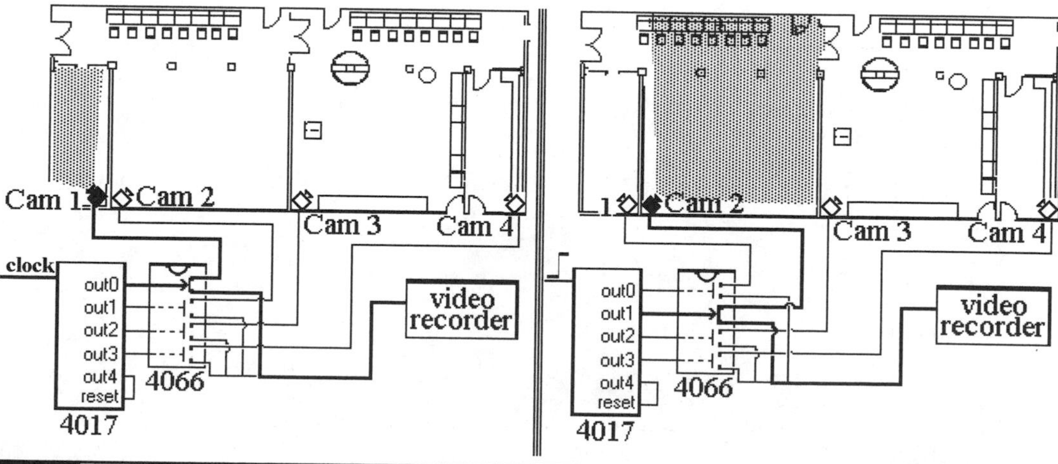

Figure L42-2

8. How are the three inputs conditioned so they are always connected to either high or low?

   a. Clock________________________________

   b. Reset________________________________

   c. Enable_______________________________

9. Look at the schematic. The "enable" is conditioned to low through the 10-kΩ resistor. Start the circuit from the 4046. In the middle of this standard rolldown, connect the enable to V+. Note what happens. Leave it attached for 30 seconds, and then trigger the rolldown. Disconnect the enable from V+ so it is low again. The rolldown should pick up again in the middle. Describe what the enable does.

   ____

   ____

10. In the following system diagram, describe the processor for the video security system shown in Figure L42-2 (see Table L42-1).

TABLE L42-1 System Diagram

Input	Processor	Out
Video camera signal	Three items are necessary for this processor.	Video recorder
	1. A circuit that generates clock signals.	
	2. The 4017 does what? ________	
	3. The 4066* does what? ________	

* The 4066 is the equivalent of four high-quality NPN transistors in a single package. What does it do here in the circuit?

SECTION 13

Running a Seven-Segment Display

THE SEVEN-SEGMENT DISPLAY depends on two major chips. I introduce the "slave" 4511 first so you can understand how it is controlled. Then the 4516 "master" is introduced. You will investigate and play with most of the possibilities of each chip as they are presented. At the end of this chapter, you will have the prototype of the DigiDice and be able to explain it. Better than that, you will be able to control it. Remember, *binary*'s the word.

Lesson 43
Introducing the Seven-Segment LED

Here you have a real number output. It's just a set of LEDs set up to represent numbers.

Here is the system diagram from the beginning of Part Three (see Table L43-1).

Remember that there is to be a number output too. In this chapter we will add the number readout. Everything revolves around creating a number output using a seven-segment LED display. This simple device is created by using individual LEDs in rectangular shape, set into a package that creates a number, as shown in Figure L43-1.

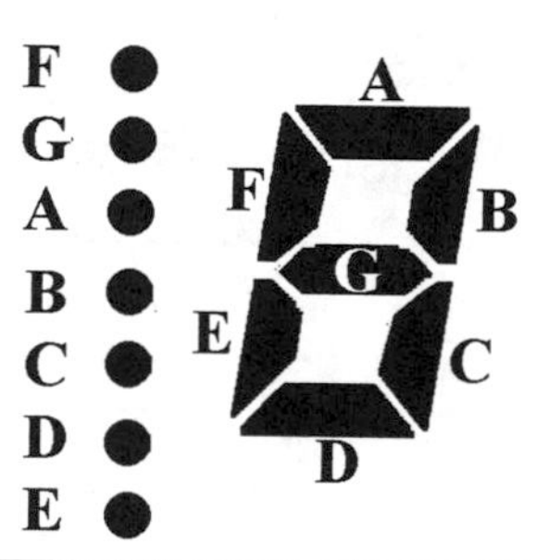

Figure L43-1

So you might have some seven-segment displays lying around. You try them out. They might work and they might not. That is because there are two major types of seven-segment displays: the CC and the CA. The type that is used here is referred to as a common cathode (CC). A proper schematic diagram for the CC is shown in Figure L43-2.

TABLE L43-1 System Factors

Input	Processor	Output
Push button	**1.** Rolldown (4046 IC) controlling 2	**1.** Fast cycling through six LEDs.
	2. Walking ring resets at 6 (4017 IC)	**2.** Fast cycling through numbers 0 through 6.
	3. Binary counting decimal (4511 IC)	**3.** Cycling slows steadily to a complete stop, randomly at points 1 through 6.

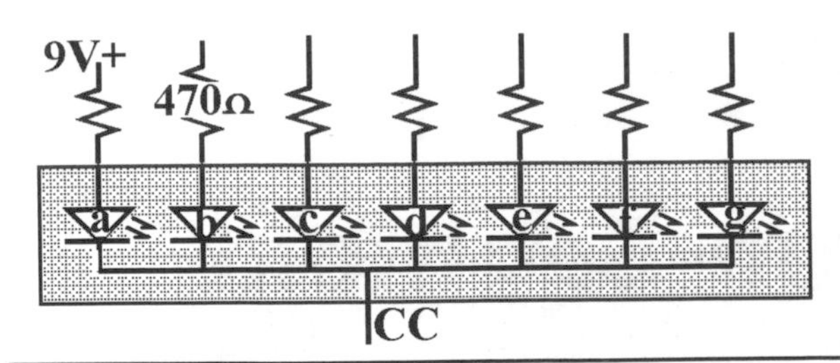

Figure L43-2

This type of seven-segment LED display is referred to as a common cathode because all of the LEDs share the same ground line. You have to add resistors on the anode (positive) side for each LED. The alternate type of seven-segment display is shown in Figure L43-3. Note that it is almost identical, yet completely opposite, of the CC layout.

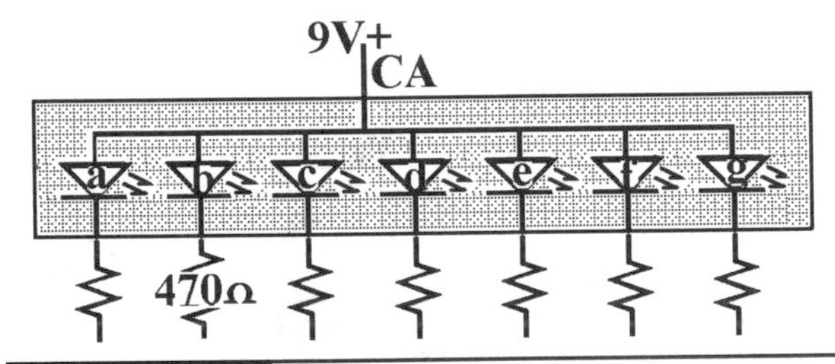

Figure L43-3

Just as with the CC, you still have to add resistors. This is done on the cathode (negative) side for each LED in this package. Just a further note: The two package styles are *not* interchangeable. A CC will not substitute for a CA (common anode) package in any circuit because it is the same as placing an individual LED in backward. It just doesn't work.

Right now, you need to map out your seven-segment display. Set the seven-segment display into your breadboard as shown in Figure L43-4.

Be sure to use a 470-ohm resistor to cut the voltage to the individual LEDs as you map out the pins. The center pin of each side is a common ground connection. Only one needs to be connected to ground.

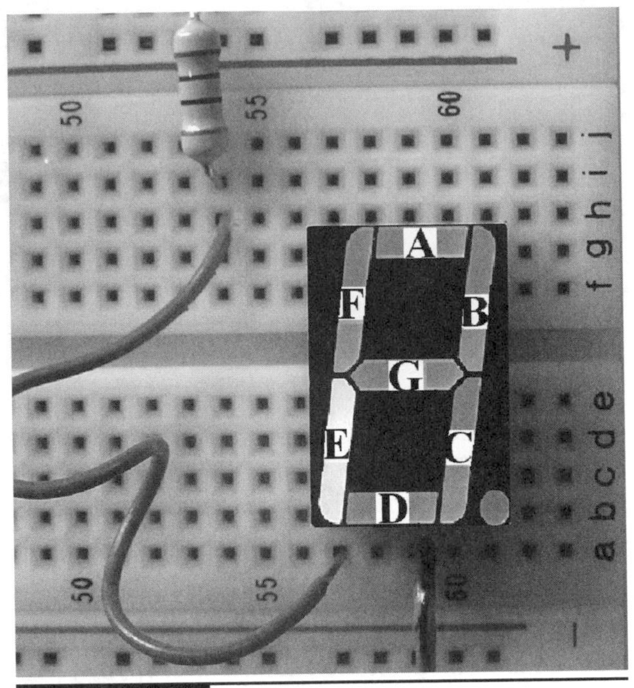

Figure L43-4

Now use a single wire to determine which pin powers which LED. Label your drawing. You will need this information to wire up your next circuit.

How would you create the number 7?

Lesson 44
Control the Seven-Segment Display Using the 4511 BCD

1. Getting numbers to present onto the seven-segment display has to be easier than rewiring all the time. Electronics is about control. Here is a chip that controls the seven-segment display.
2. Several IC processors are available to change inputs into number outputs.
3. The 4511 BCD CMOS IC used here is a basic binary-counting decimal processor.

4511 Data Sheet

The system diagram and pin assignments for the 4511 BCD are shown in Figure L44-1.

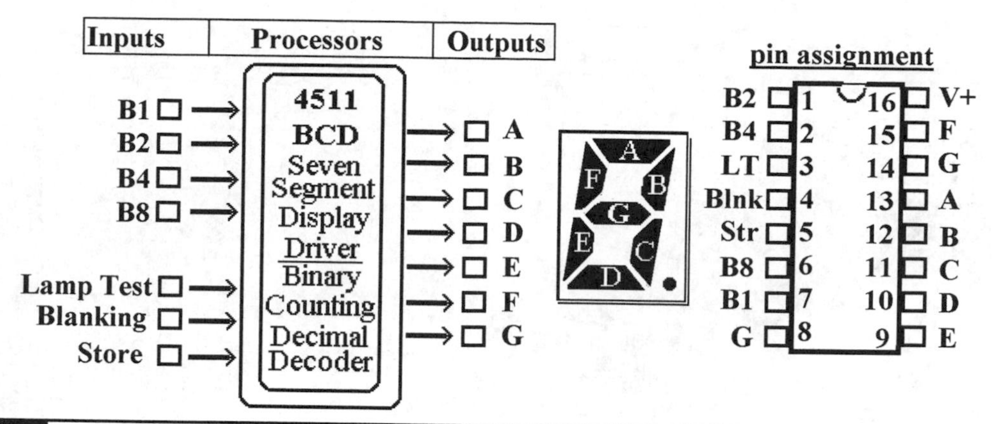

Figure L44-1

Basic Operation

The 4511 BCD accepts a binary input code and converts it to be displayed on a seven-segment, common cathode LED (all segments share the same connection to ground).

The binary code placed on B1, B2, B4, and B8 inputs is translated as a code for display on a CC seven-segment LED. For instance, if B8 is low, B4 is high, B2 is high, and B1 is low [0110], this input combination will make outputs c, d, e, f, and g high, while outputs a and b remain low, creating a decimal six.

In normal operation the lamp test and blanking are held high while store is connected to ground.

- **Lamp test:** If lamp test is grounded, all of the lettered outputs will go high.
- **Blanking:** If blanking input is made low, all lettered outputs go low, turning the display off.

Note as well that any input representing a number bigger than decimal nine will blank the display. Even though this 4-bit binary word can count from 0 (0000) to 15 (1111), the 4511 is designed to operate a seven-segment LED that cannot count above nine. Any binary word above 0101 is unreadable. The 4511 will blank, turning off all outputs to the display.

If the store input is made high, the value of the input code at that instant is held internally and can be used later.

The 4511 BCD system layout is described in the Table L44-1.

TABLE L44-1 4511 BCD System Layout

Input	Processor	Output
A binary word provides high signals at the binary inputs.	Binary code is converted for display by the binary-counting decimal processor.	Decimal value of binary input shown on the seven-segment LED.

Breadboarding the Seven-Segment Display and 4511 Display Driver

Set the chip and seven-segment display near the end of your breadboard as shown in Figure L44-2.

Now build this circuit from the schematic in Figure L44-3.

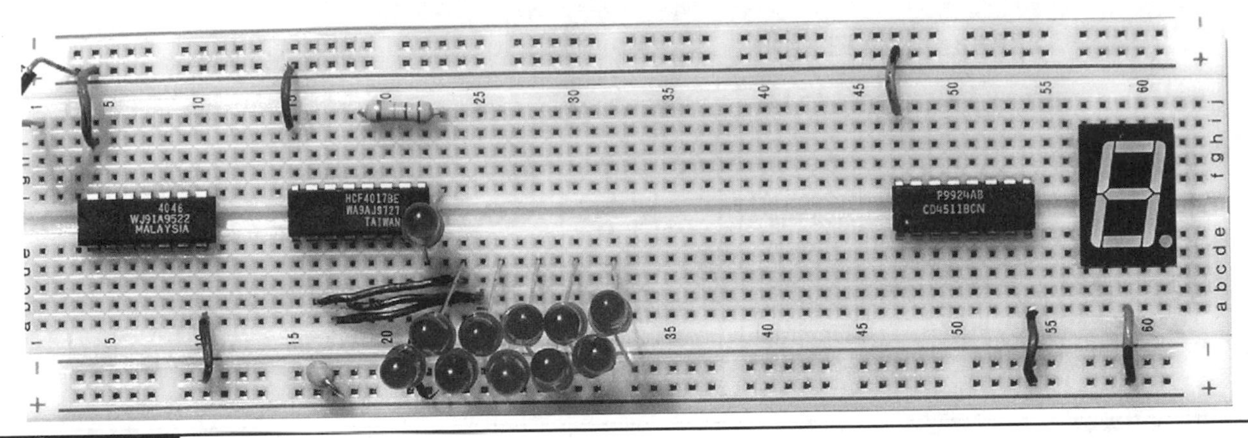

Figure L44-2

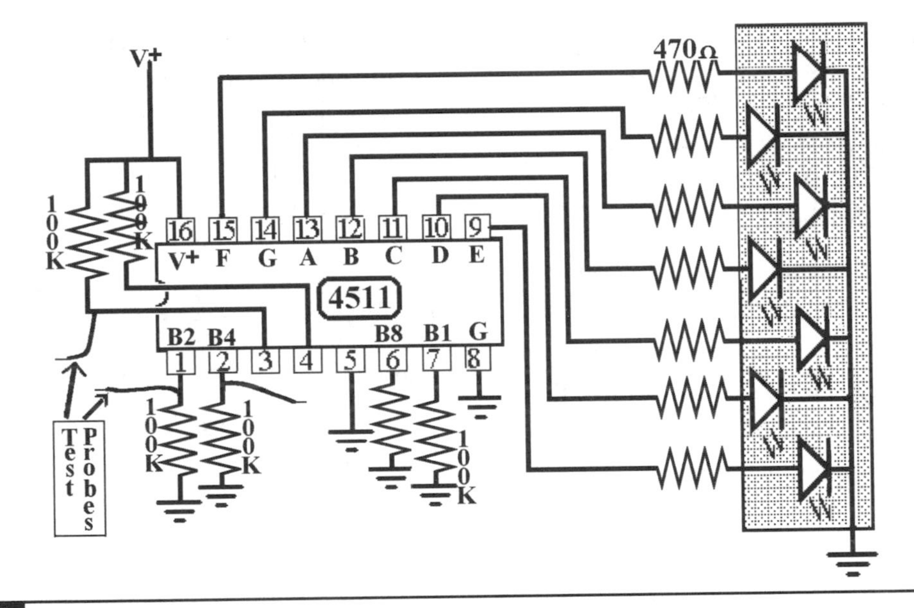

Figure L44-3

The wire probes are only needed for testing in this lesson.

Building this circuit on the breadboard can prove to be challenging. The picture in Figure L44-4 is provided to help guide you. There is a correct technique for breadboarding, and then there is a mess.

- Note that the resistors have had their legs clipped shorter on one side. Then they are placed upright.
- Insulated wires are used to make the long connections. Those wires are a bit longer than

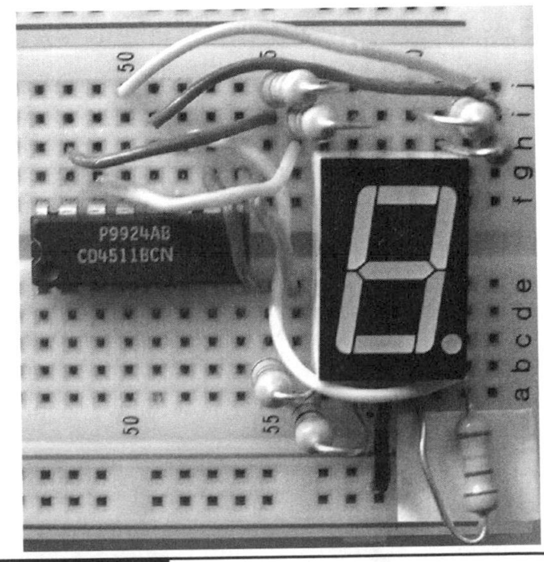

Figure L44-4

the actual connection. That relieves any strain that might cause the wire to spring out of its hole.

- The wires are gently bent. Sharp corners actually encourage the solid wire to break under the insulation. Such breaks are frustrating and hard to find.

What to Expect

When you attach power to this circuit, the seven-segment display should read ZERO. That is, segments A, B, C, D, E, and F are lit. Not the crossbar, G. If you have a backward "6," you have switched the outputs "F" and "G."

1. Touch the wire probes connected from inputs B1, B2, B4, and B8 separately to V+. This "injects" a high signal into the various binary inputs.
2. Touch both B2 and B4 to voltage at the same time.
3. Try different combinations.
4. What happens when you try to inject a high at B8 and B2 at the same time? It should blank the display.

Troubleshooting

Run through the PCCP. The most common problems in this setup are as follows:

- Crossed wires.
- Reversed display.
- Individually burned-out LEDs on the display.
- Blanking at pin 4 and lamp test at pin 5 are connected to V+. Either connected to ground would keep this circuit from working.

Exercise: Control the Seven-Segment Display Using the 4511 BCD

The inputs B8, B4, B2, and B1 represent binary inputs.

Together, all four binary inputs create a 4-bit binary word. Figure L44-5 presents an excellent representation of how the binary word input relates to the display driver outputs.

1. Use the wire probe to connect B1 (pin 7) to voltage. What is the readout of the display?

   ____

2. Use the wire probe to connect B2 (pin 1) to voltage. What is the readout of the display?

   ____

3. Use the two separate wire probes to connect both B1 and B2 to voltage. What is displayed?

   ____

4. Work through Table L44-2.

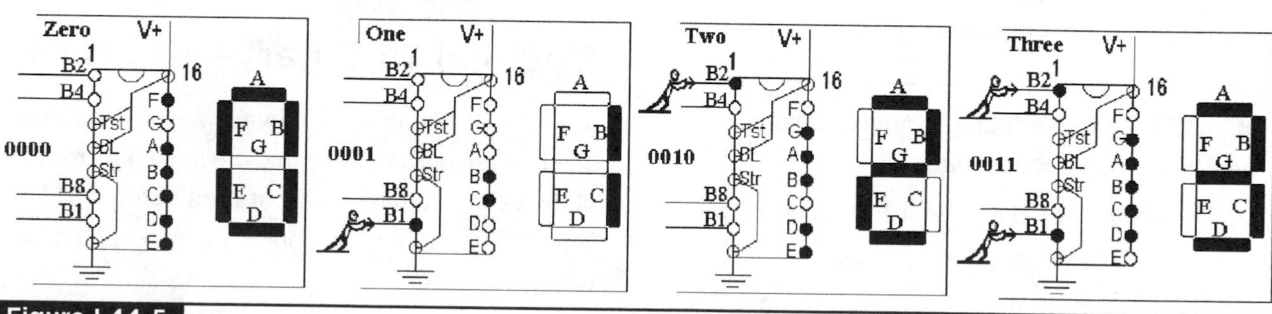

Figure L44-5

TABLE L44-2 Wire Probe

Number	B8	B4	B2	B1	Binary Word	Seven-Segment Display Output
Zero	Low	Low	Low	Low	0000	A B C D E F
One	Low	Low	Low	High	0001	C and D
Two						
Three						
Four						
Five						
Six						
Seven						
Eight						
Nine						
Ten						

The highest binary input that the 4511 recognizes is when inputs B8 and B1 are both high.

That means the highest it can count is 9. Anything over a binary input equal to "decimal 9" will cause it to blank out.

5. Why are inputs 1, 2, 6, and 7 connected to ground through 100kΩ resistors?

____

____

6. What is the maximum count of a 4-bit binary word? ______________________

7. What is the highest number that the 4511 seven-segment driver is able to display?

____

8. BCD stands for binary-counting decimal: Binary in—decimal out.

What is the binary word for decimal 7?

____

9. Besides the binary word input, there are three other inputs to the 4511. What are they (from the 4511 data sheet)?

a. ______ at pin______ is conditioned ______ in normal operation. What is its purpose? ______________________

b. ______ at pin______ is conditioned ______ in normal operation. What is its purpose?______________________

c. ______ at pin______ is conditioned ______ in normal operation. What is its purpose?______________________

10. Test blanking and lamp test separately. Simply connect them to ground momentarily.

Lesson 45
Decimal to Binary—The 4516

The 4516's output is the binary word input to the 4511. It changes the decimal clock signal from the 4046 VCO to a binary output used by the 4511. This lesson explains the operation and application of the 4516 BCD IC.

Data Sheet: 4516 Binary Counting Decimal—Up/Down Zero to 15 Counter

The system diagram and pin assignment for the 4516 are shown in Figure L45-1.

Basic Operation

For each clock signal input, the binary word output changes by one. The binary word shows as a high signal on corresponding outputs. For example, for the decimal number "7," the output B8 is low, while B4, B2, and B1 are high [0111].

In normal operation the inputs carry-in, reset, and preload are held low.

The binary word increments one count on the clock signal at pin 15. The movement is dependent on the setting of the up/down control. The output appears as the binary word at the outputs B8, B4, B2, B1.

Up/Down Control

When the U/D control is set to high, the count proceeds upward. When decimal 15 [1111] is reached, the count cycles back to zero [0000] and carryout is triggered. When the U/D control is set to low, the count proceeds backward. When decimal zero [0000] is reached, the count cycles backward to 15 [1111], and carryout is triggered.

Carry-In

The carry-in must be held low to allow counting. Holding the carry-in high stops the count, making it function like the enable switch of the 4511.

When multiple stages are used, the chips share the clock signal, as shown in Figure L45-2. Carryout of each stage connects to the carry-in of the next level.

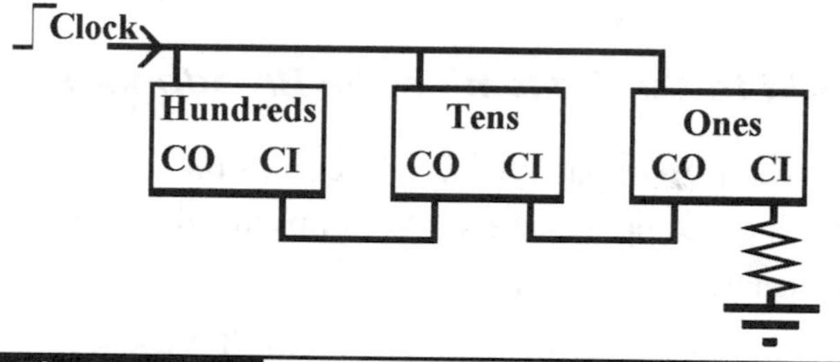

Figure L45-2

Reset

By momentarily connecting reset to V+, the count returns to zero [0000]. The reset must be returned to ground to allow counting to continue.

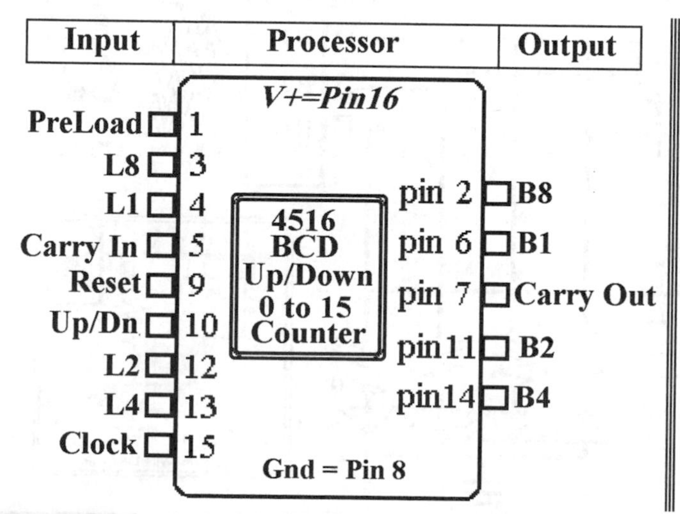

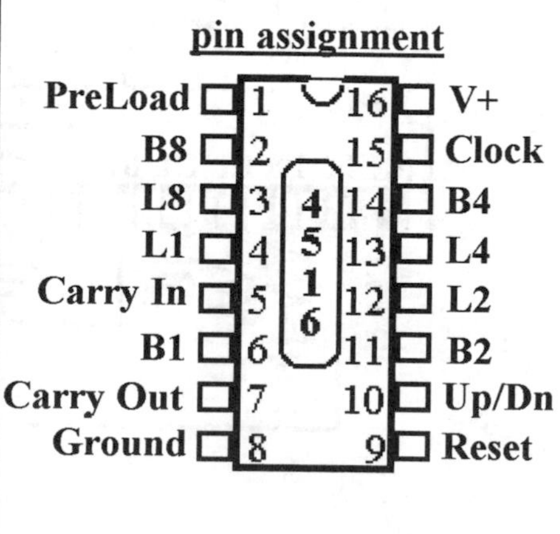

Figure L45-1

Preload

Any number can be preloaded into the 4516 by presetting the appropriate binary word to voltage through the preload inputs [L8, L4, L2, L1]. These preloaded inputs are loaded on to their corresponding binary outputs [B8, B4, B2, B1] by bringing the preload at pin 1 high for a moment. As with reset, the preload must be returned to ground to allow counting to continue.

NOTE **Pin 15 should be in a low state and there should be no clock inputs to pin 15 when preloading is done.**

Add to the System on the Breadboard

Leave a small patch of five free lines on one side for the last part of the system introduced in the next lesson.

As the schematic shows in Figure L45-3, the 4516 uses seven more 100 kΩ resistors and more wire. You also add a temporary display of the binary word using four extra LEDs.

1. Remove the 100 kΩ resistors from the 4511's inputs.
2. Preload control at pin 1, and the preload presets L8, L4, L2, and L1 are attached to ground through 100 kΩ resistors.
3. Carry-in is connected to ground, because it is an unused input.
4. Carryout can be left open. It is an output used to trigger another stage in counting.
5. Reset is attached to ground through a 100 kΩ resistor.
6. Start with the up/down control connected by a 100 kΩ to V+.

What to Expect

Here's what to expect before you start to play. Use the clock signal from the carryout of the 4017 as the INPUT clock signal to the 4516. The number will advance one for every run down all 10 LEDs. Note that this wiring is specific to where you are now. It is subject to being played with by curious fingers and nimble minds.

Also, at this point, the up/down control at pin 9 can be connected to V+ (counting up) or ground (counting down) through a 100 kΩ resistor.

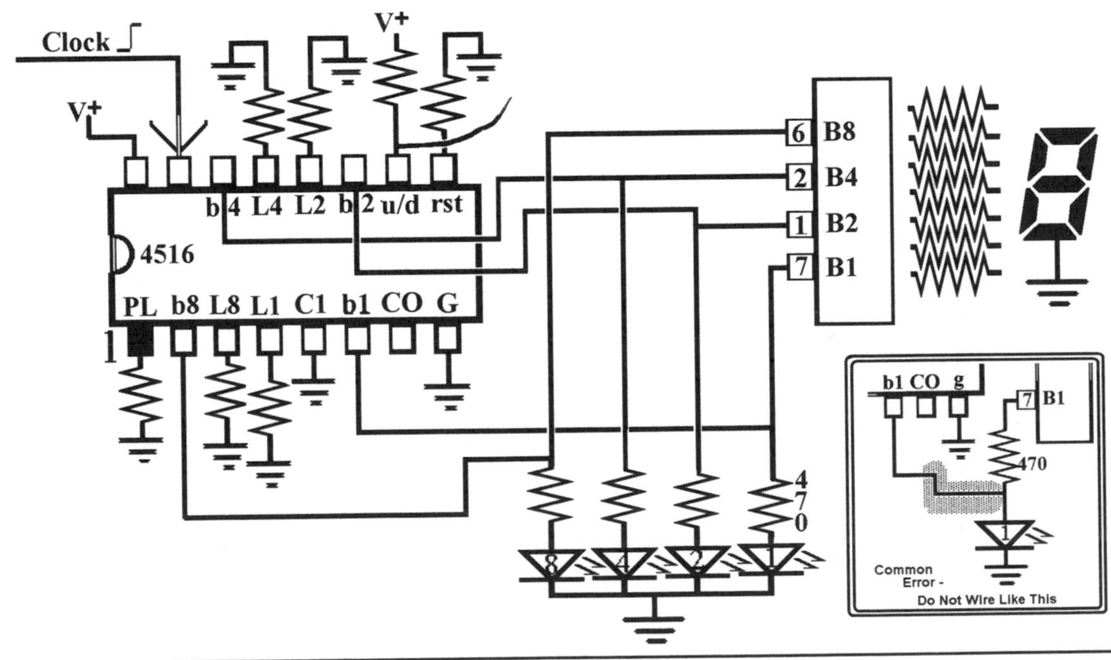

Figure L45-3 The insert shows the most common error—do not wire like this!

Exercise: Decimal to Binary—The 4516

A: Initial Setup

The 4516 acts as the interpreter. It changes the decimal clock signal source into the binary input used by the 4511 display driver.

The outputs of the 4516, B8, B4, B2, and B1 are *wired directly* to corresponding inputs of the 4511. The four LEDs display the 4516's binary word. It matches the decimal number displayed on the seven-segment LED. Now place a 10-megohm resistor from pin 12 of the 4046 to ground. This sets the minimum frequency output of the 4046 to about 1 Hz. This way, you don't have to continually trigger the roll down.

Make sure that the 4516's up/down counter (pin 10) is conditioned to V+ so it counts upward.

1. The shortened "DCB" stands for what?

   D____________ C______________

   B____________

2. The shortened "BCD" stands for what?

   B____________ C______________

   D____________

B: Inputs Inputs Everywhere

Be careful while you are exploring. You can accidentally connect V+ directly to ground. That shorts out the power supply. Even though it looks like everything turns off, a short circuit can damage your power supply.

- **Up/down counter.** First, try out the up/down counter at pin 10. Use a long wire as shown in Figure L45-4 to connect the up/down control to ground.
- **The reset.** Use a wire probe and momentarily touch the reset input to voltage as shown in Figure L45-5.
- **The preload.** This is easy to do. Don't remove the resistors. They allow flexibility. Pop a short wire into both L2 and L4 as shown in Figure L45-6, and connect them to V+. Connect a long wire to preload. Touch the preload wire to V+, just for a moment. Your display should now show the number "6." When the preload is set to high, the values preset on the load inputs are dumped onto their matching binary count outputs.

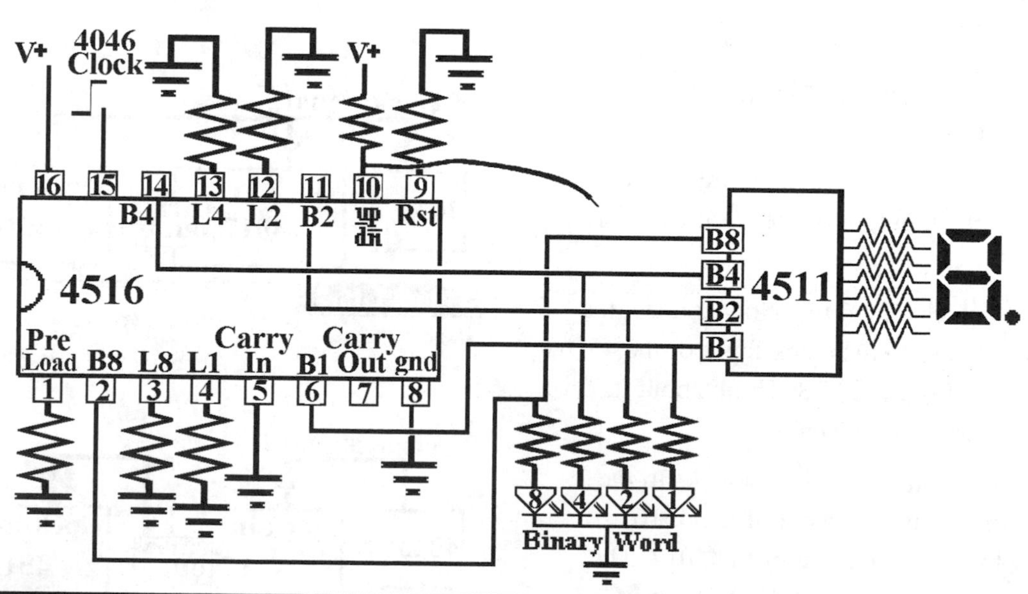

Figure L45-4

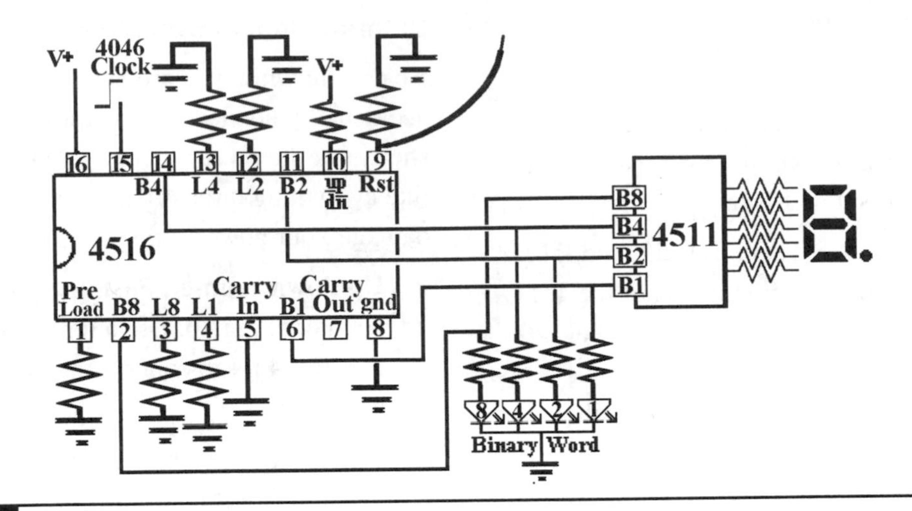

Figure L45-5

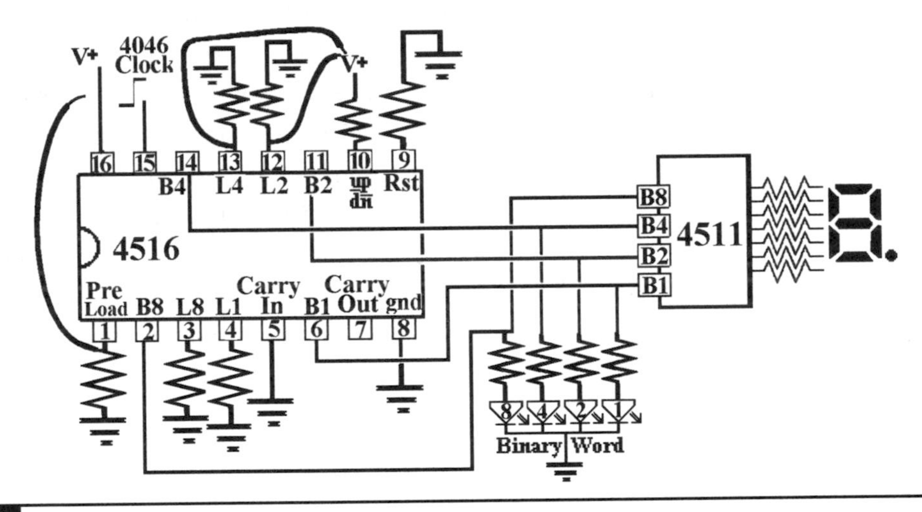

Figure L45-6

C: Now for Some Serious Playtime! Exploring the Possibilities

Right now, both the 4017 and 4516 are connected directly to the clock output of the 4046 as shown in Figure L45-7.

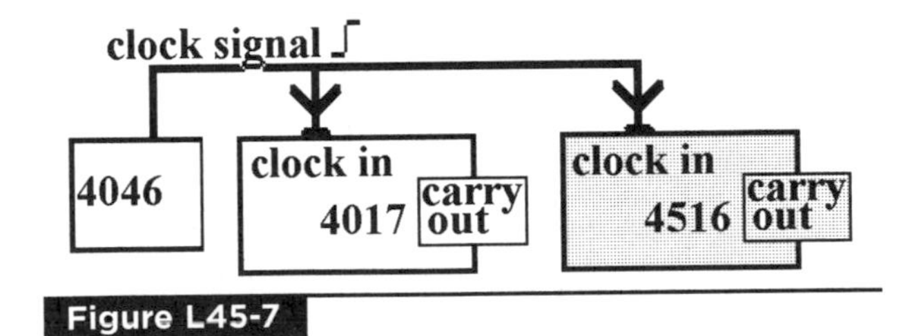

Figure L45-7

1. The 4046 clocks the 4017. Now connect the 4017's carryout to the clock input of the 4516 as shown in Figure L45-8. Think about it. Try to explain what is happening.
2. This is even better. The 4046 clocks to the 4516. Connect the carryout of the 4516 directly to the clock input of the 4017. The setup is shown in Figure L45-9. What is happening?

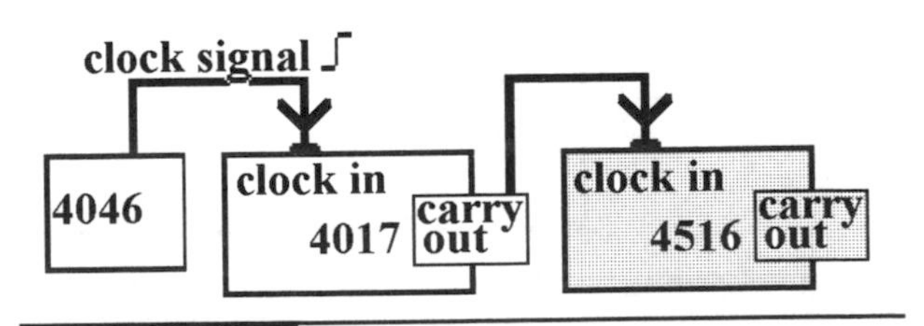

Figure L45-8

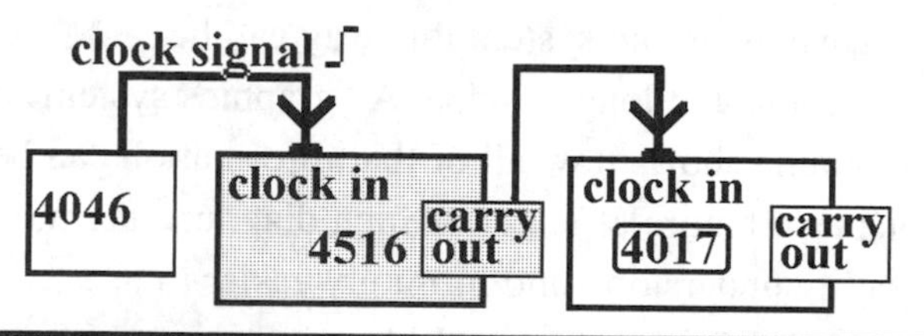

Figure L45-9

3. Will the 4516 still count zero to nine in this setup when you play with the reset or enable of the 4017? Try it out.
4. Remember the carryouts of the 4017? It is high for the first five numbers, and then goes low for the second five numbers. Try the setup shown in Figure L45-10. Connect the 4017 carryout to the up/down counter of the 4516. Predict what's going to happen. Were you right?

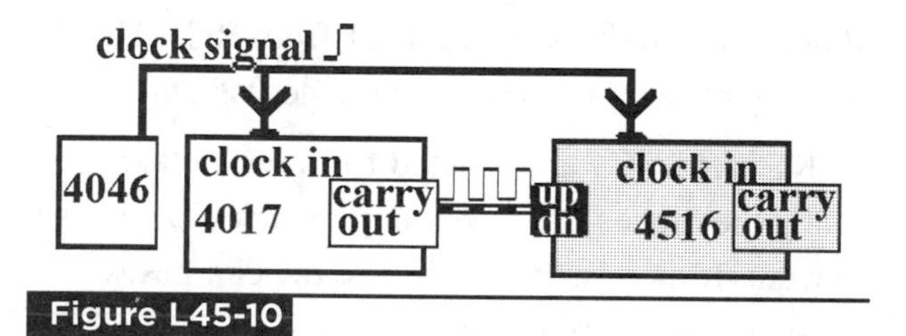

Figure L45-10

5. Any other ideas?

D. Setting Up the DigiDice

Figure L45-11 is the graphic system diagram that you can use to set up your prototype for DigiDice. Notice that some particular features have been incorporated into the system.

The 4046 clock feeds to both 4017 and 4516.

When "Out 6" of the 4017 goes high, this signal is used to trigger the following:

1. The reset of the 4017 (the first six LEDs light up in sequence)
2. The preload (pin 1) of the 4516
   - L1 (pin 4) must be set directly to voltage.
   - This will load the number "1" when it is triggered, effectively skipping "0."
   - Resulting in numbers displayed 1-2-3-4-5-6-1-2-3 . . .

At this time, you can remove both the 10-megohm resistor from the 4046's pin 12 and the LEDs used to display the binary word. The minimum frequency will return to full stop, and you will have room for one last addition to the entire system.

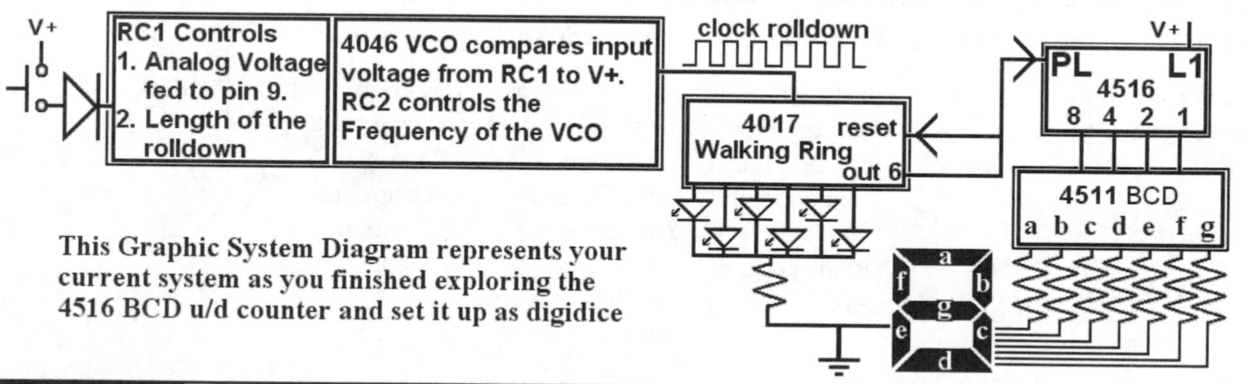

Figure L45-11

Lesson 46
The Displays Automatically Fade Out

Of course, your mother never reminded you, "Can't you remember to turn off the lights after you leave the room?" With the next step, you'll never have to worry about that with your circuit.

Hey! If you didn't do it yet, do it now. Remove those LEDs used to show the binary word.

Power consumption is important if you use batteries to power the system. CMOS electronics are popular because the systems use so little power. In comparison, the LEDs gobble energy in this system. In this lesson we apply a simple RC to the system that automatically cuts the power to the LEDs.

A basic system diagram is shown in Table L46-1. It includes the RC that automatically turns off the LEDs.

A detailed system diagram for DigiDice deals with the input/processor/output of each piece of the system (see Table L46-2).

Displaying the system this way can be confusing and long-winded. A "graphics system diagram" shows how all of that information can be displayed more efficiently. Such diagrams are more readily understood and infinitely easier to use when designing and troubleshooting systems. Figure L46-1 is the graphic system diagram showing how RC3 is set into the system.

But even a graphic system diagram does not show the full schematic. For that matter, at this point, I'm not going to show the full schematic either. All that needs to be shown is the specific schematic of interest and the notation of how it is connected into the system. Figure L46-2 is the schematic showing how RC3 controls the NPN transistor.

Everything used here should be fairly obvious, except for the extra diode. Why is the extra diode there? The diodes separate RC1 from RC3. Here is a complete explanation why it is necessary.

RC1 and RC3 are isolated from each other using the diodes to prevent any reverse flow of the current. If there were no diodes, the charges would be shared between the capacitors of RC1 and RC3. This is shown in Figure L46-3.

TABLE L46-1 Basic System Diagram

Input	Processor 1	Processor 2	Processor 3A	Processor 4	Output 1
A contact switch like a PBNO	RC1 controls the timing of the rolldown	4046 VCO RC2 controls max and min frequency of the clock signal	4017 Walking ring 1-10 counter. Seventh LED (out 6) is connected to reset and preload controls on the 4516	Timed off as a power saver. RC3 controls the transistor that cuts power to the LEDs	6 LEDs
			Processor 3B		**Output 2**
			4516DCB(P) 4511BCD		Seven-segment display counting 1-2-3-4-5-6

TABLE L46-2 Detailed System Diagram

Input	Processor	Output
First Section, Including the 4046		
Input	**Processor**	**Output**
Push button closes Voltage filling RC1	Falling voltage controlled by the RC1 drains from max to minimum. The voltage at pin 9 is compared to V+. This controls speed of voltage-controlled oscillator.	Length of rolldown controlled by RC1. Rolldown clock signal.
Second Section 4017		
Input	**Processor**	**Output**
Clock rolldown from the output of the 4046 Reset	1 of 10 counter reset at out 6.	Walking ring provides high to out 0 to out 6. Out 6 is used to reset 4017 to 0. Out 6 is used as controlling input to 4510.
Second Section LEDs		
Input	**Processor**	**Output**
Walking ring provides high to 1 of 6 LEDs in sequence.		Light
Third Section Starting with 4516		
Input	**Processor**	**Output**
Clock rolldown from 4046 High input from 4017 out 6 triggers preload of L1 U/D counter set up or down	Decimal counting binary	Advancing binary code
Third Section 4511		
Input	**Processor**	**Output**
Binary word Binary-counting Decimal	Converter for seven-segment LED	Decimal value of binary input shown on the seven-segment LED
Fourth Section Timed Off		
Input	**Processor**	**Output**
Voltage filling RC3	Transistor opens, then slowly closes the connection to ground as controlled by RC3	Displays fade-out1-2-3-4-5-6

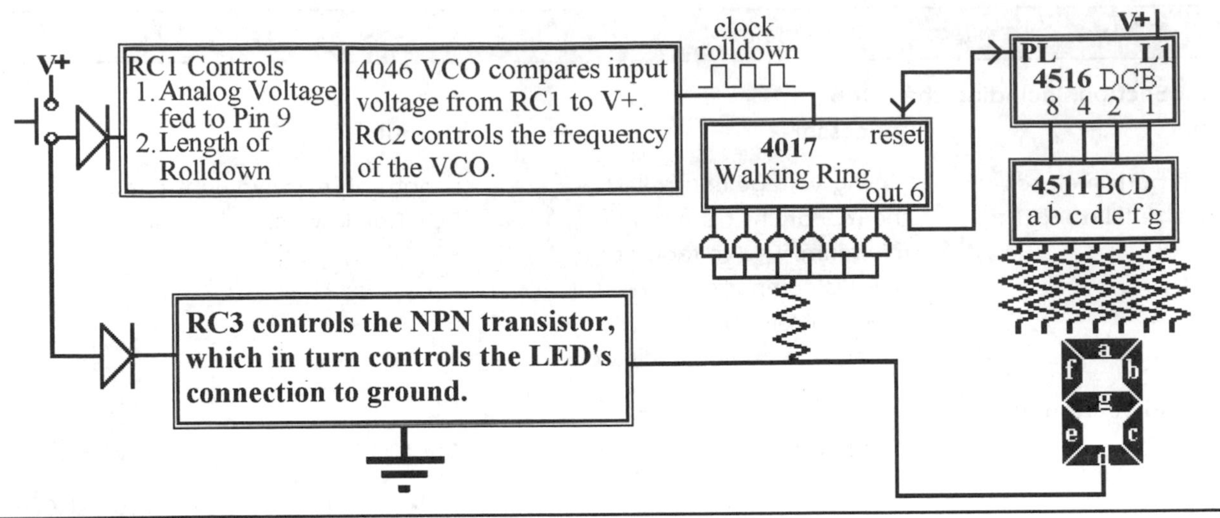

Figure L46-1

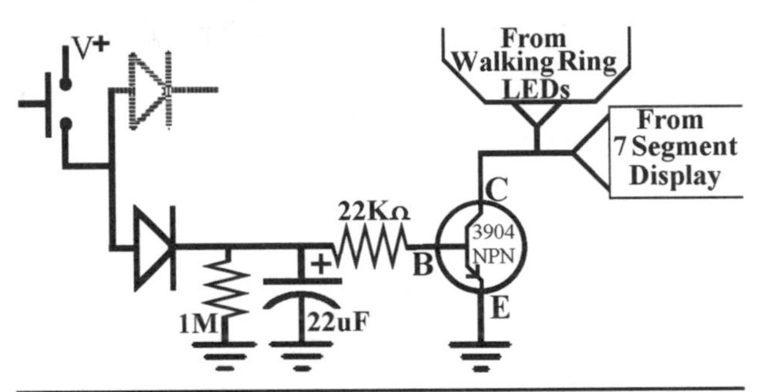

Figure L46-2

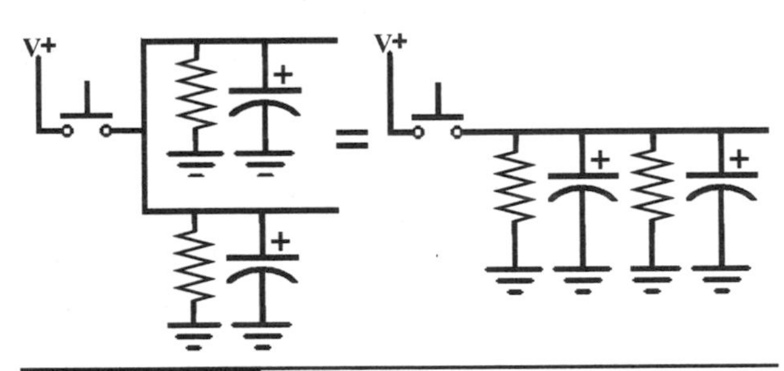

Figure L46-3

The resistors would act together, too. RC1 and RC3 would combine and act as a single RC.

If there were only one diode, the unisolated capacitor shares its charge first. This is demonstrated in Figure L46-4. The charge in the 1-μF capacitor of RC1 would push through the lower diode and add to the charge of the 22-microfarad capacitor of RC3. The reverse would also occur; if the bottom diode were not present, the charge stored in the 22-μF capacitor of RC3 would push through the upper diode and add to the charge of the 1-μF capacitor of RC1.

The complete and proper setup is shown in Figure L46-5. Each RC acts independently. The RC1 rolldown should finish 5 to 10 seconds before the fadeout really finishes. If the LEDs fade out before the rolldown is completed, adjust the timing of either RC1 to be shorter or the timing of RC3 to be longer.

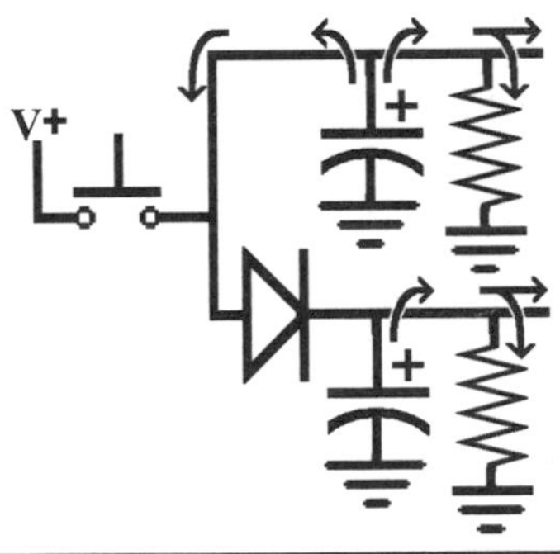

Figure L46-4

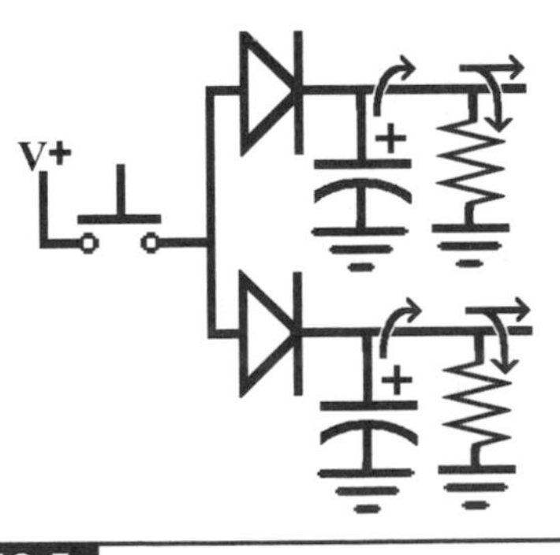

Figure L46-5

SECTION 14

Define, Design, and Make Your Own Project

Imagination is more important than knowledge.
—Einstein

IT HAS BEEN MY EXPERIENCE that when people are given instructions to make something for themselves, they will complain. Either they are given far too much detail or what they perceive as not nearly enough. All this from the same audience. What I have done here is give examples. The more complex the example, the more I choose to talk in terms of concept, not detail. I do this on purpose. This is your project. If I gave you all of the details, it becomes my project and you copying my project.

Remember the earlier comment about Lego. That was a serious statement. Parts of one fit perfectly into the parts of another. That is one of the beauties of digital electronics. Think about it. You have learned enough that you could almost build the controller for an arena's scoreboard. You can develop your own applications using just a portion of the system. Next time you go to the mall, carefully look at those fancy gadgets that eat your quarters and tell you your fortune. What? You say you can't do that? Then why not?

Lesson 47

Defining and Designing Your Project

Ideas to keep in mind:

1. In an ideal world, the sky is the limit.
2. Murphy was probably right. Murphy's first law states, "If anything can go wrong, it will." Do an Internet search of "Murphy's law" + Technology.
3. In reality you will never have:
   - Enough time
   - Enough money
   - The right equipment for the job
4. Your expectations for this project should be reasonable.
   - Five printed circuit boards are provided in the kit for this unit. This provides you with maximum flexibility.
   - The biggest limitation that exists for this project is your inability to create your own enclosure.
   - The most reasonable and inexpensive premade enclosures are old plastic VHS cassette cases.
   - These are often available with plastic lining that allows you to insert labels, signs, and directions.

Notes Regarding Possibilities

This is digital electronics. The processors can be mixed and matched, or not used at all.

1. 4046 VCO:
   - Can be set to a specific frequency
   - Rolls down
   - Can roll up (think about it)
   - Can set minimum frequency
   - Can set maximum frequency
   - Has frequency range
2. 4017 walking ring:
   - Dependent on a clock signal, but any clock signal will do.
   - Don't underestimate reset or enable.
   - Can count 0 to 9, and reset or freeze on any predetermined number.
   - What about the carryout? High for half the count, low for the other half.
   - What about using two 4017s, counting 0 to 99 in rows?
   - The outputs are not limited to LEDs, as shown in Figure L47-1. They can control transistors to give power.

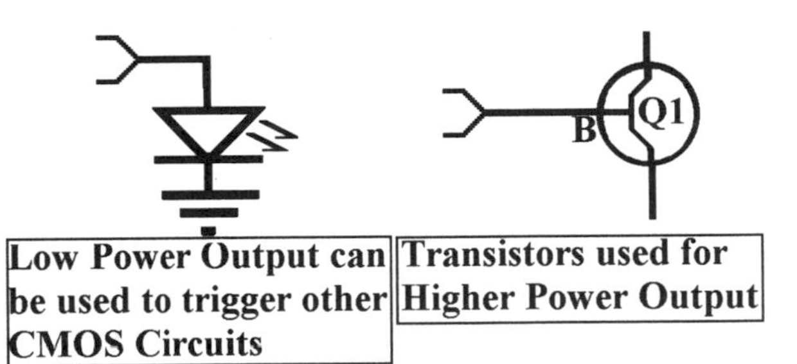

Figure L47-1

3. 4516/4511 BCD unit:
   - Like the 4017, it is dependent on a clock signal.
   - With a 1-Hz signal, you could make a binary clock.
   - The 4516 has more inputs than outputs.
   - The U/D control takes any digital input. It is not dependent on a clock signal.
   - What about having two 4516s counting 0 to 99 in digits?

Timing

There are only three separate RC timers right now. More are possible. RC1 and RC3 work together. If RC1 is longer than RC3, the displays fade before the rolldown is complete. Use Table L47-1 to set the timing of the RCs to your desires.

TABLE L47-1 Set Timing of RCs

Resistor	Capacitor	Time
20 MΩ	10 μF	200 seconds
10 MΩ	10 μF	100 seconds
4.7 MΩ	10 μF	50 seconds
20 MΩ	1 μF	20 seconds
10 MΩ	1 μF	10 seconds
4.7 MΩ	1 μF	5 seconds

Examples

Each of five examples presented has a unique twist in its application. New portions are explained in detail. Some of the design components, by their very nature, are identical between projects. When that happens, that portion is mentioned briefly.

The Ray Gun

The ray gun displayed in Figure L47-2 is a wonderful little project.

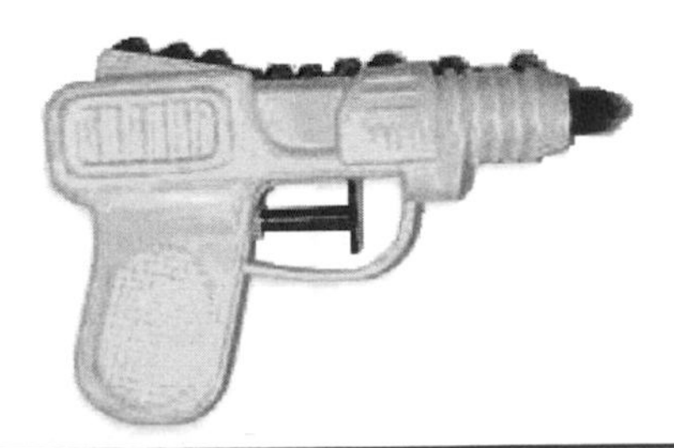

Figure L47-2

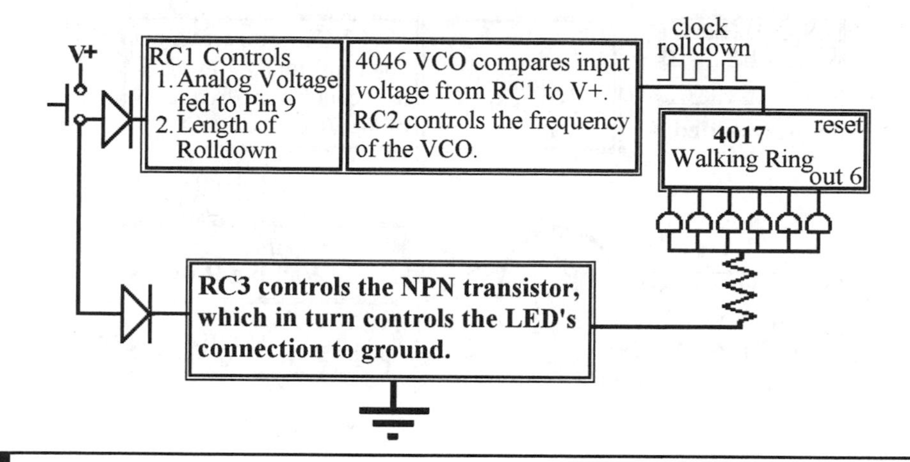

Figure L47-3

The entire system has been enclosed in a squirt gun. The unique idea here is that only a portion of the larger system has been applied as shown in the graphics system diagram of Figure L47-3.

The same system can be applied to a push button fortune teller or a 10-LED roulette wheel.

The "Whatever" Detector

The only unique quality here is the switch and the name you give it. In fact, this is the same system, but applied as a fake metering device. Fill in the word of your choice for "whatever." The beauty is that it actually changes frequency from person to person. Did you notice that you could get the circuit to work by just touching across the bottom of the push button? Try it. The packaging can be fancy, as shown in Figure L47-4, or you could use a simple box.

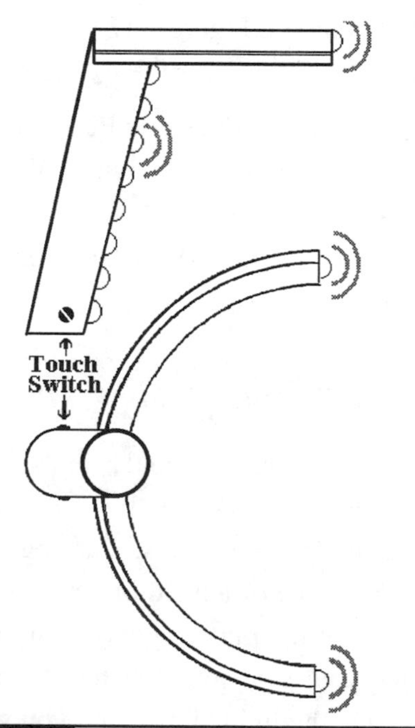

Figure L47-4

You could moisten your palm and pick up the "Ghost Detector," claiming all sorts of things as you point it toward a dark corner. Give it to your friend and watch as it goes dead. A little knowledge isn't dangerous. It's fun.

An Animated Sign

The animated sign shown in Figure L47-5 could be used for a variety of shapes.

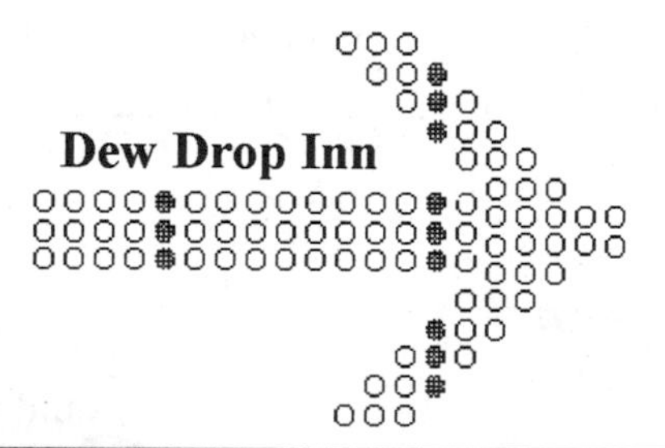

Figure L47-5

Both of the things that make this unique are shown in the graphics system diagram of Figure L47-6.

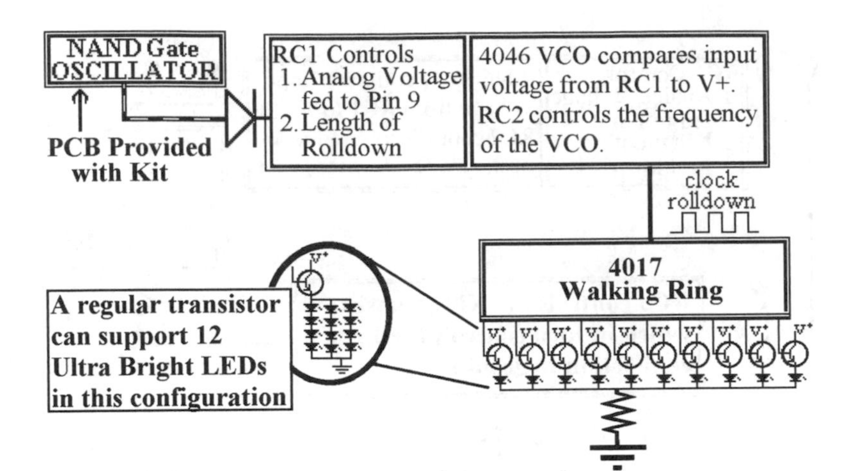

Figure L47-6

First is the subsystem that triggers the 4046. It is a 4011 oscillator set to pulse every 20 seconds. The maximum frequency is set to 20 Hertz. It rolls down almost to a complete stop before it kicks the oscillator high again. The schematic is shown in Figure L47-7.

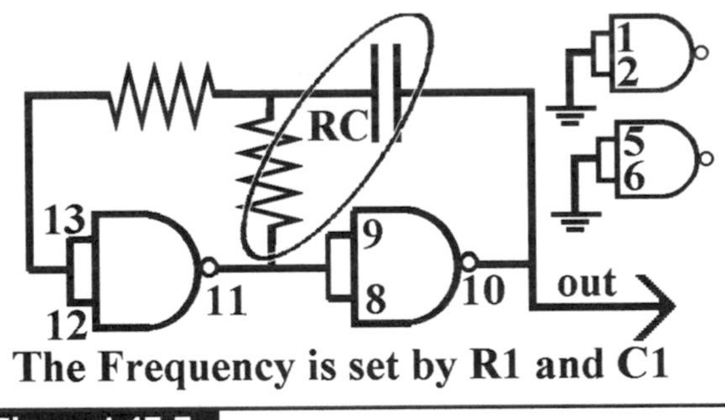

Figure L47-7

Second is the necessity for ultrabright LEDs to make the output visible from a distance. This requires the output of the 4017 to use transistors to amplify the power.

The IQ Meter

This is another wonderful gadget. It could just as easily be called a lottery number generator. You secretly control the first digit by a hidden touch switch made from two pinheads. Shown in Figure L47-8, when you hold it, you're guaranteed a number above 100.

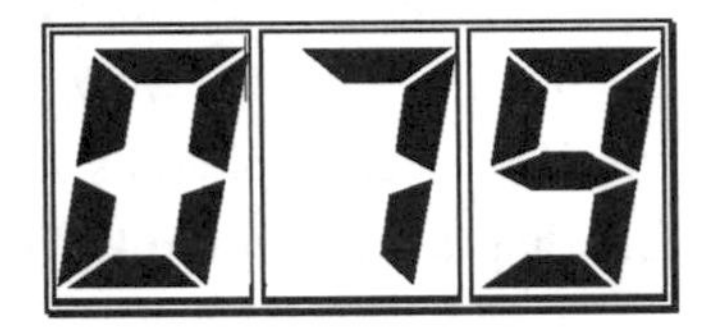

Figure L47-8

Everybody else gets a number under 100. It doesn't use the walking ring at all. To do it properly, though, does require two sets of the number display circuits and three seven-segment LEDs. Directions to make your own printed circuit boards are provided at www.mhprofessional.com/computingdownload. Figure L47-9 shows the graphic system diagram for the IQ meter.

Modify the dual NAND gate PCB, included in the kit, so it is a single-touch switch. You can figure out the wiring to get a high output to show 1 and a low output to show 0 on the seven-segment display.

Love Meter Fortune Teller

This wonderful gadget is popular in the malls. It takes two inputs to work. You pay a quarter and do one of several different things. You might put your palm on the outline of a hand or pass your hand over the hand of the plastic gypsy. You might touch one pad and your girlfriend touches the other; then you kiss. Then the lights flash, and you get your

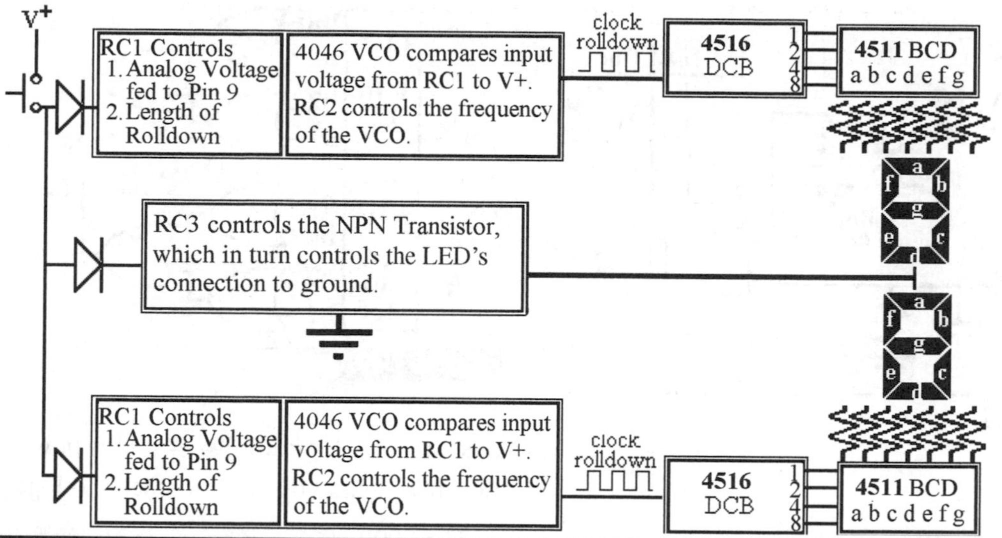

Figure L47-9

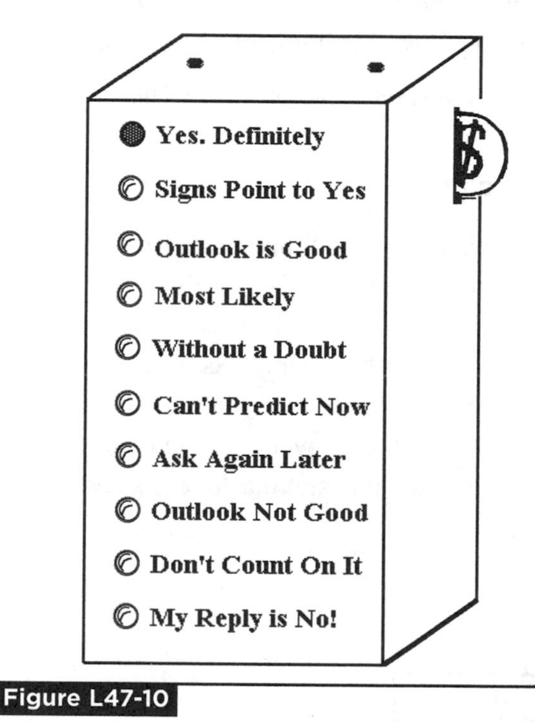

Figure L47-10

fortune told. The outcomes shown in Figure L47-10 come directly from The Magic 8-Ball toy made by Tyco Toys.

The graphics system diagram for the fortune teller is shown in Figure L47-11.

The schematic for the dual input trigger is shown in Figure L47-12. It is important to give plenty of time for the coin input to time out, so that by the time they trigger the second input, the first has not timed out.

Event Counter and Trigger

This device will count how many times something happens and then trigger an output. With two 4516 chips, the output can be set to trigger at any number between 0 and 255. The graphics system

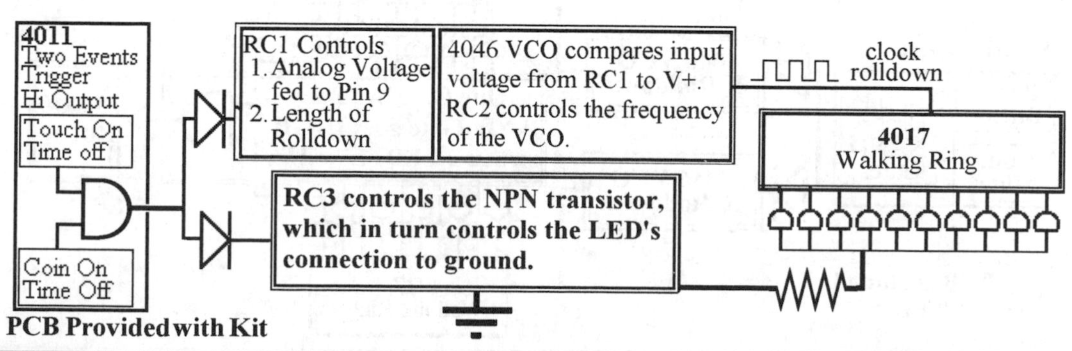

Figure L47-11

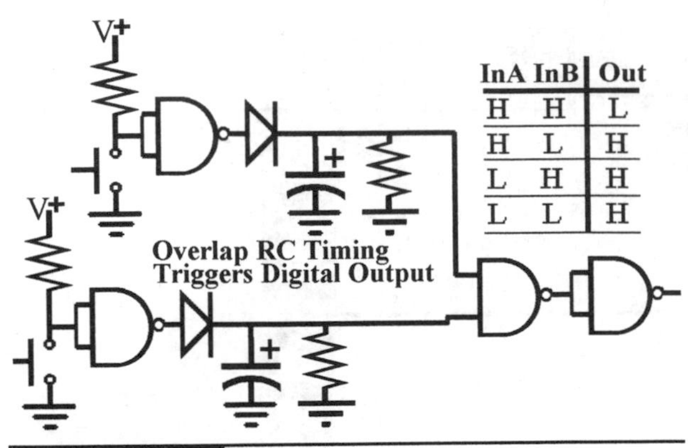

Figure L47-12

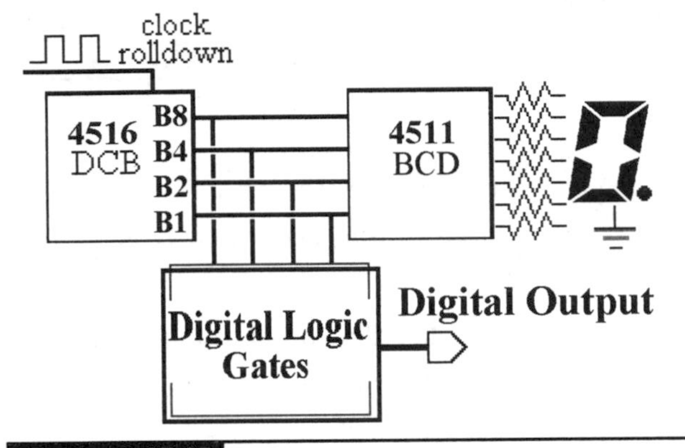

Figure L47-13

diagram in Figure L47-13 is set up for a single numbering system.

The 4516 can be preset to any number from 0 to 15. You don't have to feel confined to what the display can show. Set the up/down control to count backward. The output is triggered when everything

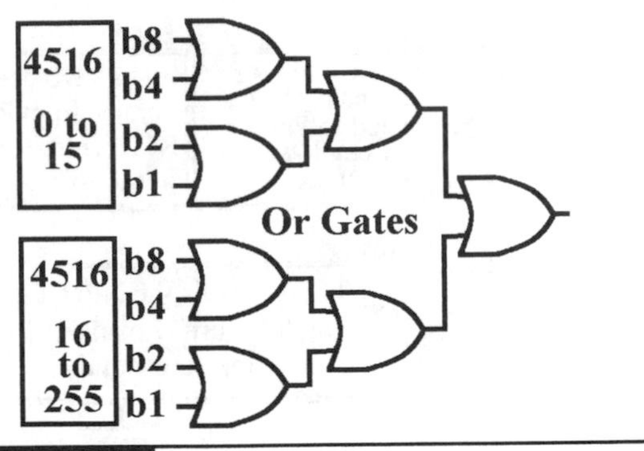

Figure L47-14

reaches zero. Figure L47-14 shows how to apply two 4071 quad OR gates to check to logic output.

The 0000 0000 can be made by the following:

- Counting up and rolling over the top number
- Counting down until you hit the bottom
- Triggering reset by an outside event

Slot Machine

It is possible to do this. What if I get two rows of walking rings and trigger them both? When the LEDs match, it can work like a slot machine and trigger a payout. What a simple concept. Move forward with this one only if you have confidence. It can work, but it takes work. The graphics system diagram for a two-line slot machine is shown in Figure L47-15.

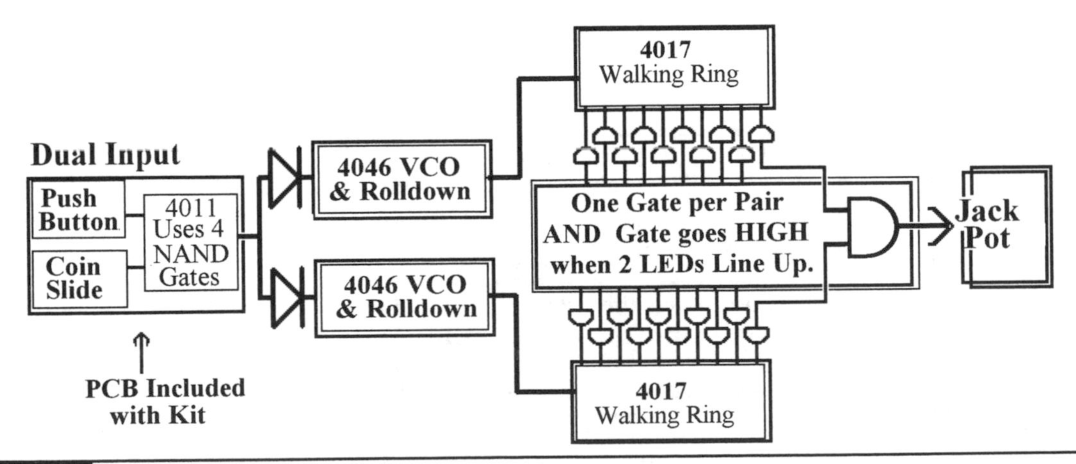

Figure L47-15

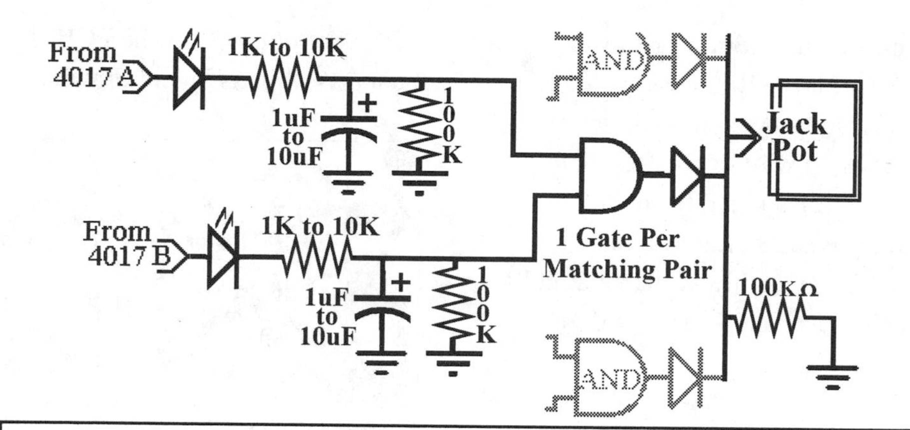

Figure L47-16

It appears obvious that you can compare the outputs using an AND gate. The 4081 dual input quad AND gate is ideal. But consider that a high output from the two walking rings might cross each other for a millisecond. That is enough to trigger the AND gate falsely. It is important to slow down the comparison to prevent false triggering. Such a circuit is displayed in Figure L47-16. This ensures that each LED in the pair must be high for a minimum of half a second before the AND gate input moves above the triggering voltage.

What about the payout? I'll be honest. Check your phone directory for any shop that repairs vending machines. Unless you are a skilled machinist and have time, you will save yourself many headaches by purchasing a used coin return mechanism. Usually, one clock signal in kicks one quarter out. It would be fairly safe to say the output from the 4081 will be a clean clock signal. Have fun.

One Last Word

Be creative. Be patient with yourself. Breadboard your prototype before you start building onto your printed circuit board.

Lesson 48
Your Project: If You Can Define It, You Can Make It!

Lesson 48 includes the following for building each subsystem of your project:

- The printed circuit board layout
- The schematic laid out onto the PCB
- The parts placement onto the PCB

The possibilities for creating different applications with the subunits of this circuit should be limited only by your imagination. But I have to apologize. There are real limitations to this course. I have strived to provide you with a wide range of printed circuit boards. You can control all of the various inputs for each processor. Be reasonable and realistic in your applications.

If you choose to expand your project beyond the components provided, they should be available at any electronics components store, or order them from www.abra-electronics.com. If you really want to play, there are instructions for creating your own printed circuit boards at www.mhprofessional.com/computingdownload.

WIRE TYPE: A comment is necessary regarding the type of wire that you are using as you start to build your project. Up to this point, you have been using 22- to 24-gauge solid-core wire for prototyping on your breadboard. Solid wire is great for prototyping. But it breaks easily. Bend a piece back and forth 10 times. If it didn't already break, it will soon. As you build your subsystems, use stranded wire to connect them. Stranded wire will bend back and forth without straining itself. But it is difficult to use on breadboards. This wire is readily available at dollar stores, but is disguised as "telephone wire."

Inputs

There are four different types of input switches.

- **The contact switch:** Recall that basic mechanical switches were discussed in some detail in Part Two.
- **The touch switch:** The initial switch using the zener diode is sensitive enough to react to skin's resistance. This is not digital. Connect the push button's wires to pinheads instead.
- **Two Input Digital:** A coin or other action at the SW1 in Figure L48-1 will turn the output of the first gate high for about 20 seconds. You would then have that time to activate the second NAND gate using the SW2.

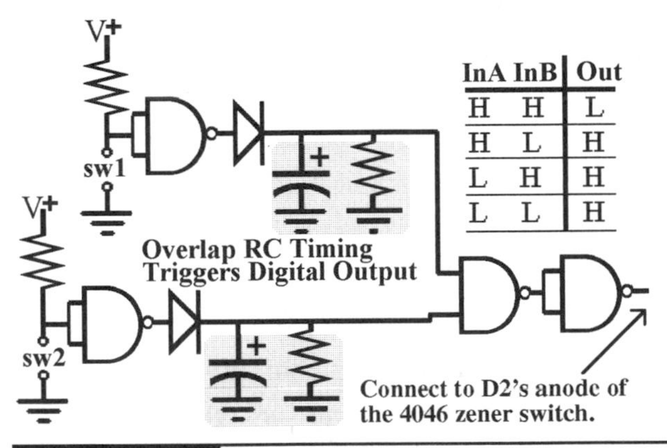

InA	InB	Out
H	H	L
H	L	H
L	H	H
L	L	H

Figure L48-1

The bottom view of the PCB is shown next to the top view parts layout drawing in Figure L48-2.

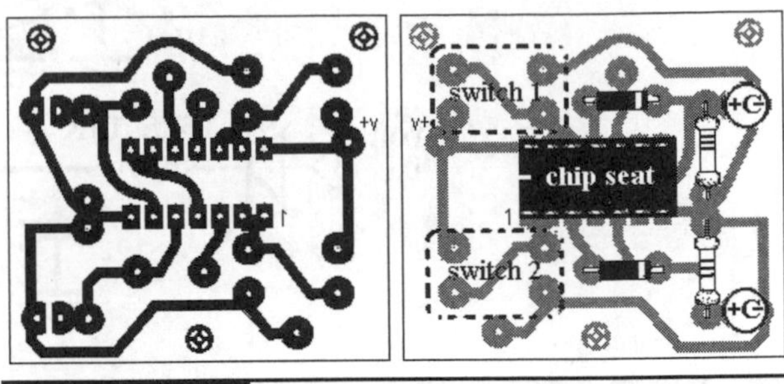

Figure L48-2

Many of the possible variations for these input switches are discussed in depth in Part Two. Use the same layouts and component values, depending on the type of switch you want to build.

- **The Self-Kicking Oscillator:** The schematic for the self-kicking oscillator is shown in Figure L48-3. Remember that the frequency is set by the RC. The inputs of the unused gates have been set to ground.

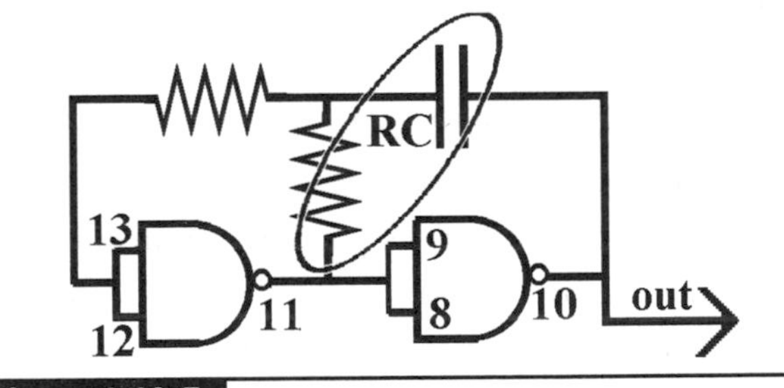

Figure L48-3

The timing for the rolldown can be adjusted, but the oscillation timing should be longer than the rolldown timing if you want the rolldown to stop completely before it gets kicked again. The schematic is shown in Figure L48-4.

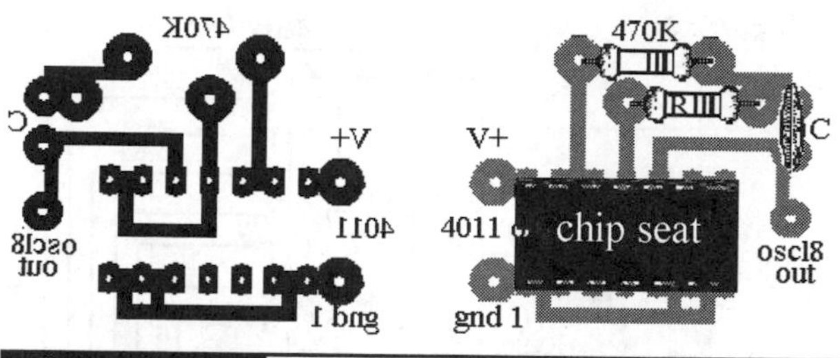

Figure L48-4

Processors

There are many processor systems and subsystems here. Each has its major input and a vairety of minor inputs. The potential combinations are limited only by your imagination.

VCO and Timed Off

The heart of the system's rolldown is the VCO's reaction to the voltage drop of RC1. To save space, three subsystems have been grouped together on one printed circuit board. Naturally, RC1 and the 4046 VCO would inhabit the same board. I've included the Timed Off circuit on this board as well. Figure L48-5 shows the individual schematics that exist on this one PCB.

This does make for a more complex PCB layout, as you can see in Figure L48-6. Note that the components that make up the Timed Off portion have been marked with an *.

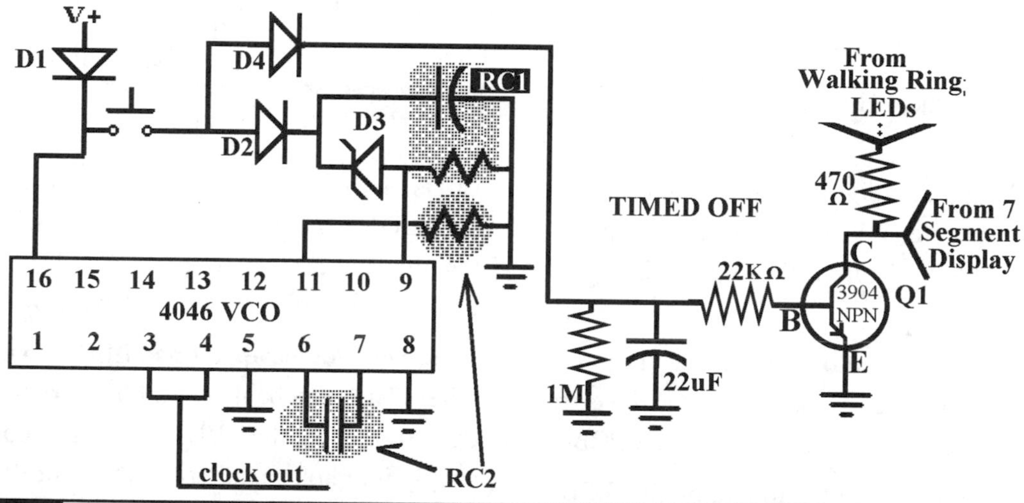

Figure L48-5

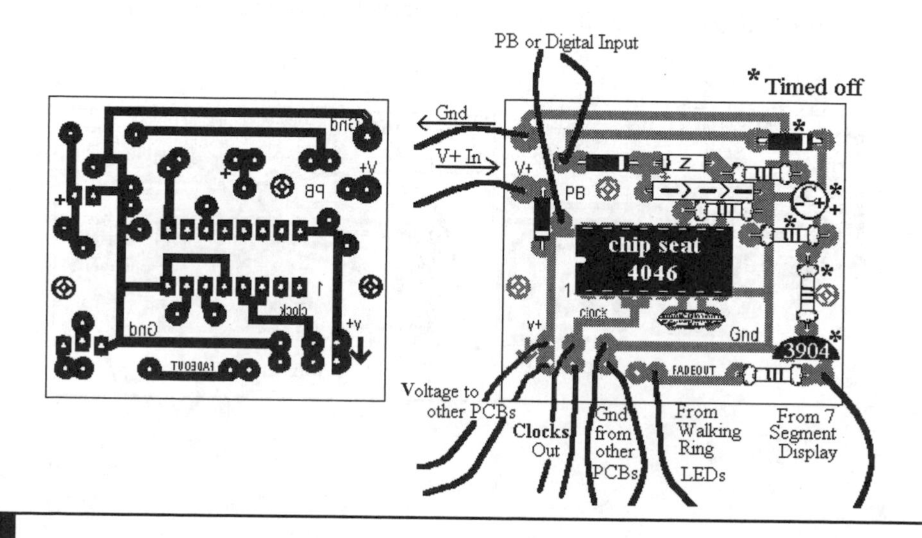

Figure L48-6

Because the PCB is fairly crowded, another view in Figure L48-7 is provided. This top view identifies each component by value and relation.

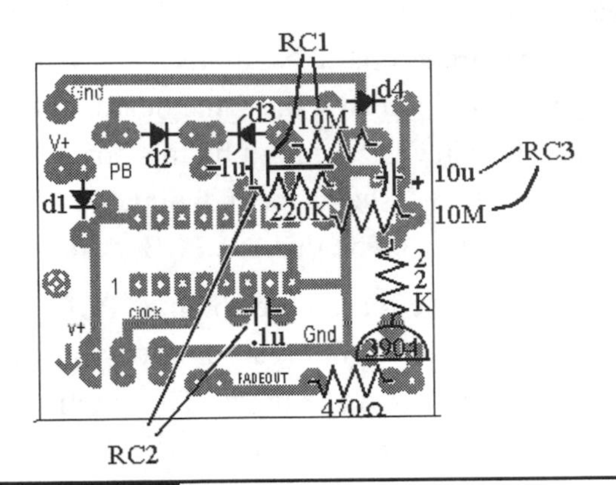

Figure L48-7

The Walking Ring

The schematic of the 4017 walking ring in Figure L48-8 says it all.

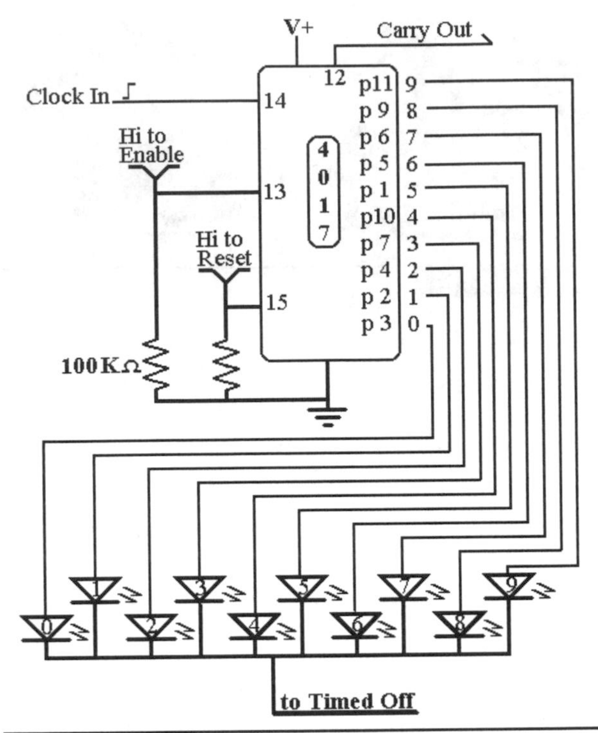

Figure L48-8

Remember that any digital high can be used to control the reset or enable. Otherwise, these inputs need to be held low. Figure L48-9 shows the PCB and parts layout.

Even thought this is not as complex as the previous PCB, Figure L48-10 emphasizes the extra control inputs available on the walking ring PCB.

The LEDs are not expected to live on this PCB. Remember to use stranded wire to connect the LEDs. Also, there is no ground connection for the LEDs on the 4017 PCB. That is because each of the LED's negative legs (cathode) is wired together, as shown in Figure L48-11, and that one connection leads back to the Timed Off circuit.

So now you have 10 or so LEDs floating on wires. They can be neatly mounted onto an enclosure using the LED collars. The collars fit snugly into holes made by a 7-mm or 9/32" drill bit. Figure L48-12 shows how to use these marvelous little items.

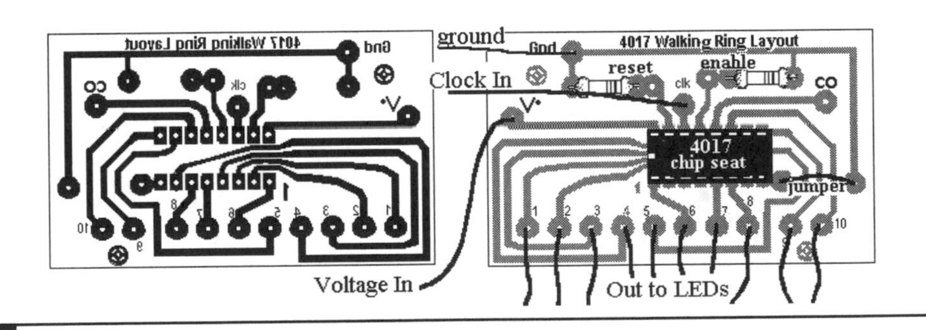

Figure L48-9

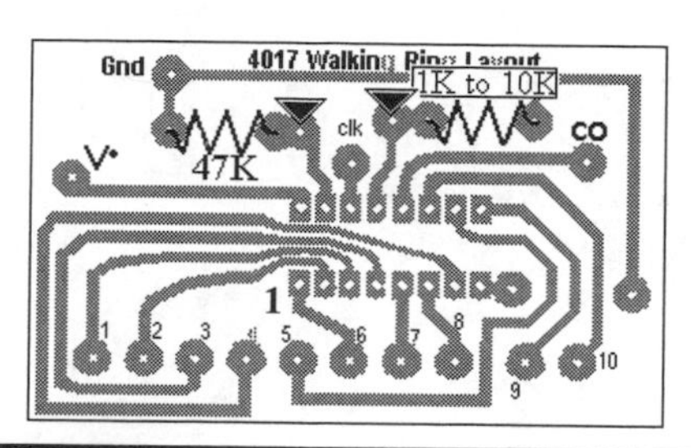

Figure L48-10

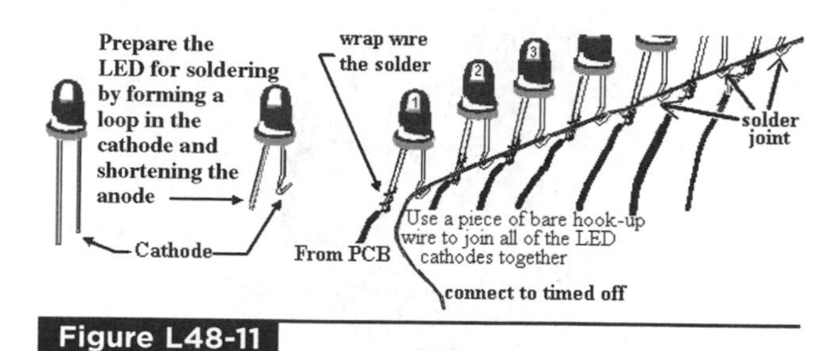

Figure L48-11

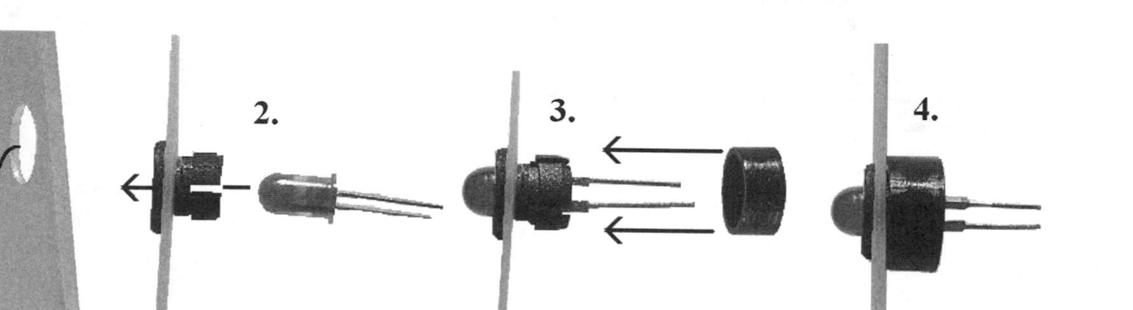

Figure L48-12

4516/4511 and Number Display

Examine the schematic in Figure L48-13.

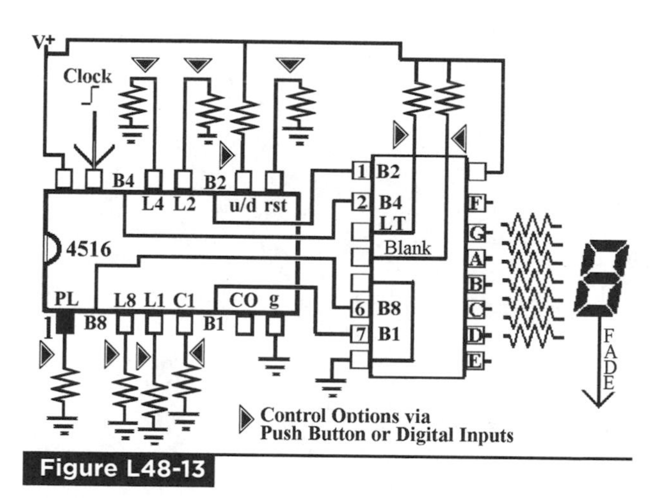

Figure L48-13

The printed circuit board layout and parts placement are all displayed nicely in Figure L48-14.

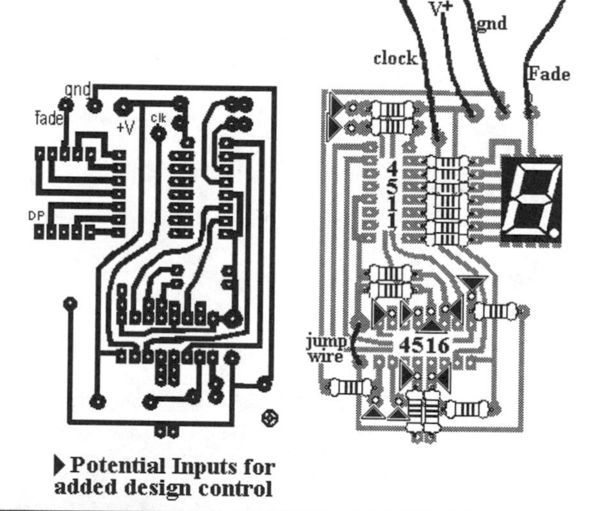

Figure L48-14

Inputs, Inputs Everywhere

Consider all of the preloads as a single input. They have to act together, and are controlled by pin 1 of the 4516. Also, ignore carry-in unless another 4516 feeds to it. So there are really only five inputs on this subsystem.

Remember to use stranded wire for all of the connections to the other subsystems. All of the resistors are marked either 100 kilo-ohms or 470 ohms in Figure L48-15. These are necessary to condition all of the many inputs that exist.

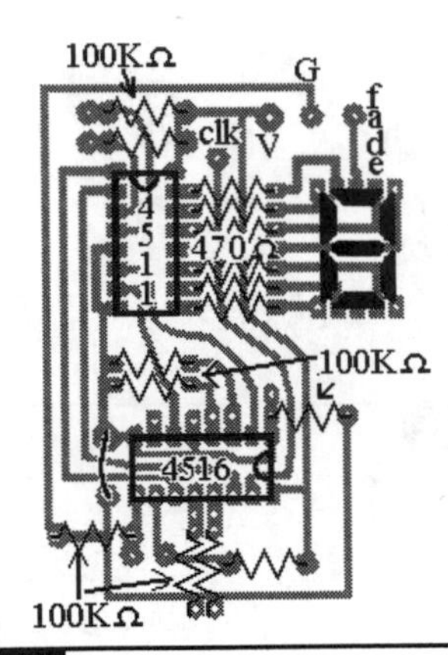

Figure L48-15

That's really all there is to it. The rest is up to you.

PART FOUR

Amplifiers: What They Are and How to Use Them

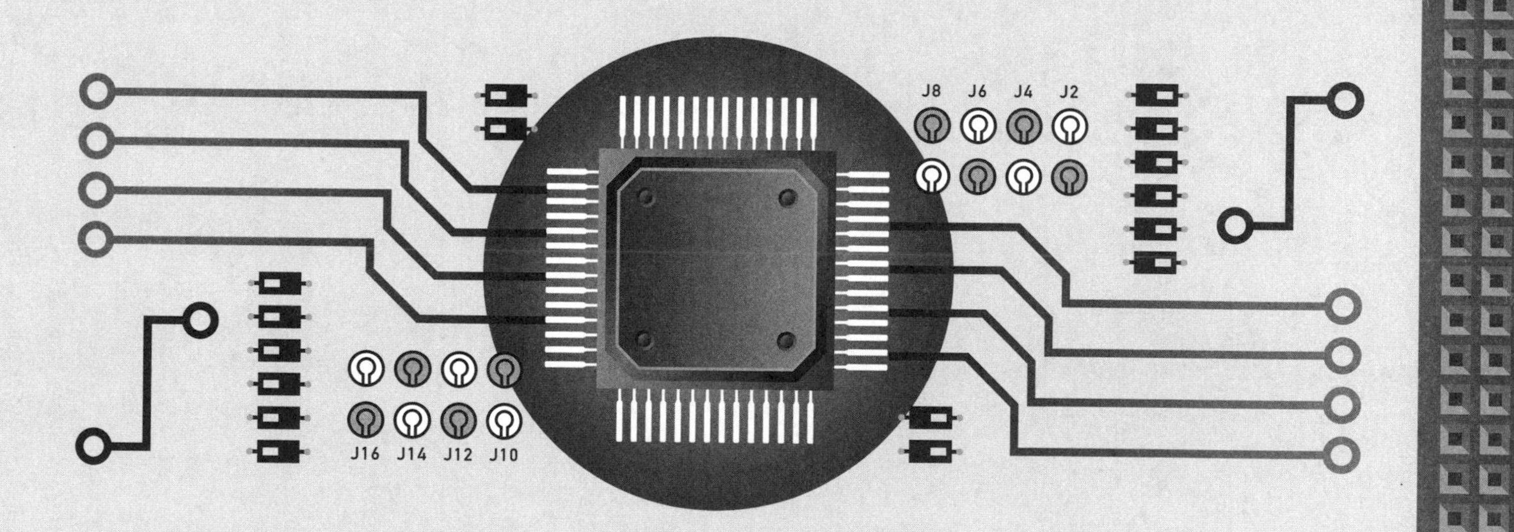

Pump Up the Volume

Where would we be without amplifiers? The telegraph system was invented in 1844 long before amplifiers existed. The figure below shows an early input key. The signal had to be relayed down the line from each station because it weakened with distance. Imagine our world without amplifiers. It would be a quiet place.

The Parts Bin on the following page has the complete parts list used in Part Four.

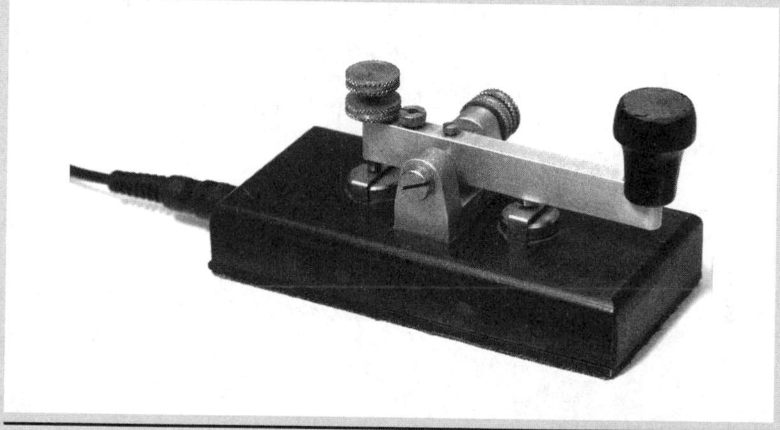

Imagine our world without amplifiers. It would be a quiet place.

PARTS BIN FOR PART FOUR

Description	Type	Quantity
2N-3906 PNP transistor	TO-92 case	2
2N-3904 NPN transistor	TO-92 case	2
LED	5 mm	1
Electret microphone	Mic	1
10 Ω	Resistor	1
100 Ω	Resistor	2
100 kΩ	Resistor	2
1 kΩ	Resistor	2
4.7 μF	Capacitor	1
470 μF	Capacitor	1
1000 μF	Capacitor	1
8 ohm	Speakers	2
100,000 ohm 1/2 watt	Potentiometer	1
LM 741 op amp	IC	1
Audio transformer	Transformer	1
Alligator clips (red and black)	Hardware	1 each
1/8″ plug	Hardware	1
1″ heat shrink 1/8″ d	Hardware	2
Speaker wire	Hardware	4 meter 12′
4PDT	Switch	1
Door phone	PCB	1

- Not all components will be consumed by project work.

SECTION 15

What Is an Amplifier?

Water is always a great analogy. A small amount of power on the handle allows a very large amount of power to flow from the source through the amplifier.

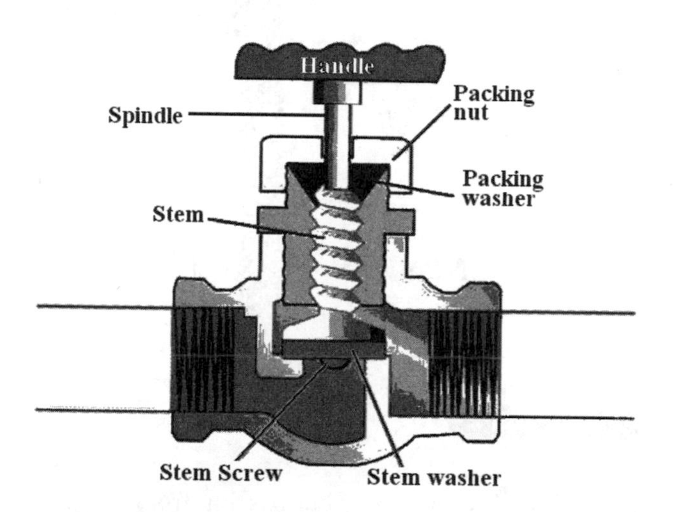

Lesson 49
Transistors as Amplifiers and Defining Current

So, what is an amplifier? It takes a small signal and makes it bigger. Signals are made of both voltage and current.

The definition from dictionary.com is very simple.

> **am·pli·fy**—verb. 1. To make larger or more powerful; increase. 2. To add to, as by illustrations; make complete. 3. To exaggerate. 4. Electronics—To produce amplification of or to amplify an electrical signal.
>
> **am·pli·fi·er**—noun. 1. One that amplifies, enlarges, or extends. 2. Electronics—A device, especially one using transistors or electron tubes, that produces amplification of an electrical signal.

Review

We've applied the transistors as amplifiers in a few ways. You may not have recognized it at the time. Let's review what you did.

Remember the Night Light

Its action is shown in Figure L49-1. The signal given to the base of the NPN transistor is amplified.

The voltage divider created by the potentiometer and the LDR control the voltage to the base of the transistor. Remember, the voltage at that point is dependent on the following:

1. The setting of the potentiometer
2. The amount of light on the LDR

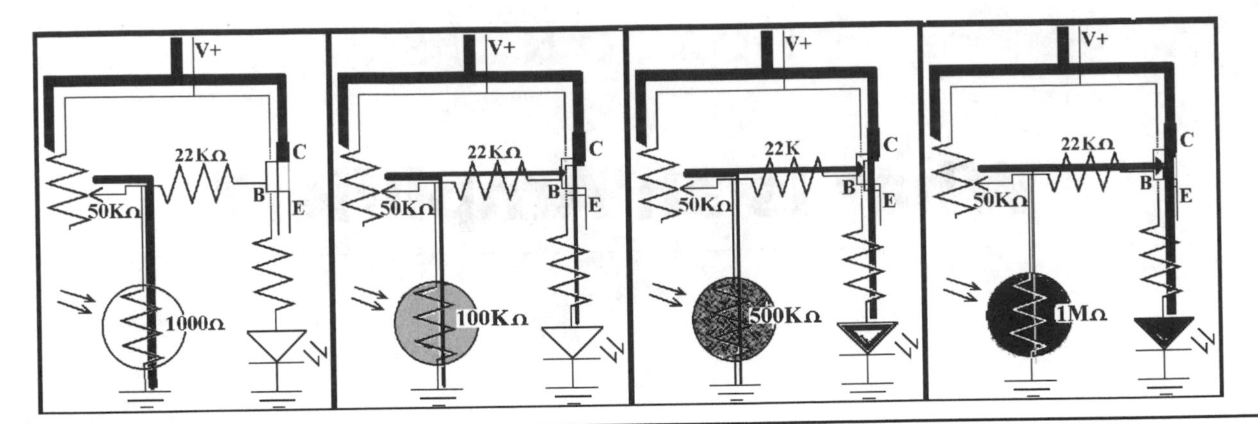

Figure L49-1

The signal to the base of the transistor has little power, not even enough to turn on an LED, because so little current gets through the 100-kilo-ohm potentiometer set at midrange.

The action of how the signal to the base is amplified by the NPN transistor is demonstrated in Figure L49-2. This action acts as a valve controlling the power source to the LEDs:

- The smaller signal controls the transistor's valve action.
- The resulting amplified signal from C to E is a much more powerful version of the original small signal.

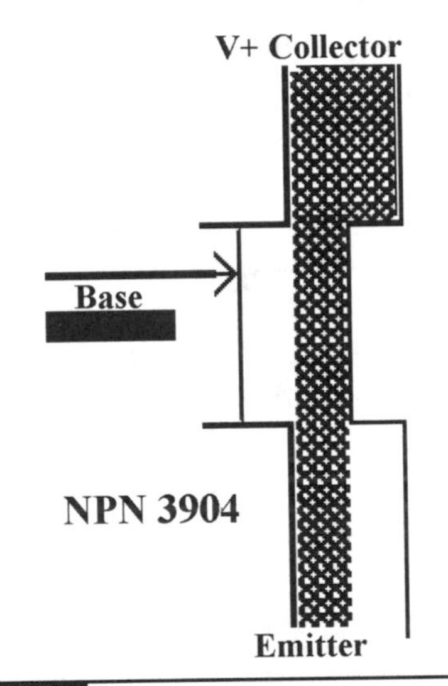

Figure L49-2

Audio for the NAND Gate Oscillator

Remember when you first connected your speaker directly to the NAND gate output? The signal was so quiet that you had to place your ear right on top of the speaker to hear it … and that was in a quiet room. The amount of current passing from V+ to pin 10 when it was low was enough to light an LED, but not enough to really shake the speaker. This is the schematic in the left side of Figure L49-3.

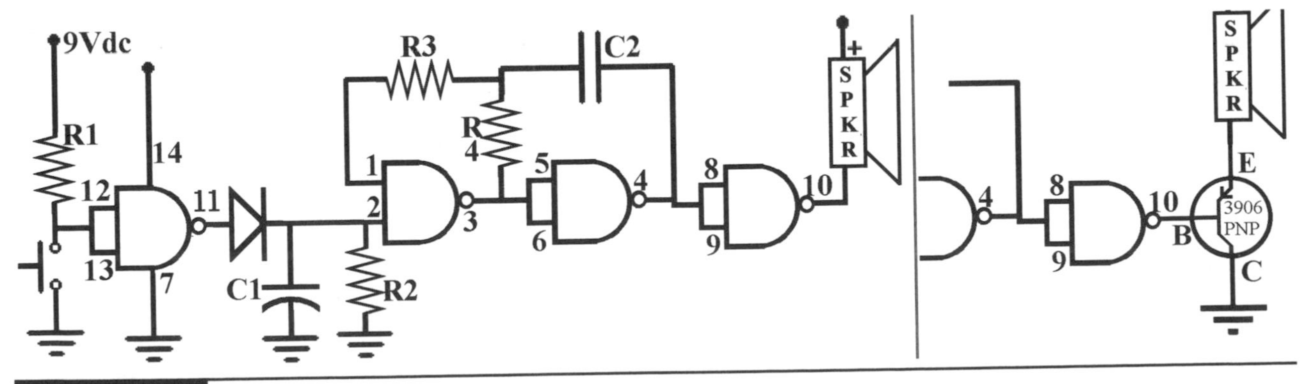

Figure L49-3

That small signal to the base was amplified by the PNP transistor, shown inserted into the schematic in the right side of Figure L49-3. The small signal was used to control the transistor valve action that controlled the power directly from the voltage source.

The weak signal to the base was amplified by the PNP transistor, providing plenty of power to the speaker coil to make an amazingly annoying loud sound.

"Wait a minute ...," you say. "Didn't the digital gate provide 9 volts, and the transistor gave 9 volts as well, so why was the transistor output more powerful?"

Calculating the Current: Amperage

The difference was not in the voltage, the push behind the electrons, but in the quantity of electrons being pushed, the current.

Figure L49-4 starts a very practical way of picturing *current in an electrical system.*

Figure L49-4

In electronics, voltage is the push behind the flow, or the force.

For a creek, gravity is the push, the force behind the flow, as water runs downhill.

A creek, by definition, has little current. Current is measured by how quickly water passes by in liters per second or cubic feet per second. This creek has a current of 5 ft^3/s.

cur·rent (noun). 1. A steady, smooth onward movement: a current of air from a fan; a current of spoken words. Synonyms—flow. 2. The part of a body of liquid or gas that has a continuous onward movement: *rowed out into the river's swift current*. 3. A general tendency, movement, or course. Synonyms—tendency. 4. Symbol I, Impedance (a) A flow of electric charge. (b) The amount of electric charge flowing past a specified circuit point per unit time. (dictionary.com)

Now think of a stream like the one pictured here in Figure L49-5. It has a modest current, but you could walk across it.

Figure L49-5

Consider that the downhill "slope" is the same as the creek. The force is the same, but the big difference is the "current," the amount of water. It is more powerful.

This stream has a current of 100 ft^3/s.

This river pictured in Figure L49-6 has a very large current. Even though it has less "slope" as the creek and stream shown, it has more water flowing by a single point, every second. If you tried to swim across, you would be swept away.

Figure L49-6

Obviously, the river is the most powerful. The force is the same, but the difference here is the amount of water. The river has a current of 20,000 ft^3/s.

In electricity, current is electrons passing through a wire.

The standard unit of current is an ampere (Amp or A, for short): 1 Amp = 1 Coulomb/second (1 Coulomb = 10^{18} electrons).

Current is the general term. *Amperage* is the unit of current. The common abbreviation for current is *I* (impedance). People new to electronics say, "It's such a pile of new words. 'I' looks odd." But think of the word *stampede* with a herd of electrons pounding through the wires. Now that's an "impede." There's a wild one shown in Figure L49-7.

Figure L49-7

So how much is a coulomb? How much current is moving through an electronic system? How could we look at the electrical equivalent to a creek in Figure L49-8 and figure the current used by an LED?

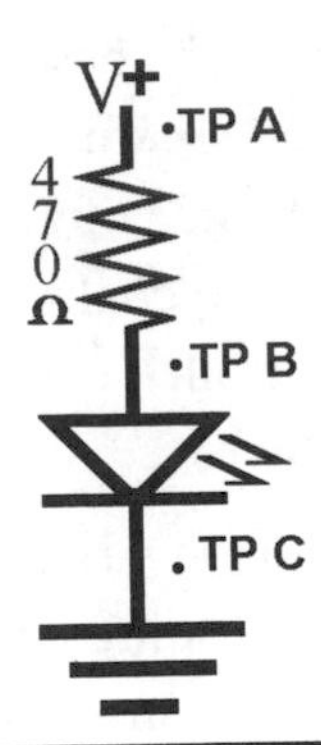

Figure L49-8

The current is a quantity we cannot measure directly with the tools at hand. But we can figure the amperage by using Ohm's law. Ohm's law simply states that V = IR, where George Ohm calculated that the amount of pressure in a circuit is directly related to the amount of current passing through the load.

Unit	Measuring
V = Voltage	Voltage is the unit of force.
I = Amperage	Amperage is the unit of current (impedance).
R = Ω	Ohm is the unit of resistance.

Here's how Ohm's law works. Think of a hose. A regular garden hose will do. The hose in Figure L49-9 has a fixed diameter, so the resistance is not going to change.

The higher the pressure, the more water flows through it. The lower the pressure, the smaller the current becomes.

But if you change the resistance as demonstrated in Figure L49-10 and Figure L49-11, that affects the amount of current, too.

Figure L49-9

Figure L49-10

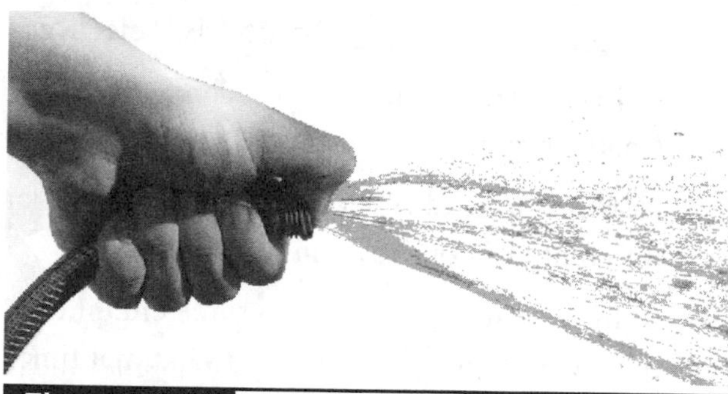

Figure L49-11

The fire hose shown in Figure L49-12 offers very low resistance because the hose is bigger. It is able to handle lots more current, but the pressure from the city supply is the same.

Figure L49-12

Back to Ohm's Law (V = IR)

This is a simple three-variable equation: $A = B \times C$. If you know two variables, you can figure the third. But how would you figure the current in the simple LED circuit shown in Figure L49-8?

To figure the current flowing through the LED's circuit, you need to know (a) the voltage used by the resistor (the pressure drop) and (b) the resistance.

Now, back to considering the basic circuit. Let's make some assumptions.

1. The power supply is a perfect 9 volts.
2. "R": The resistor's value is a perfect 470 ohms.
3. "V": Connect power and measure the voltage drop across the resistor—from TP a to TP b. The voltage drop across the resistor is 7.23 volts. V = IR so:

$$\frac{V}{R} = I$$

$$7.23\ V = I \times 470\ \Omega$$

$$I = 0.0153\ A$$

So we just figured out how much current passed through the resistor. Does that tell us how much passes through the LED? Yes, it does. Consider the following:

> **A simple statement. Unless some is added or taken away:**
>
> - **The current in a hose is the same along the entire hose.**
> - **The current in a creek is the same along the entire creek.**
> - **The current in a simple circuit is the same throughout the entire circuit.**
>
> **For a simple circuit, the amount of current is the same throughout the system. The current (I) passing through the resistor is the same amount of current through the LED as well.**

Exercise: Transistors as Amplifiers and Defining Current

1. Define amplifier as it relates to electronics.

____

2. Match these units to the terms they represent.

V, A, W, hose width, force, coulombs per second, gravity, ft³/sec, I, R, current

Pressure	Current	Resistance

3. Define the two combined factors that make a current.

a. _______________

b. _______________

4. What is a basic unit used to show current in the following?

a. Water system

____

b. Electrical system

____

5. Write without scientific notation how many electronics there are in a 1-coulomb charge.

____

6. There are 0.0153 amperes in the LED circuit. Write out exactly how many electrons pass a single point in the wire in 1 second.

_______________ = 1.53 10^{16} electrons

Recognize that this is considered a very small current.

7. Now it is time to observe the actual effect that different resistors have on the current.

In Lesson 5, you used the same circuit as the one shown in Figure L49-13. At that time, you observed what happened as you changed resistors. Now you get to understand what happened.

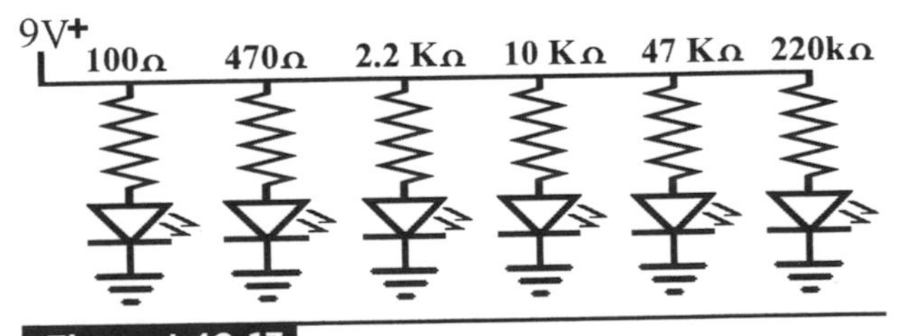

Figure L49-13

Measure the resistors used here with your DMM. Do all figuring here to three significant figures. This means that in terms of accuracy, all numbers have three digits. For example, a 10-kilo-ohm resistor has two significant figures because the color coding only shows two digits. On the DMM it might show 9.96 kilo-ohms. That shows three significant figures.

You can copy most of the information from the exercise you did for Lesson 5 to Table L49-1.

8. Think of your garden hose.

a. If you turn the hose on, the pressure comes from where?____________________

b. Pressure in an electronic system is called what? ____________________

c. If you leave the pressure the same and squeeze the hose, this increases what?

____

d. As you increase the resistance, what happens to the current?____________________

e. If voltage in a system remains constant, current will decrease if resistance is what?____________________

f. Conversely, the current will increase when resistance is what?____________________

Lesson 50
Defining Work, Force, and Power

What are we amplifying? Here you learn how work, force, and power are defined and measured in electronics. Amplifiers by their very nature are analog devices. The transistors you've worked with are analog. They responded to any voltage to the base in varying degrees. Remember, NPN was turned on by more voltage to the base, and the PNP was turned off as the voltage increased. Here you will be introduced to the two types of amplifiers that match the opposite actions of the transistors you've already dealt with.

What Is Force?

Force is the amount of energy exerted. In Figure L50-1, Atlas is exerting a force to hold up the sky. But by definition, no work is happening because there is no movement. If an object is not moved, no work is done, no matter how great the force.

TABLE L49-1 Information from the Exercise

Resistor in Order	DMM Resistor Value	Voltage Drop Across Resistor	Current in System Amps = $\frac{V_{drop}}{\Omega}$	Voltage Drop Across the LED
100 Ω	______Ω	______Volts	0.______ A	______Volts
470 Ω	______Ω	______Volts	0.______ A	______Volts
2.2 kΩ	______Ω	______Volts	0.______ A	______Volts
10 kΩ	______Ω	______Volts	0.______ A	______Volts
47 kΩ	______Ω	______Volts	0.______ A	______Volts
220 kΩ	______Ω	______Volts	0.______ A	______Volts

Figure L50-1

Force is measured in newtons. Roughly, 1 newton will move about 100 grams of mass upward against the force of gravity as demonstrated in Figure L50-2. Actually, it is 98 grams, but 100 is easier to remember.

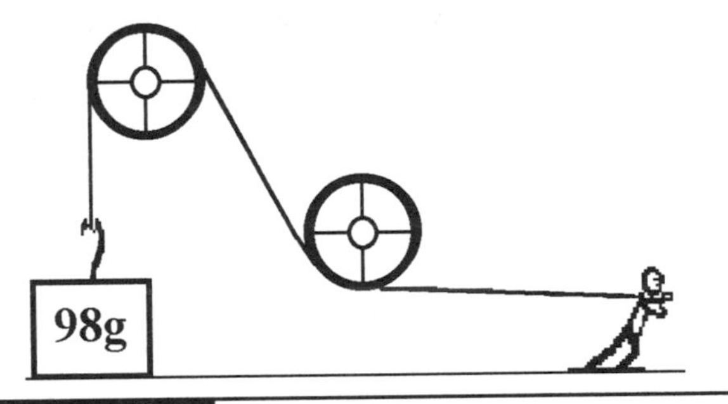

Figure L50-2

The force in electricity is measured in volts. The matter being moved is electrons. This is shown in Figure L50-3.

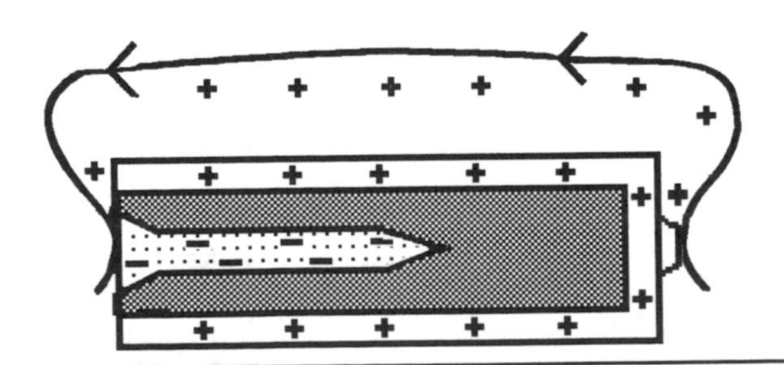

Figure L50-3

What Is Work?

Work is measured as the amount of force exerted on an object through a distance. There has to be both force and distance for work to be done. Work is measured in standard units of newton-meters or joules: 1 N-m = 1 J. The terms are interchangeable.

The distance can be done via a bottle rocket or a snail as shown in Figure L50-4. The speed doesn't matter.

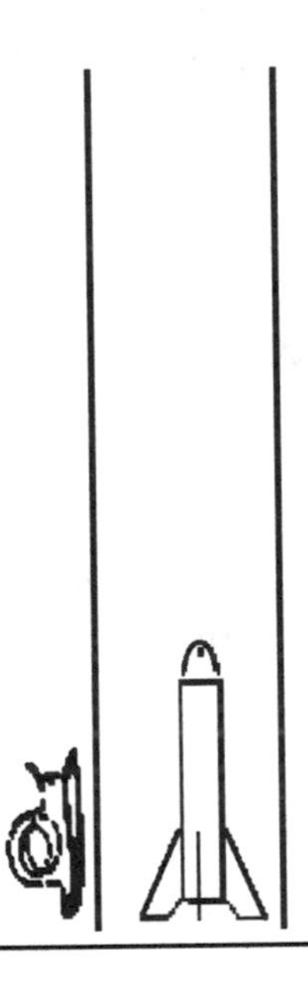

Figure L50-4

Watt Is Power?

Power is the *rate* that *work* is *done*. The standard unit of power in electricity is watts. One watt is the standard power unit defined by the force needed to move 98 g upward against gravity 1 meter in 1 second. This is graphically shown in Figure L50-5.

Two items can have the same force exerted on them, but can have different power. The bottle rocket may do 1 joule of work in 0.01 second. That works out to 100 watts. The snail may take 1,000 seconds to do the same work. It has a power of 0.001W.

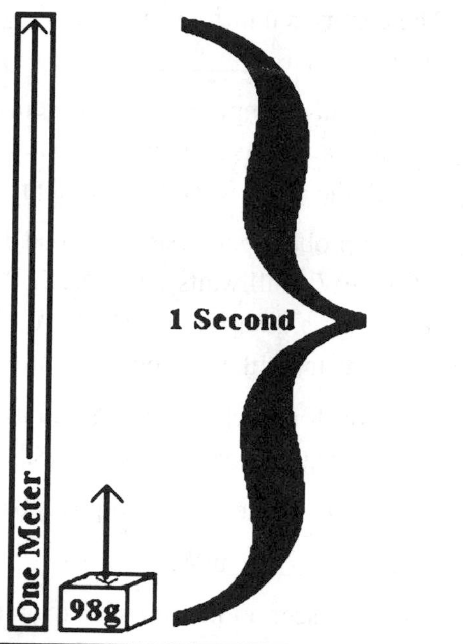

Figure L50-5

But look at the power in different water systems. The force of gravity pulls water over the edge of the cliff in both of the waterfalls shown in Figure L50-6. The current of both waterfalls has the same force of gravity on them. It is 180 feet at each of these falls. But power is defined by the current multiplied by the force. By definition, the current includes both quantity and rate. I would stand under only one of them.

Figure L50-6

Multiply the current by the voltage, and that gives power. In any system, power is determined by how fast work is done.

- In these natural water systems:

  Power = current × force

  Current = ft³/s = the amount of water passing by and how fast it is flowing

  Force = slope = the amount of push behind the current, the pressure

- In electrical systems:

  Power = current × force

  Force = voltage

  Current = amperage measured as a standard where 1 A = 1 C/s

  Remember that 1 C is a standard mass of electrons.

  Power is measured in watts.

  1 W = 1 A × 1 V

The LED is a low-power light output. How much power is used by the LED? Consider the basic circuit shown here in Figure L50-7.

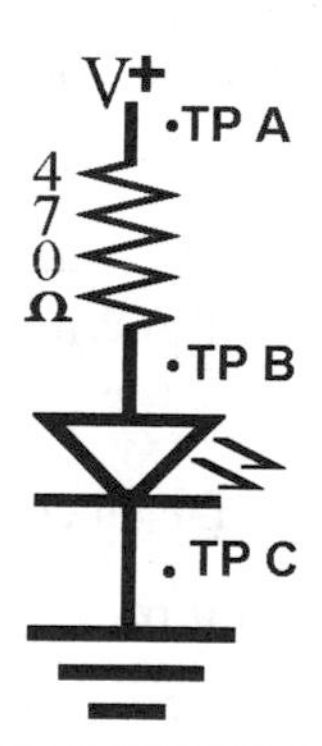

Figure L50-7

To figure the power used by the LED, you need to know both voltage and current.

Now this becomes a three-step process. But they are three simple steps.

1. You need to measure the voltage drops across the LED (TP B to TP C).
2. Then you have to calculate the current moving through the resistor. V = IR, where voltage is now the drop specifically across the resistor.

TP A to TP B. There is no way to measure the resistance of the LED to otherwise help figure out the current. Remember that the current moving through one component in series is the same current for all components along that line.

3. Then you calculate the power.

Voltage drop across the LED multiplied by the current in the system equals power used by the LED.

Your numbers should be close to the sample provided. Plug these numbers into the formula for power:

Watts = Volts × Amperes

0.0270 W = 1.77 V × 0.0153 A

Oooh! 27 milliwatts. Brilliant.

Exercise: Defining Work, Force, and Power

1. Force is the amount of energy exerted. It is measured in what units? ________________
2. Roughly 1 N moves ______ grams upward against the force of gravity (specifically it is ______ g).

   Work = Force × Distance
3. Work is the amount of _______ exerted on an object over a __________.
4. If an object is not moved (or the distance equals zero), how much work is done?
5. The unit used to measure work is ______________.

   Another name for the same unit is joules.

   1 joule = 1 ___________.
6. Both are equal to the amount of work done when moving ____ g __m upward.
7. Power is the __________ that work is being done.
8. The common unit for power is __________, abbreviated as _______.
9. Provide an example where two items have different amounts of power output even though the same force is exerted on them.

   You probably already know that 1 W = 1,000 mW (milliwatts). The prefix "milli" represents a thousandth (0.001). You've seen it in chemistry (mL) and physics (mm).

   Power for electricity is figured using the formula watts = volts × amperes.
10. The LED sample power output is 0.027 watt.

    That is ________mW.
11. You have seen in the previous exercise how increasing resistance dampens current. The power output is directly proportionate to the current available. If the supply voltage remains constant and the resistance increases, the current decreases.

    Don't breadboard the setup in Figure L50-8. You did this same set of resistors early on, and you can copy the values directly from the exercise sheet you did in Lesson 5.

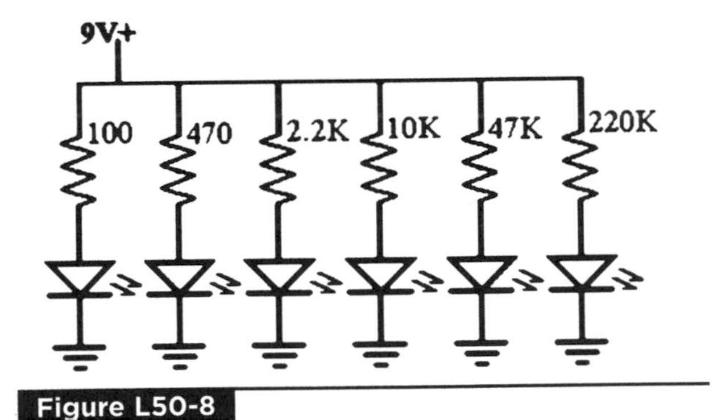

Figure L50-8

Calculate the power available to the LEDs for each setup (Table L50-1).

TABLE L50-1 Power Available

R Value Measure	V + to Gnd	Current in System $\frac{\Omega}{\text{Ohms}}$ = Amp	Voltage Drop Across the LED	LED Brightness	Power Used by LED
Copy	Copy	Calculate	V_{total} - VR = V_{LED}	Copy	W = (V) (I)
100 Ω	___Volts	0.____ A	____Volts	______	_0._______W
470 Ω	___Volts	0.____ A	____Volts	Normal	_0._______W
2.2 kΩ	___Volts	0.____ A	____Volts	______	_0._______W
10 kΩ	___Volts	0.____ A	____Volts	______	_0._______W
47 kΩ	___Volts	0.____ A	____Volts	______	_0._______W
220 kΩ	___Volts	0.____ A	____Volts	______	_0._______W

12. John's car has a 150-watt amp working off 12 volts. His parents have a 22-watt home system that runs off 120 volts. Which system is more powerful?

    (Hint: What is the color of George Washington's white horse?)

    ____

13. What is the current in a 100-watt light bulb working off 120 volts? ___________

    People used to say the standard lighting needed to read was the light provided by a 100-watt incandescent light. The same amount of light can be provided by five ultrabright white LEDs at 50 mA each.

14. Compare LEDs for power usage (see Table L50-2).

15. Use Ohm's law to calculate the amount of current passing from C to E in Figure L50-9.

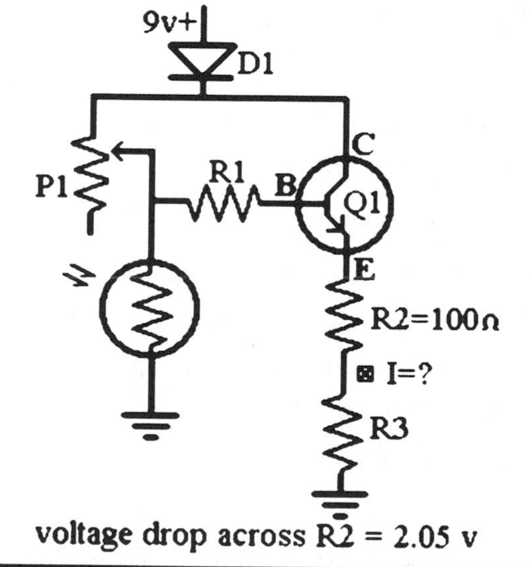

Figure L50-9

TABLE L50-2 Power Usage

Type	Voltage	Current	Brightness	Power
Regular red diffuse	1.8 v	30 mA	2,000 mcd	
Regular green diffuse	2.1 v	30 mA	2,000 mcd	
Regular yellow diffuse	2.3 v	30 mA	2,000 mcd	
High-intensity yellow	2.1 v	50 mA	6,500 mcd	
Ultrabright red	2.4 v	50 mA	8,000 mcd	

16. How much power is available at the point between R2 and R3? _______mW

17. Is this enough power to light up an LED? Yes / No

Lesson 51
What Do I Have to Gain?: Definitions

As you turn the volume control on your radio, the potentiometer's changing resistance isn't just using up some of the power to the speakers. Its actions are much more subtle. Amplifiers use an efficient method to control their power output. Read on and see what you have to gain.

Defining Gain

In an amplifier, the input signal is small and the output signal is big! *Gain* is the ratio that compares input to output.

- Gain = output/input
- Gain = $signal_{out}/signal_{in}$

Gain is a statement of a basic ratio. It has no standard unit because the ratio compares identical unit values, such as current.

- Gain = I_{out}/I_{in}

Gain in the NPN Transistor

Set up a modified version of the night light circuit. The schematic in Figure L51-1 shows a few changes needed to calculate the gain of the NPN 3904 transistor.

1. R1 at 47 kilo-ohms simulates the potentiometer set to midrange.
2. R2 will be various resistors simulating the LDR at different light levels. Bright light would have low resistance.

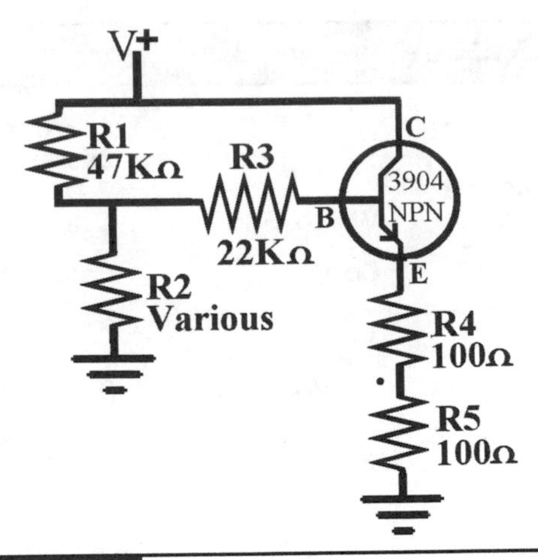

Figure L51-1

3. R3 is 22 kilo-ohms. I never did explain before what it did. It restricts current to the base (input) of the transistor.
4. In Figure L51-2, R4 and R5 make a voltage divider. Measuring the voltage drop across R4 makes it an easy task to calculate the current available at the emitter (output).

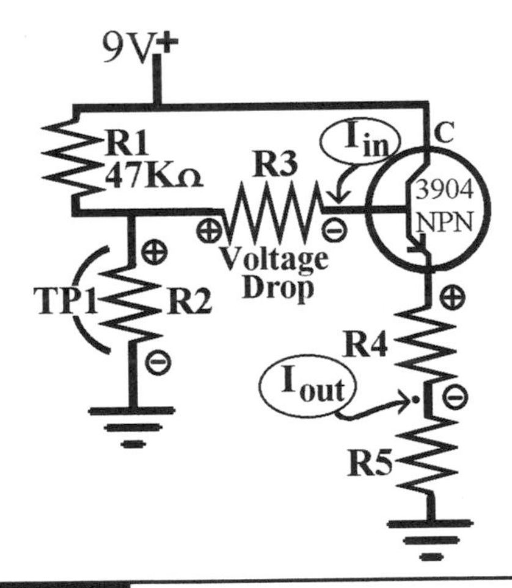

Figure L51-2

How would we calculate gain for this setup?

This is a matter of comparing the current input and current output.

They get set into a ratio, and there is your gain.

Gain = I_{out}/I_{in}

Table L51-1 was done with a 12-volt wall adapter as a power supply. The readings here will not necessarily match your readings.

TABLE L51-1 Ratio and Gain

	R2	VTP1	V_{drop} R3	I_{in} (base) V/Ω = A	V_{drop} R4	I_{in} @ R4 V/Ω = A	Gain = I_{out}/I_{in}	Power Used by R4 W = V x A	Power @ E P_{R4} + P_{R5} = P_{total}
1	1 M	7.54 V	2.83v	2.83 V/ 22,000 Ω = 0.129 mA	2.05v	2.05 V/ 100 Ω = 20.5 mA	20.5 mA/ 0.129 mA Gain = 159	2.05 V 20.5 mA = 0.042 W	0.082 W = 82.0 mW
2	470 K	7.39 V	2.73v	0.124 mA	2.0v	20 mA	161	0.040 W	80.0 mW
3	100 K	6.23 V	2.27v	0.105 mA	1.64v	16.4 mA	159	26.9 mW	53.2 mW
4	47 K	5.10 V	1.83v	0.083 mA	1.29v	12.9 mA	155	16.6 mW	33.2 mW
5	10 K	2.26 V	0.718v	0.033 mA	0.433v	4.33 mA	131	1.87 mW	3.74 mW
6	4.7 K	1.26 V	0.299v	0.013 mA	0.15v2	1.52 mA	116	0.231 mW	0.462 mW

For figuring the power output of the transistor, the following considerations have been made:

1. Since R4 = R5, they both use the same amount of power.
2. The voltage drop across both loads is identical.
3. The current on the same line is unchanged.
4. They use all the power available from the transistor.
5. Therefore $P_{R4} + P_{R5} = P_{Emitter}$

Remember that the LED used 27 mW of power. Based on the numbers developed in this sample, the LED begins to fade when there is a bit less than 50 kilo-ohms on R2 because from that point on, there is not enough power.

Exercise: What Do I Have to Gain: Definitions

1. Test the quality of your transistors using the DMM diode test. Record your readings in Table L51-2.

   For the circuit-test DMR 2900 multimeter, you first set it to "Continuity" and then press the DC/AC button. The symbol in the top-left corner changes from the "beeper" icon to a diode symbol. Many multimeters simply use the continuity tester to give the reading.

2. Before we move into measuring a complex circuit, let me prove to you that when you calculate the current in one portion of a circuit, that current is the same for the remainder of that circuit. You don't need to make the simple schematic shown in Figure L51-3. All the needed information is there.

TABLE L51-2 Readings

NPN 3904	The Readouts Are ± 5%	PNP 3906	
$E_{re}d$ to B_{black}	(expect OL)	E_{red} to B_{black}	(expect 0.68)
B_{red} to C_{black}	(expect 0.68)	B_{red} to C_{black}	(expect OL)
E_{black} to $B_{re}d$	(expect 0.68)	E_{black} to B_{red}	(expect OL)
B_{black} to C_{red}	(expect OL)	$B_{blac}k$ to C_{red}	(expect 0.68)

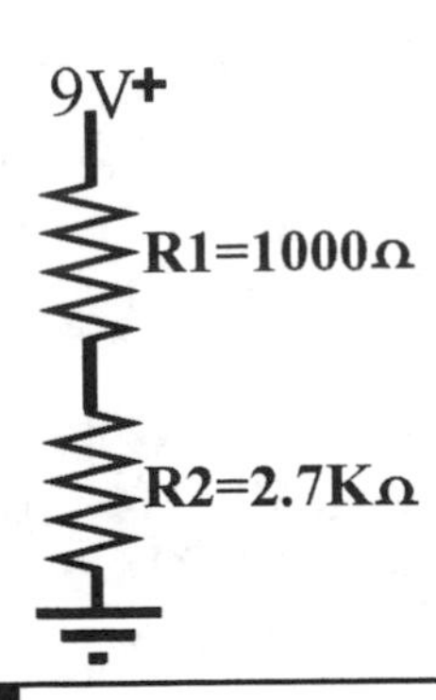

Figure L51-3

Now how much voltage is used across R1 and R2 individually?

Remember the formula to calculate this was:

$V_{drop} = V_{total}$ [R1/(R1 + R 2)]

V_{drop} across R1 = ____

V_{drop} across R2 = ____

3a. How much current is passing through R1?

Use Ohm's law: V = IR

$\frac{V}{\Omega} = A$

V_{drop} across R1/1,000 Ω = current at R1

3b. How much current is passing through R2?

V_{drop} across R2/2,700 Ω = current at R2

4. Is the current through R1 and R2 nearly identical? Does IR1 = IR2? Yes/No

5. If you want to set up a test circuit, go ahead and do the real measurements. Remember that you can use three significant figures if you measure the resistance with the DMM directly. Use only two if you don't measure the resistance but just use the color code.

6. Now, referring to the schematic in Figure L51-2, use the values shown for R2. R2 is the only resistor that changes values. R3 keeps the 22 kΩ value shown in Figure L51-1 (see Table L51-3).

Lesson 52
The World Is Analog, So Analog Is the World

Here we finally get to see how an amplifier relates to the real world. But keep in mind that this is not digital. Nothing in the real world is digital. Analog is the world.

Amplifiers are analog systems. They deal with changing voltage. Modern technology promotes the use of digital storage and transfer of information. As well, we have digital transmission of information on cable and the Internet. Good amplifiers feed digital information in and provide digital information out. But amplifiers, by and large, deal with sliding analog voltages to duplicate the sound and light as it occurs naturally.

TABLE L51-3 Readings

	R2	VTP1	V_{drop} R3	I_{in} @ B V/Ω = A	V_{drop} R4	I_{in} @ R4 V/Ω = A	Gain = I_{out}/I_{in}	Power Used by R4 W = V x A	Power @ E $P_{R4} + P_{R5} = P_{total}$
1	1 M								
2	470 K								
3	100 K								
4	47 K								
5	10 K								
6	4.7 K								

The Noninverting Amplifier

As shown in Figure L52-1, the NPN transistor works in a very direct manner. As the signal to the base increases, the valve opens. As one goes up, the other goes up in direct proportion. This is a direct relationship. The output is proportional with the same voltage as the input. More voltage in creates more voltage out. Voltage is not inverted.

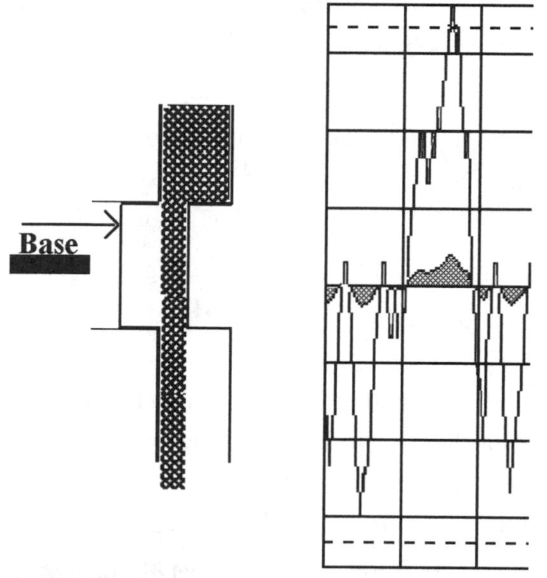

Figure L52-1

The Inverting Amplifier

Look at Figure L52-2. The PNP transistor has the opposite action of the NPN transistor. As the signal to the base increases, the valve closes. As one goes up, the other goes down in direct proportion. This is an opposing relationship. The output is proportional but opposite of the input voltage.

The voltage is inverted, but the *gain* is the same as the NPN because gain is always stated as an absolute value.

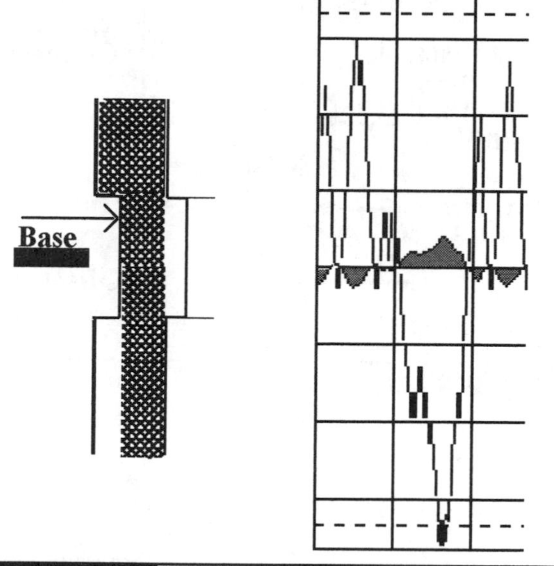

Figure L52-2

The NPN is a "noninverting" amplifier.

The PNP is an "inverting" amplifier.

Of course, in high tech, we have to use fancy words such as *noninverting*.

> **Definitions**
>
> *Inverting*—to turn upside down.
>
> *Noninverting*—not to turn upside down. To leave right side up.

The Op Amp

We are going to be using an 8-pin DIP 741 amplifier on a chip. The basic hookup diagram is shown in Figure L52-3. These are called *Op Amps*, short for operational amplifiers. There are many more amplifiers than there are CMOS 4000 series IC chips. Just look at the partial listing at www.mhprofessional.com/computingdownload. Each Op Amp has its own features regarding:

- Power input and output
- Response time

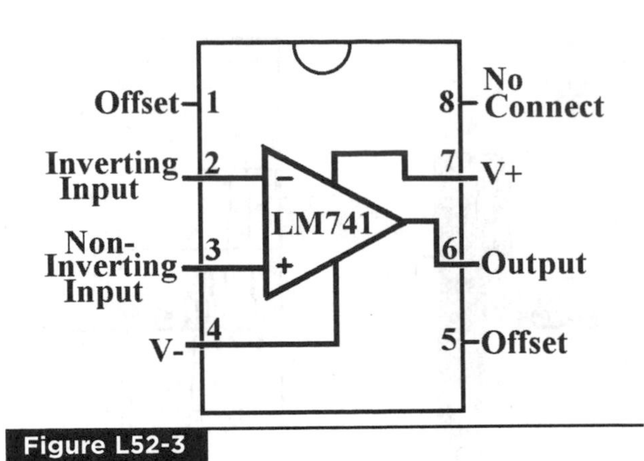

Figure L52-3

- Frequency
- Other factors

The main idea to keep in mind is how any Op Amp works. The Op Amp compares the voltage difference between the two inputs (pins 2 and 3 here) and responds to that difference.

Here we will use the LM741 because of its low power needs and simplicity.

Note that there are both inverting and noninverting inputs.

Exercise: The World Is Analog, So Analog Is the World

1. Which input pin of the LM741 has action similar to the NPN transistor? pin ______
2. That action is called what? ______________
3. Define the term *inverting amplifier.*

   ____

4. For an inverting amplifier, what is the gain when the output is − 100 mV and the input was + 10 mV? Be aware that gain is always stated as an absolute value.

   a. − 10

   b. + 10

   c. − 0.1

   d. + 0.1

5. Look at www.mhprofessional.com/computing download for a partial list of amplifier ICs that are available on today's market. Would I be right in saying that there is a limited selection of amplifiers available to work with?

   Download a full LM741 data sheet at http://cache.national.com/ds/LM/LM741.pdf. You need to have Adobe PDF reader installed to be able to access this document.

6. On the LM741 data sheet, page 1, the LM741 is manufactured in how many package styles?

   ____

7. From the LM741 data sheet, page 2, examine the numbers on your chip. Determine if you are working with a 741, 741A, or 741C.
8. From the LM741 data sheet, page 2, what supply voltage is the LM741 expecting?

   ____

9. More interesting information from the LM741 data sheet, page 2—your solder pen gets up to 400°C—how long can the LM741 endure direct heat from a solder pen? __________
10. Now turn to page 3 of the LM741 data sheet. The LM741 is considered to be a low-power Op Amp. For example, at rest the typical current draw for this Op Amp is what?

    ____

11. And more info from the LM741 data sheet, page 3—when really pushed, the LM741 consumes a whopping _______mW of power.
12. Turn to page 4 of your LM741 data sheet. How many transistors are packaged into the LM741 IC? ___________

SECTION 16

Exploring the Op Amp

THE CIRCUIT IN THIS UNIT is a "teaching" circuit. The job of amplifying can be done using a variety of methods, each one cleaner than the next. But this circuit allows the introduction of a pile of necessary concepts along the way.

Lesson 53
Alternating Current Compared with Direct Current

Here I explain the necessary analogy comparing alternating current as sound and direct current as wind.

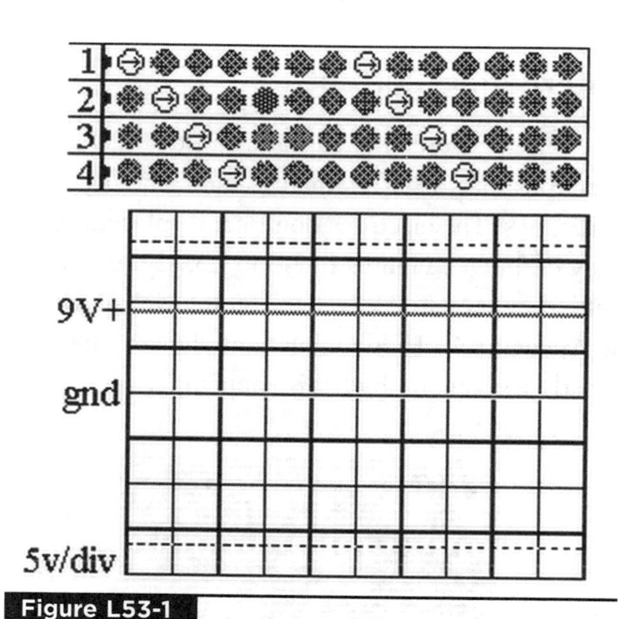

Figure L53-1

Direct Current

We always hear the terms *direct current* (DC) and *alternating current* (AC). Direct current means that the current in the system is always flowing in one direction. Electronics depends on direct current. Naturally, flowing water has been a good comparison for positive DC voltage V+. Water always flows in one direction—downhill, toward the lowest point. Direct current is easily generated by chemical reactions in batteries. Figure L53-1 shows positive movement of electrons while the scope shows the voltage is above ground. So the voltage pushes the electrons one way. However, what happens when the voltage is reversed and pushes the electrons the "other way"? Figure L53-2 shows us how that really works.

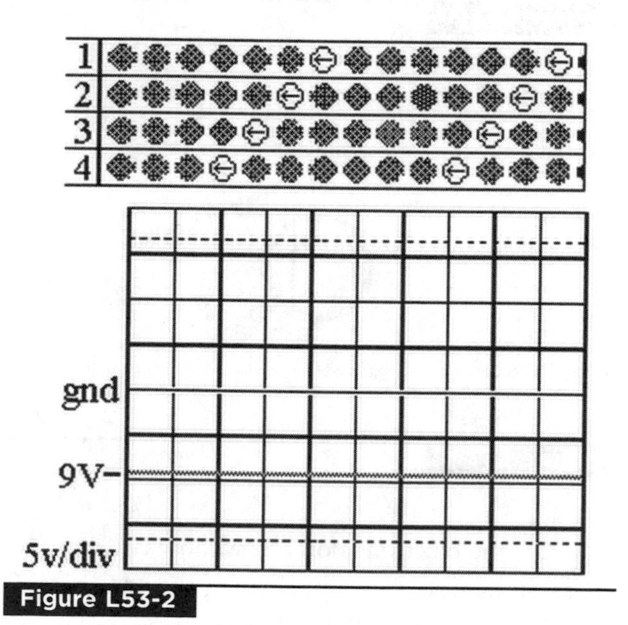

Figure L53-2

Just a note regarding the representations here of DC on a regular oscilloscope. Soundcard Scope does not respond to stable DC voltage as a regular oscilloscope would.

With direct current, you can have a positive voltage V+ or a negative voltage V–. Attach a battery backward, and you have created V–.

A more effective analogy for DC would be to think of water in a pipe or wind in a tunnel. It can move easily in both directions.

Alternating Current

Put simply, alternating current is a flow of electrons that keeps alternating and changing directions. The electrons don't get displaced, as they do in DC. Figures L53-3 to L53-5 are frame sets from animations available at the website noted in Appendix C. Those animations show far more clearly the action of what is happening.

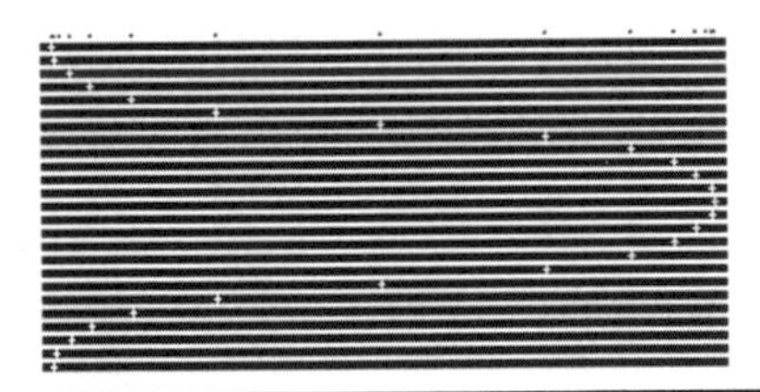

Figure L53-3

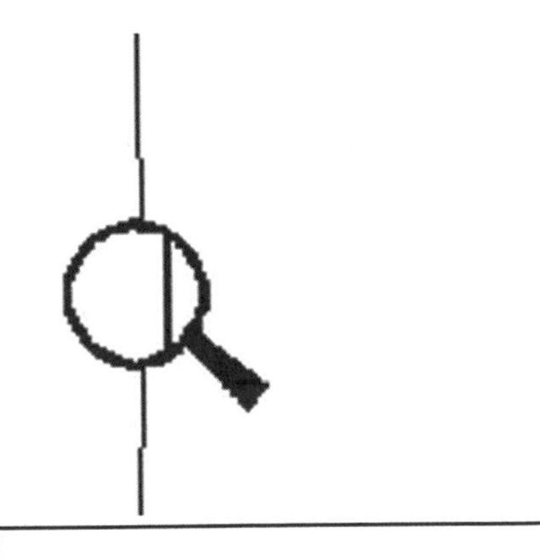

Figure L53-4

But if the electrons don't flow, how is the energy transferred? A great analogy for AC is sound waves. Air particles don't flow as sound travels through the air, but they do move; they vibrate. Sound is a pressure wave that moves through the air. A vibrating string sets one particle bumping into the next.

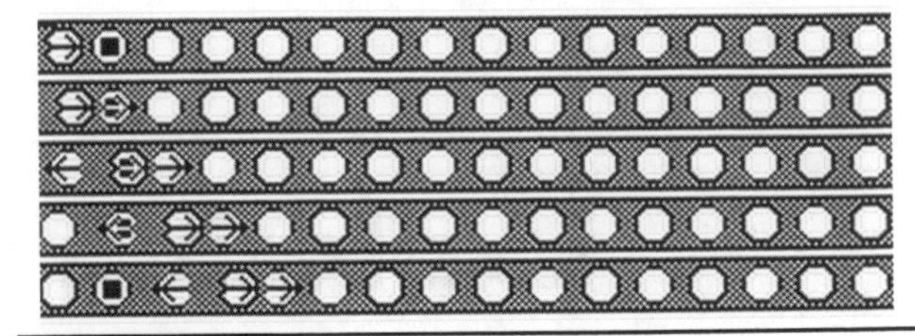

Figure L53-5

Simplified, the particles don't get displaced like they do in a wind. They vibrate in place. In fact, when sound is translated into an electronic signal, it is carried as an AC signal as shown in Figure L53-5.

Figure L53-6 represents a regularly oscillating action of electrons in an AC system. This shows a steady AC signal on the scope.

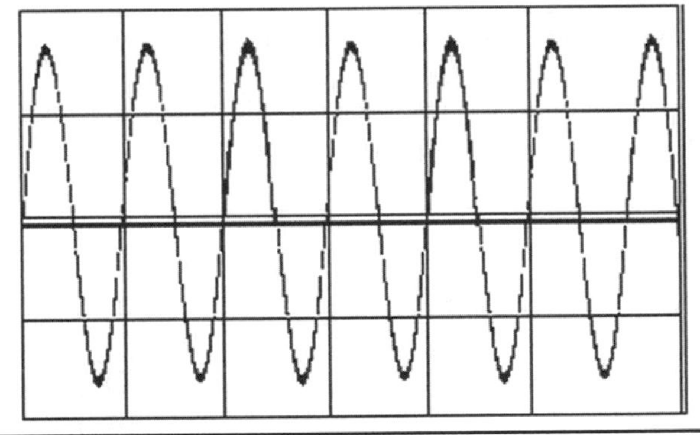

Figure L53-6

AC on the Oscilloscope

Why does a steady AC signal look like this on the oscilloscope screen? The following explains in detail the transformation of real electron movement into a sine wave:

1. This is the actual movement of the electrons, represented in Figure L53-3.
   - A lot of voltage exists as it speeds up in the center. The voltage decreases as it slows down.

- It actually has zero volts as it stops and reverses direction.

2. Figure L53-7 graphically shows the increase in the voltage of the electron as it speeds up and slows down. The voltage actually inscribes a circle.

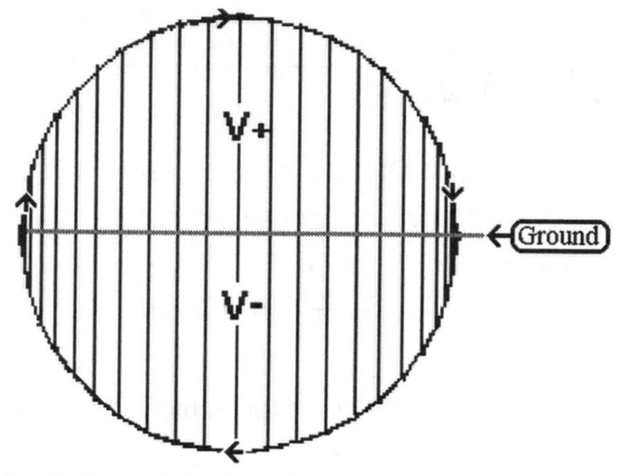

Figure L53-7

Notice two things here:

- The distance the electron moves along the base line changes as the voltage increases and decreases. Put simply, the bigger the push, the faster it moves. The smaller the push, the slower it moves.
- Each segment around the outside of the circle actually represents time.

3. It is not easy to show something that moves backward, because time moves forward. So Figure L53-8 shows the negative voltage movement happening after the positive voltage.

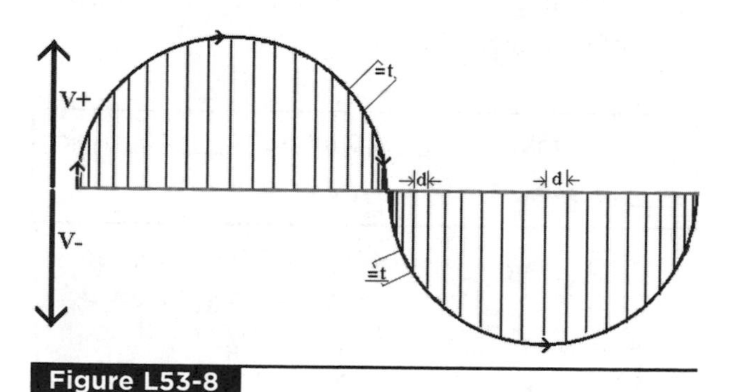

Figure L53-8

4. Time is important. How do we show time? Right now, time is chasing around on the outside of the circle and distance is marked across the horizontal ground line. But the distance the electrons move is not as important to us as timing. So the next thing to do is to impose a time measure onto the horizontal line, replacing distance as the measure. This is shown in Figure L53-9. The amount of voltage (energy or force) that defined the height remains in place. Obviously, time is marked in evenly spaced increments.

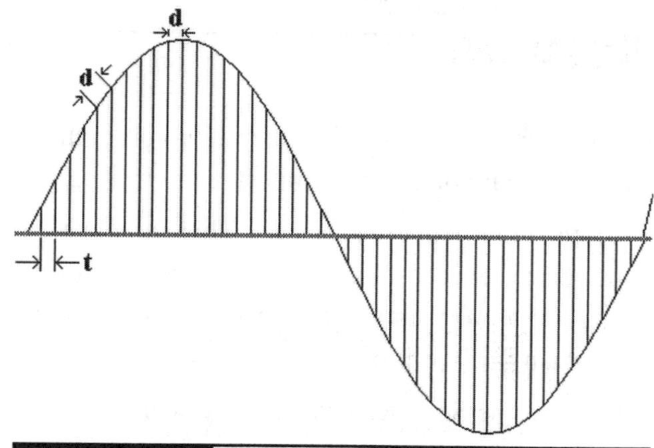

Figure L53-9

The result? The circular outline changes into the classic sine wave shape.

Note that ground is at the center of the action, the interplay between V+ and V–. As I said, our amplifier deals with sound. You will now learn how to measure with the tools at hand.

Getting Ready to Explore the Op Amp

Having a good source of a stable signal is vital when working with amplifiers. Signal generators, like the one in Figure 53-10, would be good to have available in a classroom. Otherwise, the Soundcard Scope's signal generator can create the

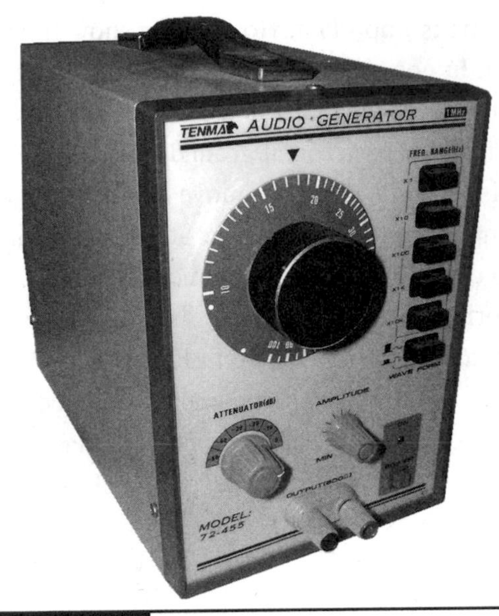

Figure L53-10

tones we need. A variety of other audio programs can record and blend these tones.

I use "Gold Wave" shareware. This complete audio editor exceeds our immediate needs, but I'd rather use a program you can grow into instead of a program that we'll outgrow quickly.

1. Open the Soundcard Scope. We will use default settings for our work here.
2. Set the Signal Generator's channel 1 to 250 Hz.
3. Use any software to record and save this frequency as a 30-second wave (*.wav) file.
4. Exit the Soundcard Scope.
5. Start up the 250-Hz wave file you just created. It does not matter what program you use.
6. Check this frequency out on the Frequency Analysis portion of the Soundcard Scope. When you are satisfied with the results here, make a 300-Hz and 1000-Hz wave file.

   Start up the 250-Hz signal again.
7. Plug the channel 1 cord into the earphone jack.

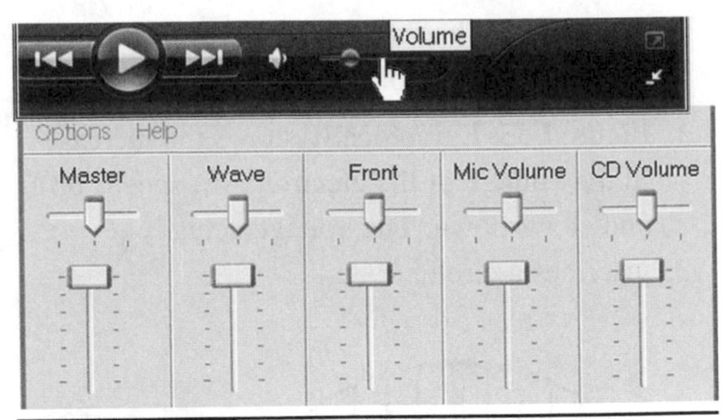

Figure L53-11

8. Use Figure 53-11 as a guide to make sure that all of the software controls are set to maximum.
9. Record the maximum output for this file. ____mVac.

   Figure 53-12 helps explain the best method for accurately adjusting the output a few millivolts at a time.

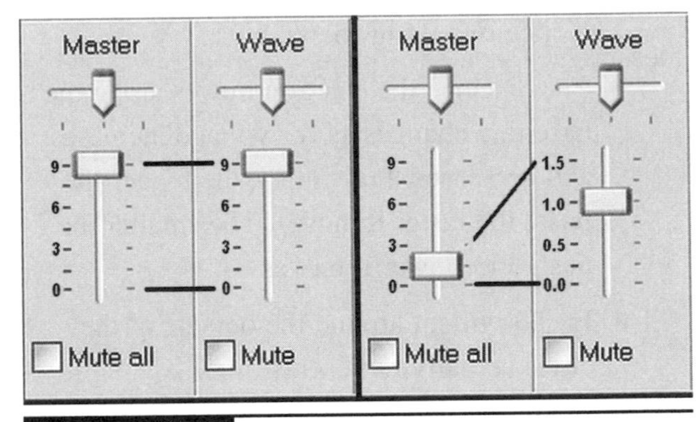

Figure L53-12

10. Now, while you are connected and the tone is running, adjust the volume controls to each of the voltage settings shown here. Check them off as you go.

___ 200 mVac	___ 100 mVac	___ 40 mVac
___ 15 mVac	___ 5 mVac	

11. Once you have that done, reset the 250 Hz to 40 mVac, and then stop the signal. Without touching the volume settings, load and start the 300 mVac signal. Record the mVac shown on the DMM. _____

12. Don't touch those dials or change any settings. Load up, start, and measure the voltage of the 1000-Hz signal. ____ mVac.

    Does the voltage remain constant as the frequency increases?

Why Doesn't DC Voltage Show Up on an AC Circuit?

Fair question, so let's check it out.

1. Okay, now you can readjust the volume back up to maximum.
2. Set your DMM to measure DC voltage. Connect the clips.

No matter how hard you try, you cannot get a measurement of DC voltage from an AC signal. Why doesn't the signal show any DC voltage? You've seen something like this before. Look at Figure 53-13.

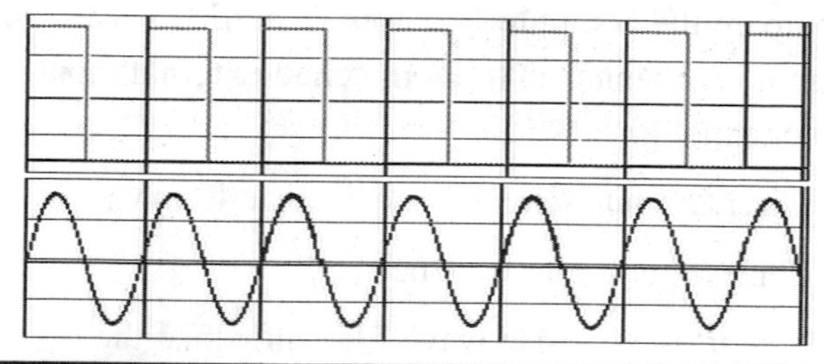

Figure L53-13

Remember when you used the DMM to record the rolldown output from the 4046? The voltage reading at pins 3 and 4 was half of the V+ fed to the IC at pin 16. The output is V+ half the time and 0.0 volts for the other half. The signal was switching quickly enough that the DMM averaged the voltage and read it as half of V+.

Consider that this is the same situation. The output is V+ half the time and V– the other half. The 200 mVac signal actually averages –.1 V for half the time, and +.1 V the other half. These two signals average out, and the DC voltage reading is ZERO.

Comparing Outies to Innies

Here are some last things to do with your recorded AC signal. Use the setup in Figure 53-14 to

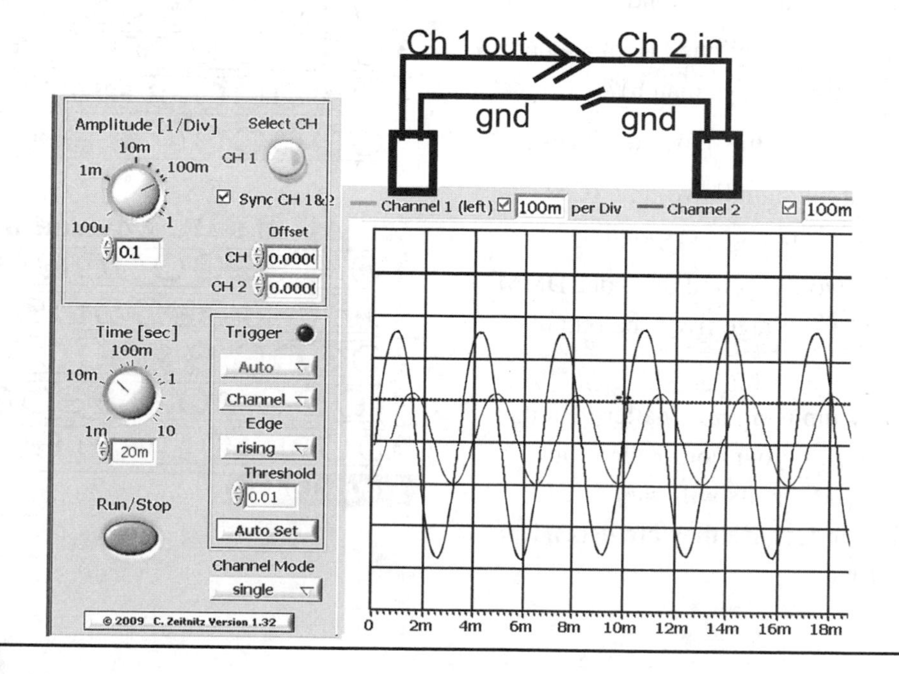

Figure L53-14

compare the computer's output (recorded wave signal on channel 1) to its matched external input on channel 2.

Ch1: Output—Headphone Jack at 40 mVac

Ch2: Input—microphone

1. Start the 250-Hz wave file and adjust the output to 40 mVac.
2. Turn on the Soundcard Scope.
3. Remove the check mark that synchronizes Ch1 and Ch2. They need to work independently.
4. Plug the channel 2 connecting cord into the microphone input.

Now you are able to compare the computer's output directly to its corresponding input. Any obvious differences you see here can be expected to show up in all of your tests.

Here's what you are looking for while you're comparing the two signals:

Compare the amplitude of output to input.

- You can expect the amplitude to drop—the question is by how much.
- Does the input even respond? The signal might be too small for the card to respond.
- Are the signals naturally aligned (lock-stepped so that the peaks and valleys match)?

Once you have observed this basic feedback:

1. Decrease the output signal's volume (Ch1) until the input on Ch2 stops responding.
2. Disconnect the two cords and use your DMM to measure the AC voltage from the output signal.

Be aware that any input signal smaller than this value will not register on your computer's input. These comparisons and results will vary from computer to computer because they are dependent on a variety of factors.

Lesson 54
AC in a DC Environment

But if alternating current goes forward and backward, how can it be used in a direct current system? How can AC be imposed onto a DC system? Relax. We can either eat on the ground, or we can pull out a table and use that raised surface. You know enough actually to set up that raised surface in electronics.

The AC Signal in a DC Environment

If AC goes forward and backward, how can we work with AC in a system that uses only V– from a single 9-volt battery?

To begin, recall the analogy comparing direct current to wind and alternating current to sound. Figure L54-1 shows how sound can be carried in the wind. The animation at the website in Appendix C shows the action much more clearly.

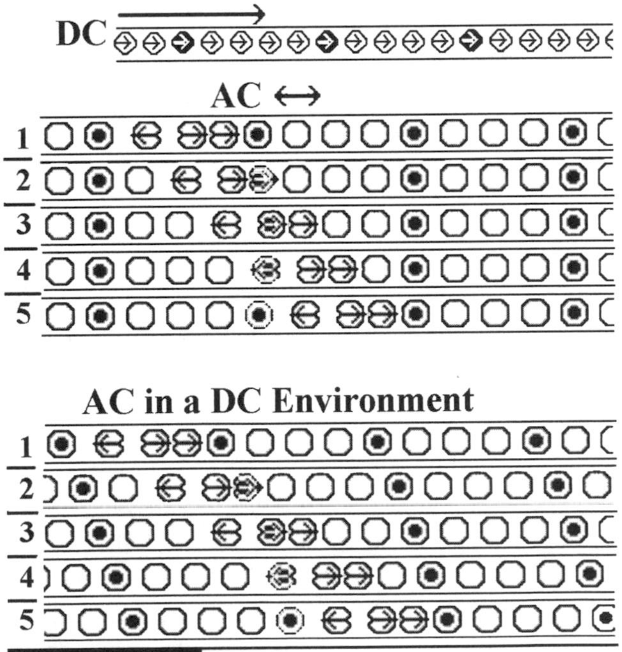

DC →

AC ↔

1
2
3
4
5

AC in a DC Environment

1
2
3
4
5

Figure L54-1

With this in mind, you can carry an AC signal in a DC environment. Here's a larger analogy to explain how. In electricity, you have above ground and below ground. As an analogy, when you go out for a picnic, you can use the ground as a convenient surface, as shown in Figure L54-2.

Figure L54-2

Or you can set up a table—and in that sense create an artificial surface above the ground as shown in Figure L54-3.

Even though the ground is a natural place to put a picnic, an artificial surface like the table is much more convenient. Adjusting the reference for the AC signal in a DC system is much more convenient, too, as you can see in Figure L54-4.

We can create an adjusted reference for an alternating current signal within a DC voltage environment by using a simple voltage divider.

If V + is 9 volts, then half of V + is 4.5 volts.

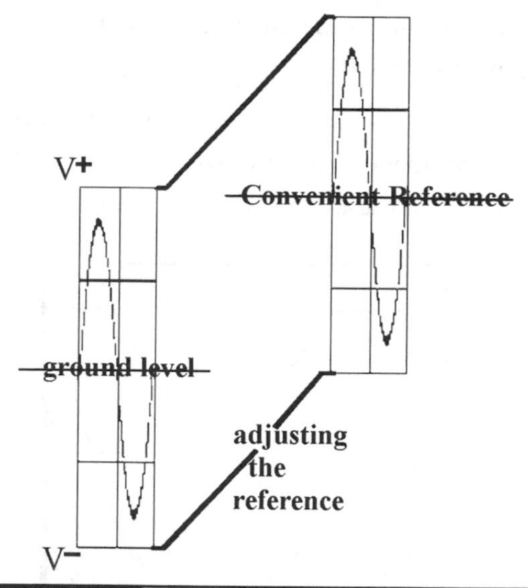

Figure L54-4

In the system shown in Figure L54-5, the adjusted reference acts identically as ground in a 4.5 VAC system.

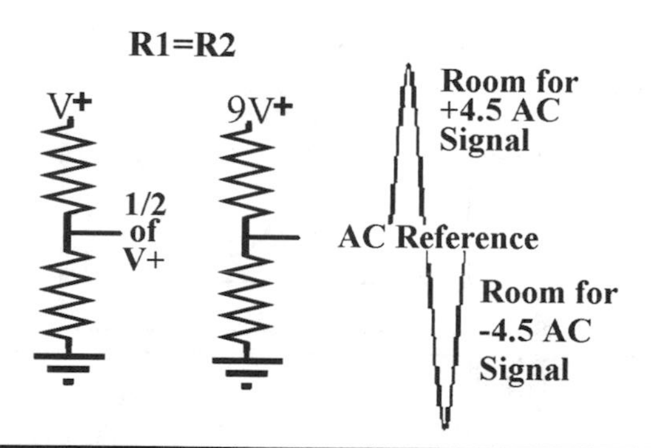

Figure L54-5

Figure L54-3

How does an AC signal move in a DC environment?

Look closely at the two drawings in Figure L54-6. The image on top shows the AC signal with ground as its natural reference.

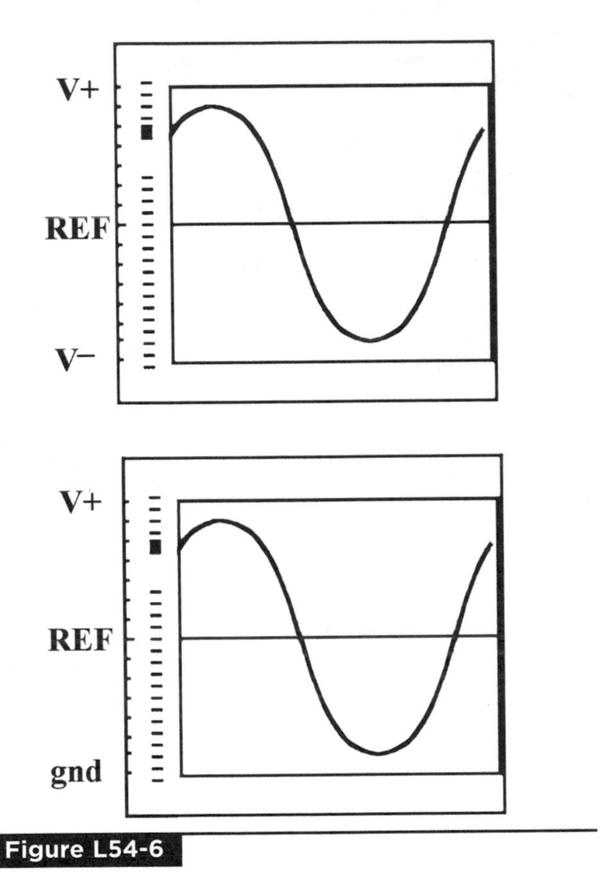

Figure L54-6

The image on the bottom shows the AC signal with an adjusted reference in a DC environment. The current does not affect the signal.

Lesson 55

Setting Up the Operational Amplifier

The Op Amps are, by far, the most versatile and valuable components in all of electronics. They are actually the heart of many digital systems. I can't pass up this opportunity to explore how they are applied and used in digital electronics.

Start Building the Circuit

As we just learned in Lesson 54, there needs to be a reference point for carrying the AC signal. Set up the two resistors as shown in Figure 55-1. These will be the voltage dividers used to set your reference point. It's best to use 5 band 1% resistors.

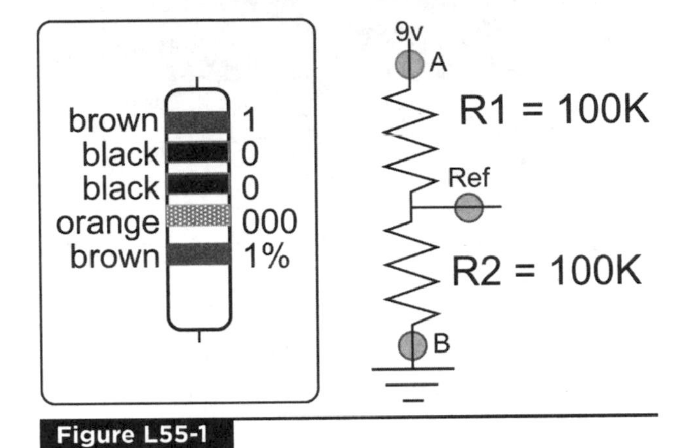

Figure L55-1

Take the following measurements:

- Total voltage ______
- The reference point voltage ______

This allows ______volts above the reference to act as the positive portion of the AC signal and _______volts below the reference to act as the _______ for the AC signal.

Now Add the Op Amp

Figure 55-2 moves us to the next step.

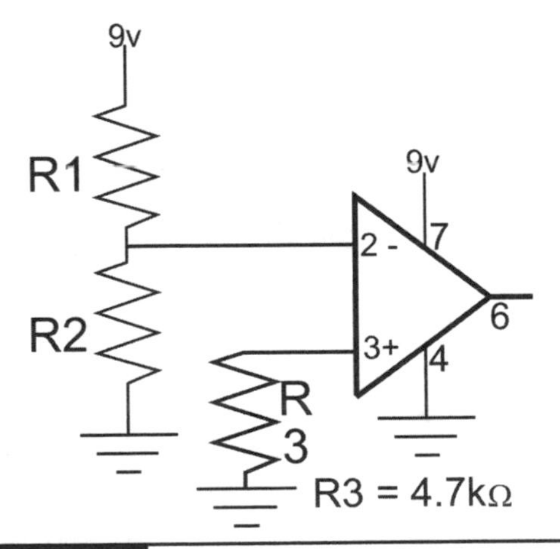

Figure L55-2

The midpoint of R1 and R2 is now simply referred to as V_{ref} (voltage reference).

Connect V_{ref} to the inverting input at pin 2 (–).

Being an input, pin 3(+) must be connected somewhere. The diagram above shows it connected to ground, injecting a Lo to the noninverting input.

The Op Amp's output at pin 6 reacts to the voltage difference between pin 2 and pin 3.

Because the *noninverting* input at pin 3 is lower voltage than V_{ref}, the output at pin 6 is also Lo.

Do your measuring.

Now shift the R3 to inject a Hi signal to pin 3. The Hi input should give a Hi output. Don't take my word for it. Check it out. Is it what you expected?

An important notice just arrived from our department of redundancy department.

*The input to the non inverting input is **not inverted** at the output.*

Moving Forward

Use Figure 55-3 as a guide and shift the V_{ref} connection from pin 2 to pin 3. Just to make sure we're on the same page, R3 should initially be set to ground, so pin 2 senses its input as Lo.

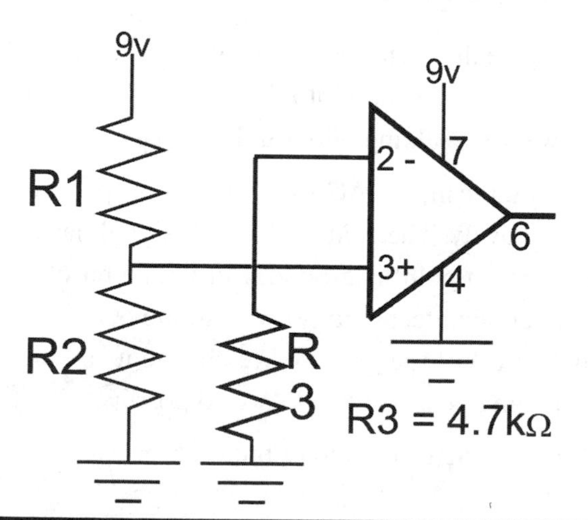

Figure L55-3

Check the output. Lo Input—Hi Output

Switch the R3 to inject a Hi input.

TAH DAH! The input is *inverted.*

The Voltage Comparator

Moving forward again with Figure 55-4, we see small but significant changes.

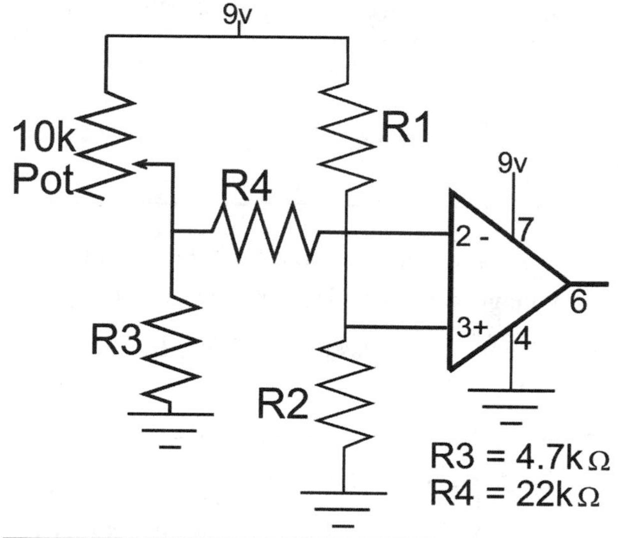

Figure L55-4

What we've seen so far is that the voltage comparator reacts to obvious voltage differences between the inverting input and V_{ref}. Using this setup, we can see how well the Op Amp compares minimal voltage differences. R3 is important. Because an Op Amp only compares voltages, current isn't important. R3 dampens the current that reaches the input.

1. Use your DMM to check V_{re}f. ______ volts
2. Attach the DMM's red probe to V_{invert} and the black probe to ground.
3. Follow the instructions in Table 55-1.

TABLE 55-1 No Matter How Small the Difference, Any Input Above V_{ref} Will Output Lo

Adjust the potentiometer so that the midpoint voltage (V_{invert}) is just a bit over V_{ref}.	Adjust the potentiometer so that the midpoint voltage (V_{invert}) is even a bit higher.
a. V_{ref} = _______	a. V_{ref} = _______
b. V_{invert} = _______	b. V_{invert} = _______
c. V_{out} = _______ (Speak Digital! Hi or Lo?)	c. V_{out} = _______

TABLE 55-2 Equal, but Opposite Reaction!

a. V_{ref} = _______	a. V_{ref} = _______
b. V_{invert} = _______	b. V_{invert} = _______
c. V_{out} = _______	c. V_{out} = _______

Do the same in Table 55-2, but with the voltage divider giving just a bit under V_{ref}.

4. When V_{invert} is less than V_{ref}, V_{out} becomes Hi or Lo?
5. When Vinvert is greater than V_{ref}, V_{out} becomes Hi or Lo?
6. Don't spend too much time on this, but try to adjust the voltage at pin 2 (V_{invert}) so that it is the same as V_{ref}.

   If you could get them to match, what voltage would you expect at pin 6, V_{out}? _____

WAIT! STOP! HOLD IT! Before we rush by, think of the power and beauty of little things, like in Figure 55-5. The voltage comparator really is the heart and workhorse of our digital world.

Figure L55-5

Really! This is your basic NOT gate. You have a reference voltage and an input to compare to it. The output responds instantly to the difference, giving a Hi output for any input less than half of V+ and a Lo output for any input greater than half of V+.

But enough of this talk. We're simply playing with adjusting a DC voltage, comparing that to another DC voltage. It's your turn again. Roll the dice and take another step forward.

Injecting an AC Signal into a DC Circuit (One Last Modification)

What we really want to be able to do is feed an AC signal related to sound into this circuit. Isn't that what we were talking about in Lesson 54?

As you recall, the AC's current moves positively and negatively. The center of an AC signal has no voltage. That point is ground. But with a bit of magic (applied technology) we can move that signal up off of the ground. We change it so that it is moving positively and "not so positively."

Get your SBB to match Figure 55-6.

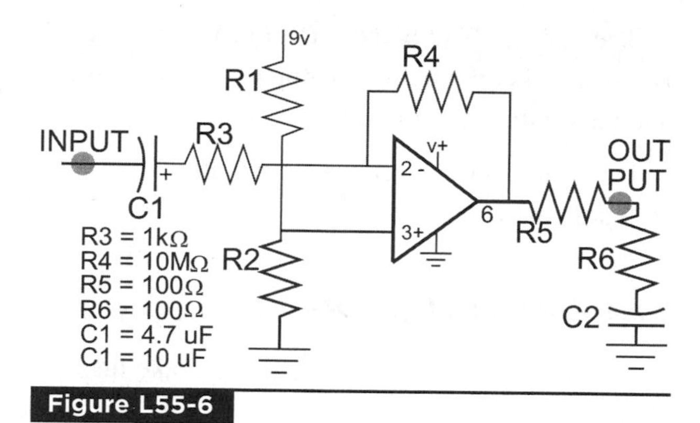

Figure L55-6

First, recognize that the capacitors are *Vital* in these positions. They are being used as audio couplers.

Audio Couplers

Capacitors have three main uses. So far, we've used them as part of RC timers. Here, we'll use the capacitor as an audio coupler. In this position, as it connects the audio signal to the circuit and it separates the AC signal (tiny voltage fluctuations) from the DC voltage:

- It isolates the Op Amp from any current input.
- The AC signal passes through the capacitor.
- The AC signal is automatically centered to the Op Amp's adjusted V_{ref}.
- Steady DC voltage does not get translated through the capacitor.

The following analogy offers the best explanation:

AC signals passing through a capacitor are similar to sound passing through a closed window, just like it is shown in Figure 55-7. Wind can't pass through a window. If the wind is steady, you don't even hear it. The wind is like the DC voltage and current. But the sound. Ahh. The sound itself is made of small vibrations; the window vibrates as well, if even just a little. And … the air on the inside of the window vibrates too, transferring the sound to other side. The sound represents the small fluctuations, variations in the voltage.

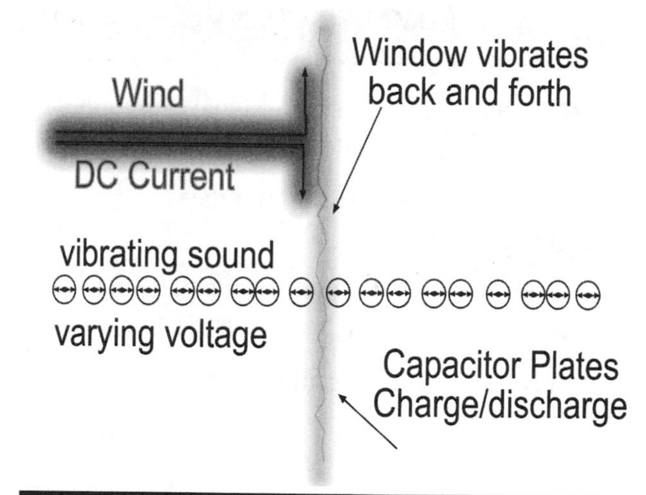

Figure L55-7

Simply stated, the active voltage changes created by the signal on one side of the capacitor affect the charge on the opposite plate of the capacitor.

Now Let's Have Some Fun

The diagram in Figure 55-8 describes what we're aiming for.

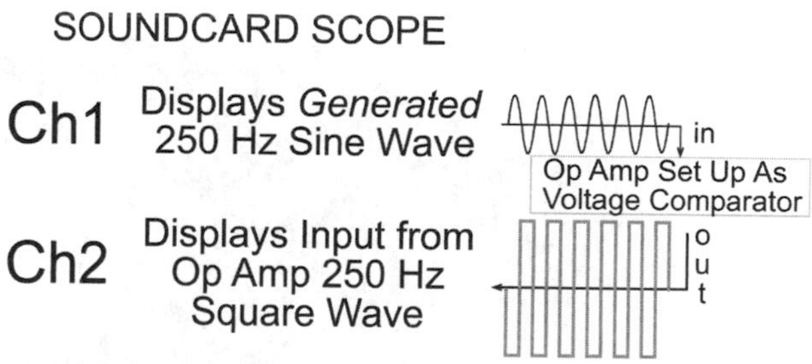

SOUNDCARD SCOPE

Ch1 Displays *Generated* 250 Hz Sine Wave — in

Op Amp Set Up As Voltage Comparator

Ch2 Displays Input from Op Amp 250 Hz Square Wave — out

Figure L55-8

Power Up Your Circuit

1. Open Soundcard Scope's Signal Generator.
2. Set CH2 to 250 Hz. Tab over to the Scope screen.
3. Clear the Sync CH1 & 2 check box.
4. Plug the CH2 connector into your computer's headphone output.

5. Using your DMM, set CH2's signal strength to 40 mVac (.040 vac).
6. Connect CH2 connector's signal clip to the circuit's "input" and the black clip to ground.
7. Plug the CH1 connector into the "microphone" input.
8. Connect the CH1 connector's signal clip to the circuit's "output," again, black to ground.

Ideally, you will see results similar to Figure 55-9.

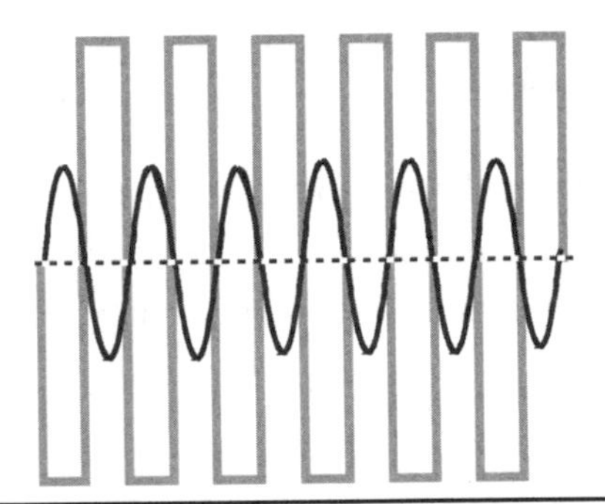

Figure L55-9

The concept here is more than just important. But before I continue, let's improve the harmony and play with the tempo.

Disconnect the power for a moment while you set up the new signal.

Take Another Step Forward

Tab over to the Signal Generator and make these changes:

1. Signal Generator
2. Channel 1 = 250 Hz
3. Channel 2 = 300 Hz
4. These two tones blend nicely. Good resonance. While you listen, go back to the scope view in Figure 55-10. Find the highlighted area at the bottom of the control screen.

They give you the ability to:

- Add the signal from each channel together.

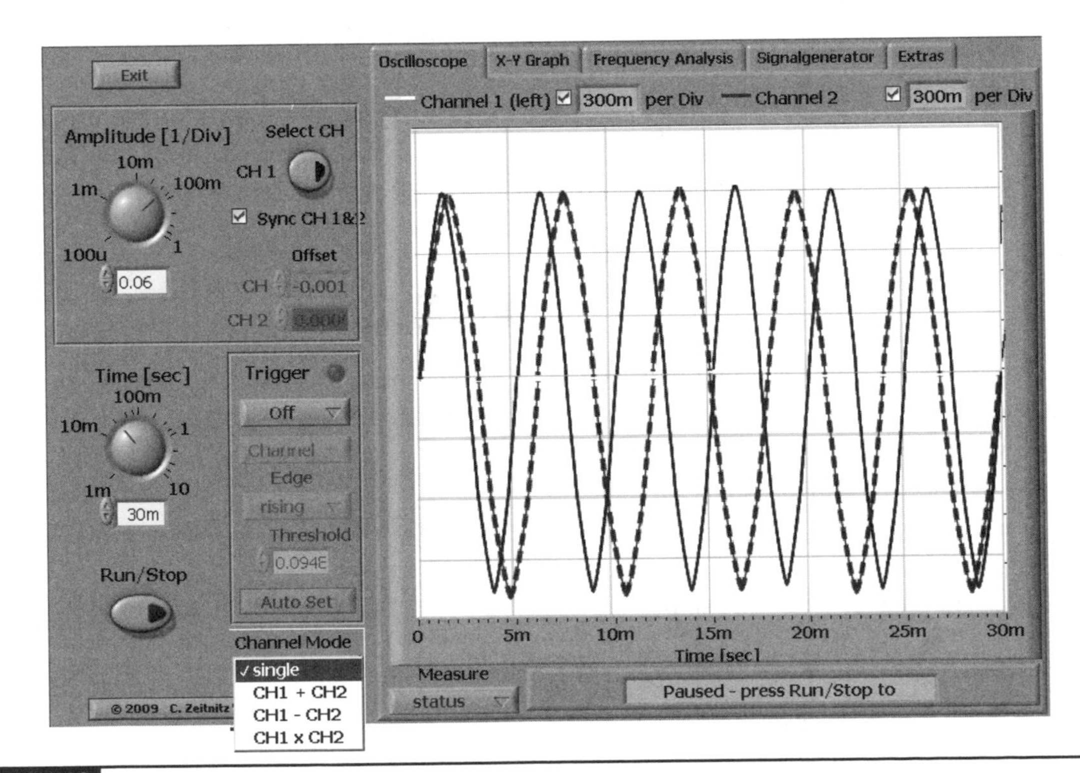

Figure L55-10

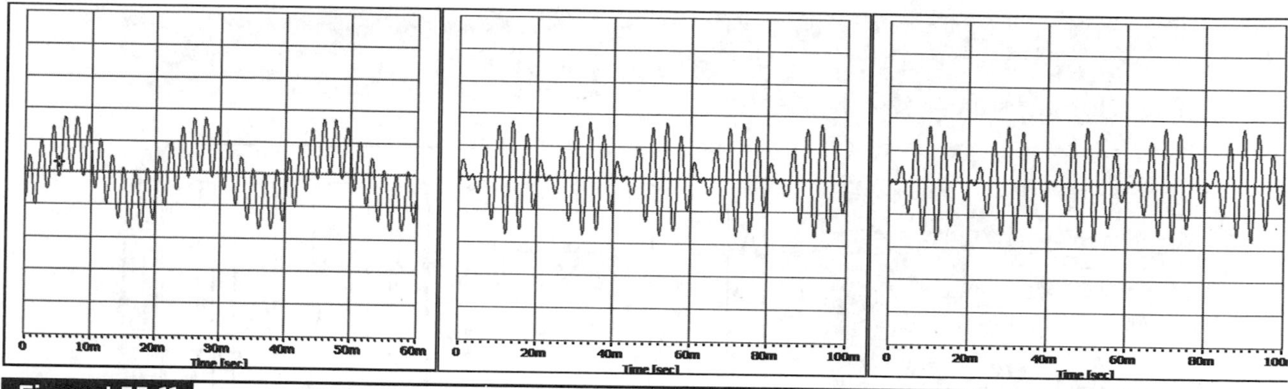

Figure L55-11

- Subtract the signals from each other: difference
- Multiply the two signals: product

Figure 55-11 provides screen shots of each function.

Now for Something Completely Different

1. Unplug the CH2 connector.
2. Plug in some earphones and listen to the tones created by the Signal Generator.

These are two separate signals.

We need to blend these two signals so they feed to your circuit from a single channel. The diagram in Figure 55-12 explains what needs to be done.

- You can create this 30-second wave file yourself, or download it (Lesson55Blended Tone.wav) from www.mhprofessional.com/ computingdownload.

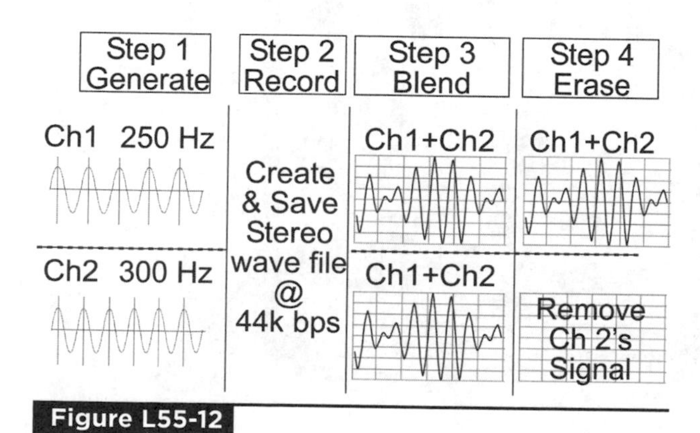

Figure L55-12

- You can't use the Soundcard Scope to play this file, but any music software will do it.
- You know you've done it properly when you hear the blended tone in just one ear when both channels are turned on.

Onwards

The circuit hasn't changed since Figure 55-6, and we're going to use exactly the same system setup as Figure 55-8.

Start the Soundcard Scope.

Make sure to clear the Sync CH1 & 2 check box.

Start the single-channel blended tone.

1. Use the CH2 connector (headphone output) to adjust the blended tone's strength to 40 mVac.
2. Connect the blended tone signal to the circuit's INPUT.
3. Connect the CH1 connector to the circuit's OUTPUT and microphone input.

Figure 55-13 compares the Soundcard Scope's signals and my real scope's signals.

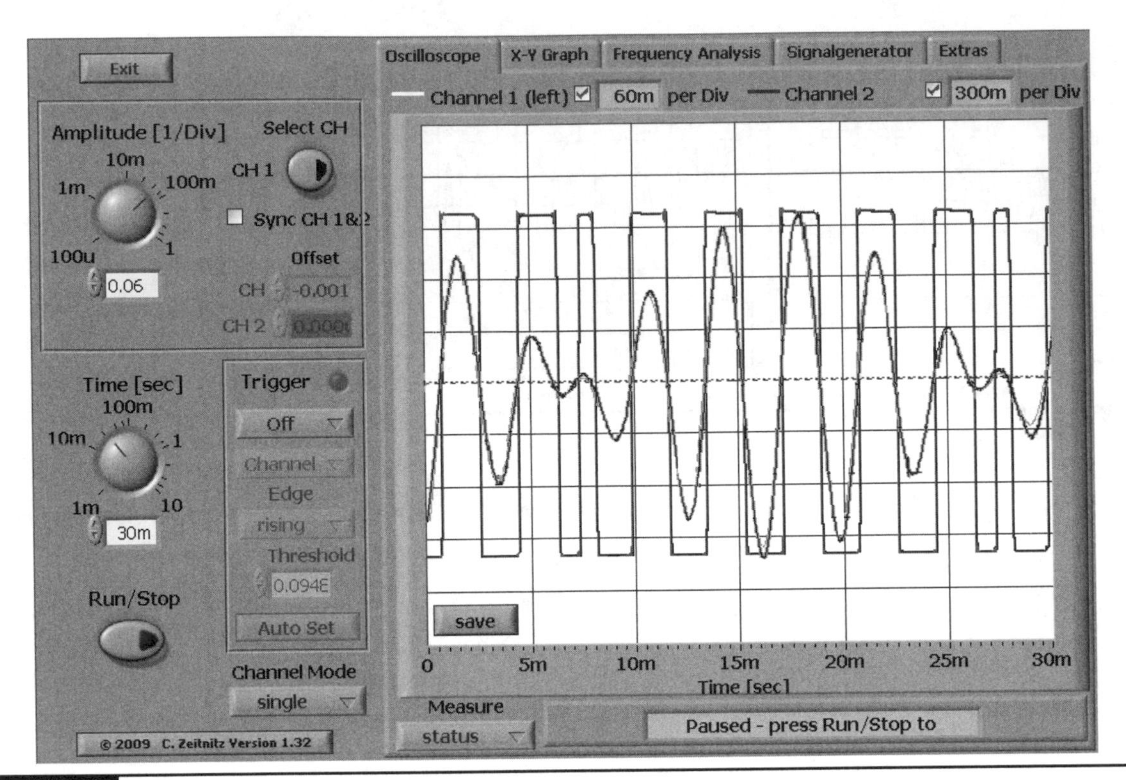

Figure L55-13

It does look pretty good, doesn't it? For the price, Soundcard Scope gives a great readout. But do you see the flaw?

Remember, this is an inverting Op Amp. Why does it show a "noninverting" output? When I checked this out on my real scope (Figure 55-14),

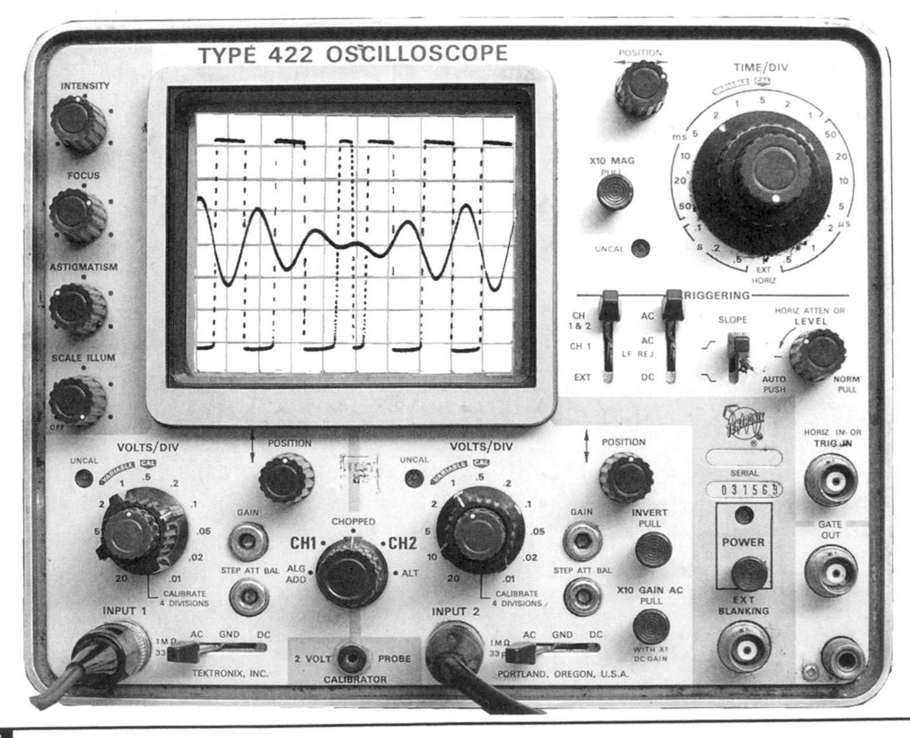

Figure L55-14

it gave me exactly what I expected. Honestly, I can't explain the error, but for our purposes, it's small. Recognize it for what it is, and move on.

Meanwhile, take a closer look at the settings on the real scope.

- Time/div =
- CH1 v/div =
- CH2 v/div =

As a voltage comparator, a millivolt's difference at the input produces an instant reaction at the output.

Lesson 56
Using Feedback to Control Gain

This lesson is dedicated to investigating and explaining the action of the potentiometer and how it acts as a volume control. The schematic is displayed in Figure L56-1.

Modify Your Breadboard (Again)

- Replace the 10 MΩ resistor with the 1 kΩ and 10 kΩ Pot in series to make a much more balanced feedback loop.
- Replace the two resistors and audio coupling capacitor at the end with the 100 Ω resistor and a 470 μF cap.

What we are using here is called "negative feedback." We will use this feedback to control the gain. You just built the Op Amp. Now you get to test it. You will set the gain through this unique feedback system. Actually, it's only a voltage divider applied in a familiar way.

This feedback loop is where things like volume and tone are controlled. Not only is it much easier to control the signal when it is small, but the components used to do it are smaller and less expensive. This is a half-watt intercom system. The potentiometer used here is rated for one-quarter watt. Using this potentiometer to control the signal to the speakers directly would create enough heat that it would burn out.

Again, refer to Figure L56-1 as we prepare to take some measurements needed as a foundation for some important concepts.

1. No power to the circuit yet.
2. Remove the potentiometer and set it to minimum resistance (0 Ω), and then set it back in.
3. Start up any of your prerecorded pure-tone frequency files.

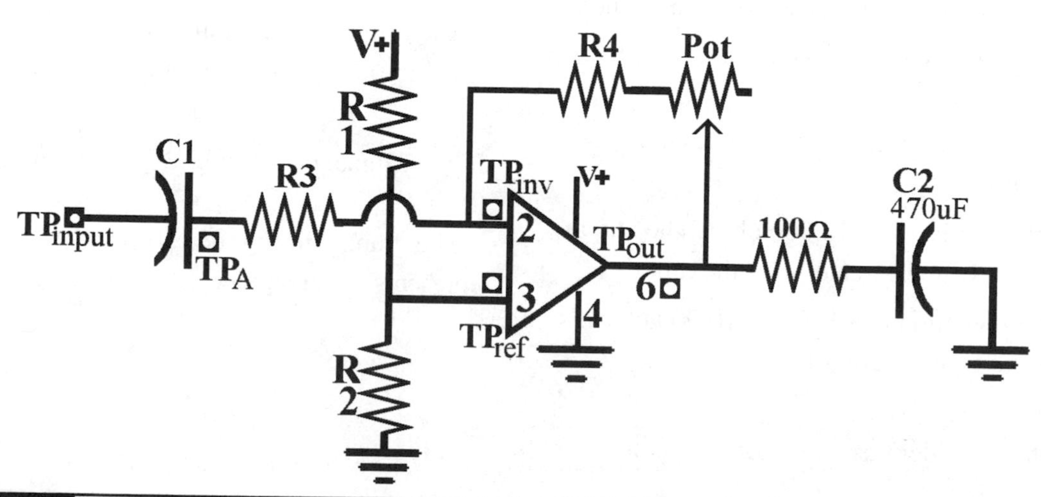

Figure L56-1

4. Use your DMM to adjust channel 1's output to a very small 10 mVac. Input = ____mVac.
5. Attach the colored clip of your CH1 connecting cord to your circuit's signal input. As usual, the black clip goes to ground.
6. Power up the circuit.
7. Directly at pin 6, check the signal's strength with your DMM. Output = _____mVac.

   Input's voltage should be identical (or at least very close) to the output's voltage. Output, Input = 1. Gain = 1.
8. Set the Pot to maximum resistance.
9. Measure the output directly at pin 6. _____mVac.

   The ouput voltage should be 11 times greater than the input (gain = 11).

   Recall that gain can be stated as a ratio of voltage to voltage, power to power, and even R_{total} to R1.

   Gain = Output/Input

What to Expect

Ideally, the input signal voltage for this circuit is between 0.010 and 0.015 V *AC*.

- When the Pot is set to 0 ohms, the output should equal the input.
- When the potentiometer is set to 0 ohms, then $\frac{R3 + Pot}{R2}$ gives a ratio of 1.
- The voltage divider has a ratio of 1. Gain should be 1.
- When the Pot is set to 10,000 ohms, the output should be 11 times greater than the input.

When the potentiometer is set to 10,000 ohms, then $\frac{R3 + Pot}{R2}$ gives a ratio of 11.

The voltage divider has a ratio of 11. Gain should be 11.

How It Works: Feedback to the Inverting Input

Here's are some questions: "Why does the volume go down when the potentiometer has lower resistance? Wouldn't less resistance mean more resistance? Wouldn't less resistance mean more current passing? Doesn't more current passing mean more volume?"

The answer is in the fact that the output signal has been inverted and a portion of the inverted signal is now being directed back onto the original signal. It is the same as adding a negative number. Think of it as subtraction. Subtraction is exactly what is being shown in the graphic representation of the signal with feedback in Figure L56-2.

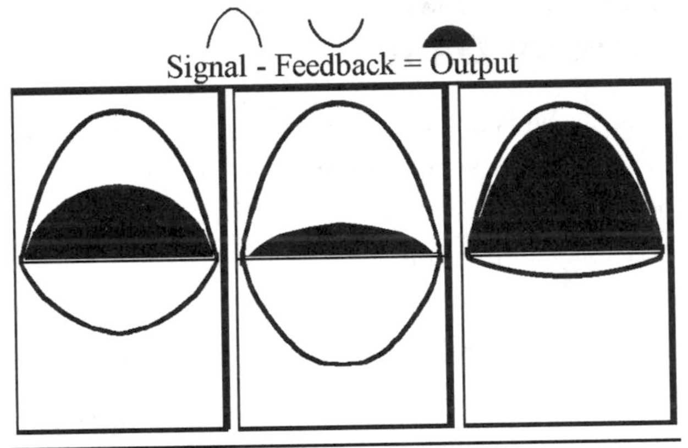

Figure L56-2

The size of the feedback signal increases as the resistance in the potentiometer decreases.

The original signal minus the larger feedback signal results in a smaller output signal.

$$Signal_{in} - Signal_{feedback} = Signal_{output}$$

The graphic in Figure L56-3 more accurately represents a real signal.

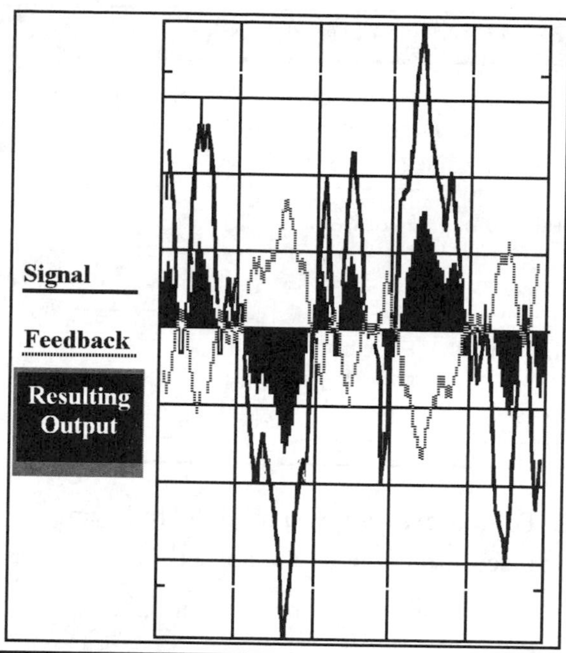

Figure L56-3

Exercise: Using Feedback to Control the Gain

1. What was the expected gain using the resistor ratios in the following?

$$\text{Gain} = \frac{\text{R4} + \text{Pot}}{\text{R3}}$$

a. ______

b. ______

c. ______

2. What was the expected gain using the voltage output compared to the voltage input?

$$\text{Gain} = \frac{V_{OUT}}{V_{IN}}$$

a. ______

b. ______

c. ______

3. Consider what would happen if R3 were burnt out and had infinite resistance (>20 MΩ).

a. Write out the expected ratio.

____

b. Think about it. Would that mean the volume would be extra loud, or not at all?

c. Would you be able to control the volume?

____

4. Consider what would happen if R3 were shorted out. In this case, R3 would have 0 ohm—no resistance.

a. State the expected ratio. ____________

b. Would you be able to control the volume?

5. Feedback—here you will need to use a bit of math to help explain the real action at the inverting input (pin 2).

TABLE 56-1 Collect This Data from the Setup in Figure 56-1

VAC between	Input = 0.0 V_{AC}*	(A) Input = 10.0 mV_{AC} R4 + Pot = 1 kΩ	(B) Input = 10.0 mV_{AC} R4 + Pot = 5 kΩ	(C) Input = 10.0 mV_{AC} R4 + Pot = 10 kΩ
TP_{IN} to TP_A				
$V_{input} = TP_A$ to TP_{REF}				
$TP_{output} = TP_{OUT}$ to TP_{REF}				
TP_{INVERT} to TP_{REF}				
TP_A to TP_{INVERT}				
Across 100 Ω resistor				

* Input from the headphones output with the volume adjusted to 0.0 volts AC. If the input to the Op Amp is unattached, static in the air will give erroneous readings. Remember what happens when inputs to a digital system are not tied either Hi or Lo? We can get the same ghosts here.

At a gain of 1, you have the least resistance in the Pot. How does the "least resistance" provide the quietest volume? Shouldn't less resistance mean more signal?

And that is *exactly* what less resistance means. But it means more signal in the feedback. More signal *subtracted* from the original input.

a. So calculate the amount of current being allowed to "feed back."

Use Ohm's law V = IR. You know the voltage, and you know the resistance (see Table L56-2).

TABLE L56-2 Calculations

Current feedback to pin 2 When R4 + Pot = 1 kΩ
Current feedback to pin 2 When R4 + Pot = 5 kΩ
Current feedback to pin 2 When R4 + Pot = 10 kΩ

b. Now calculate the current that is available at the inverting input pin 2. That is a measure of TPA to reference with R4 disconnected so there is no feedback at all. You will get a measure of the full signal. The current at the inverting input pin 2 is

____________________.

c. A little subtraction is now in order. For each of the previous three settings, calculate the real signal at the inverting input pin 2. Use the simple formula:

full signal − feedback = signal at pin 2

See Table L56-3.

TABLE L56-3 Calculations

Full signal ________	Feedback R4 + Pot = 1 kΩ ________	Signal at pin 2 ________
Full signal ________	Feedback R4 + Pot = 5 kΩ ________	________
Full signal ________	Feedback R4 + Pot = 10 kΩ ________	________

6. For this exercise, you need the following pieces from around the house:

- One long and skinny rubber band
- One short rubber band
- One large piece of cardboard
- Three tacks

Now cut the long rubber band in one spot, giving you one long strand.

Cut the short rubber band into two shorter pieces.

Gently stretch and tack the long strand across the cardboard surface. It should be tight enough to pluck.

Now tie a short piece to the center of the long one as demonstrated in Figure L56-4. Secure the end of that tack, too.

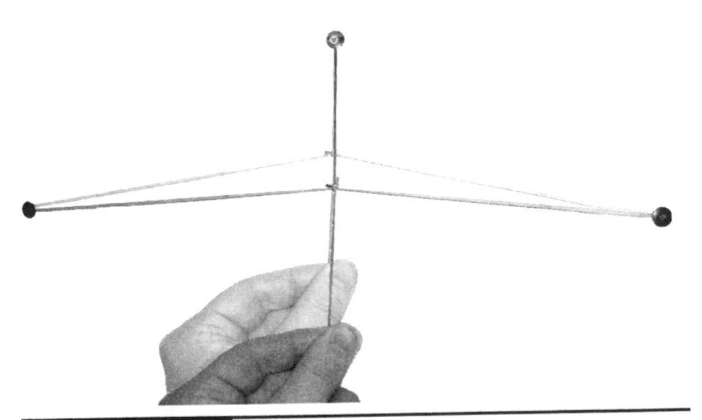

Figure L56-4

The long strand with the small connector altogether represents a signal.

Now attach the second short piece to the center of the long strand. As you tug gently on that loose strand, your input acts like the feedback; your force is being subtracted from the original force on top.

Just like this simple physical demonstration, both the original input and the "inverted" feedback combine to create the signal that is being fed into the Op Amp. The sum of these two combines to create the final output of the Op Amp at pin 6.

SECTION 17

Applying the Op Amp: Building the Intercom

HERE YOU'LL EXPLORE the individual portions of the intercom as you build the system. Consider radios. The system depends on amplifiers.

The microphone changes a voice into a small electric signal. This tiny signal from the microphone is amplified thousands of times before it is released as part of a modulated electromagnetic radio frequency.

A receiver is tuned to "resonate" to that frequency. Harmonics are a wonderful phenomenon. That weak harmonic signal is amplified in both voltage and current before it is directed to the speakers.

Lesson 57
Building a Power Amplifier Controlled by an Op Amp

So the Op Amp is now working as a preamp. The signal voltage has been increased, but there isn't much output at all. Time to crank up the power.

The preamp specifically boosts the voltage to a level that a power amplifier can use. Here we will employ transistors as the power amp. The voltage is already there. The power amplifier adds current to the voltage, effectively increasing the power.

Remember:

Power = current (I) × voltage (V) **or** $P = IV$

A huge variety of amplifier ICs are available to achieve the same purpose. But I use transistors because this is a "teaching circuit" that lends itself to explaining concepts common to all amplifiers.

The Power Amplifier

Still, many audiophiles swear that transistors are the best power amplifiers because they provide a different quality output than ICs. But before them, there were the audiophiles who swore by vacuum tubes.

Modify the Circuit

Add the components shown in Figure L57-1 to the preamp you have built. Note the rewiring of the Pot and the addition of C3 (470 μF).

The preamp created by the LM741 Op Amp effectively did three things:

1. It amplified the voltage.
2. It limited the amount of current by its very nature.
3. It still provided enough current for the transistors.

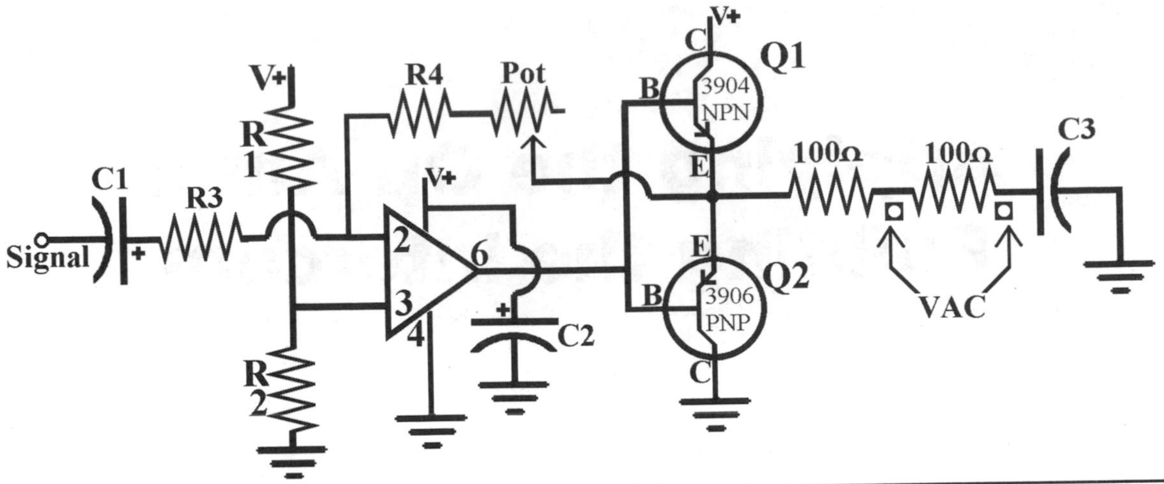

Figure L57-1

The pictures in Figure L57-2 show how the transistors act as power amplifiers. Animated versions of these drawings are available for viewing at www.mhprofessional.com/computingdownload.

The two transistors act in opposition to each other, controlling the signal output. As the valves open and close, they allow for much larger movement of current than allowed previously. The voltage and larger quantities of current create a much more powerful output than the LM741 could produce by itself.

Okay. So I have explained how the power amplifier works. Here is where you will find out how much power your amplifier is really producing. For all of the exercises, use a 1,000-Hz signal input at 10 to 15 mV.

1. Refer back to the measurements you did in Lesson 56. You took all the recordings needed and did all the calculations to get started.

   Remember:

   Power (watts) = voltage (V) × current (I)

   a. What is the VAC across R3?

   b. What is the current across R3? (V = I × R)

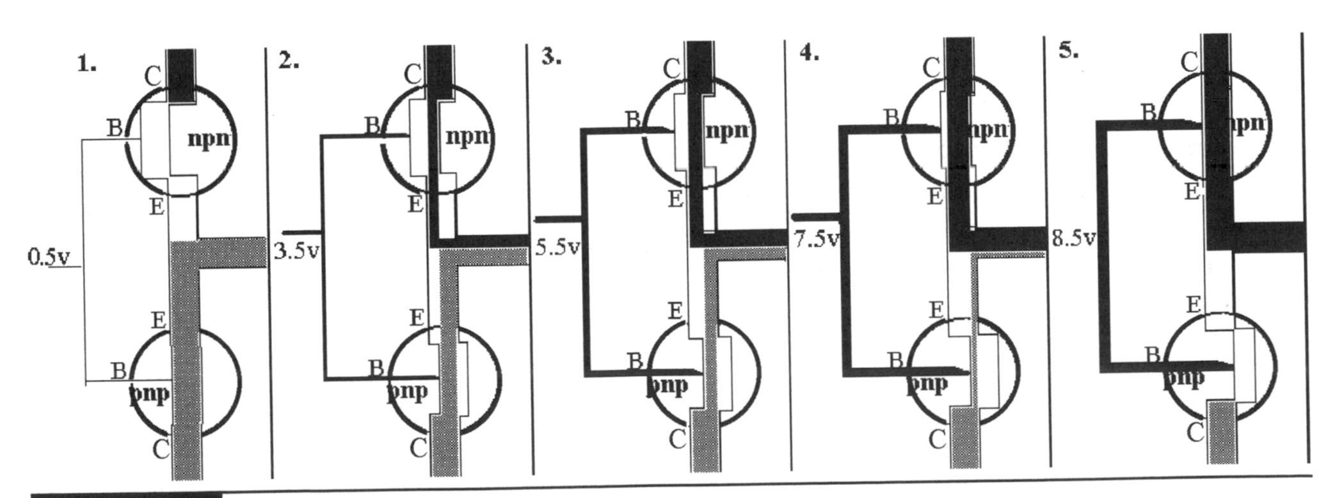

Figure L57-2

c. Figure the power across R3 when you had minimum volume (gain = 1).

Power at R3 = _______watts

d. Figure the power across R3 when you had maximum volume (gain = 11).

Power at R3 = _______watts

2. Now calculate the power output by measuring the VAC available across the 100-ohm resistor. Then measure the VAC across both 100-ohm resistors. The reading should double. What? A voltage divider in AC? With those measurements, you can now calculate the current. How does it feel knowing you are in control of all that power?

In comparing voltages used, always compare AC voltage to AC voltage. The same goes for comparing DC voltages to each other (see Table L57-1).

TABLE L57-1 Comparing Voltages

	Gain of 1 Minimum Volume	Gain of 11 Maximum Volume
VAC across the 100 Ω load	_____ V	_____ V
V/Ω = A	_____ A	_____ A
Watts = volts x amperes	_____ W	_____ W

3. In Lesson 56, without transistors, what was the power output at maximum volume? _____ W

4. In this exercise, with transistors, what was the power output at maximum volume? _____ W

C2—The Capacitor as a Buffer

What is a buffer? A buffer cushions the blow. You've been using a 1000 μF capacitor as a buffer for a while, ever since you set up the voltage regulator in Lesson 14. Refer to Figure L57-3.

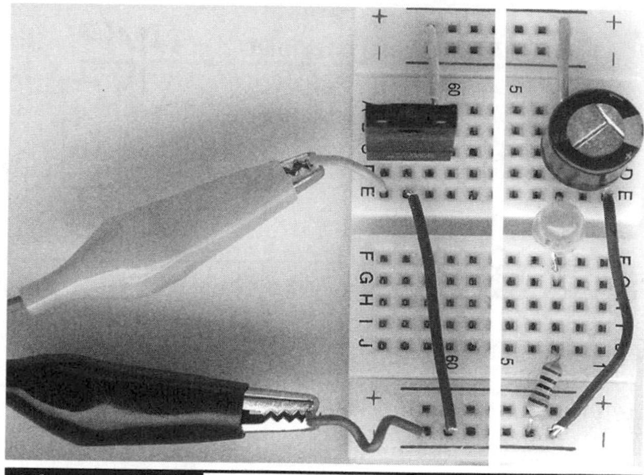

Figure L57-3

The voltage demands of the power amplifier draw quantities of power, but not steadily. This happens in quick bursts. A buffer, or reservoir in the form of a large capacitor, is necessary when using a small wall adapter or a 9-volt battery. As the power is tapped, the output fluctuates because the power supply cannot generate current quickly enough. This creates an unstable signal. The large capacitor keeps the current and voltage supply steady.

There is an interesting twist here. If you really want to hear this, you can't use our regular test frequencies. Their current draw is too stable.

- Connect the system to a 9-volt battery.
- Inject some music into your system.
- Adjust the volume to fairly loud.
- Pull out the 1000 μF cap.

The effect is quite noticeable.

The Second Audio Coupler

The second audio coupler is in place specifically to isolate the AC output from the DC voltage as shown in Figure L57-4.

In reality, if this capacitor were not in place and the speaker were connected between the output and ground, the entire signal would cease to exist. The AC output signal would be destroyed because

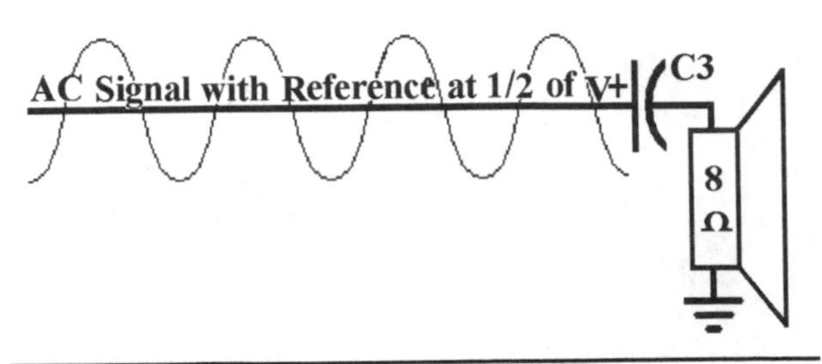

Figure L57-4

the signal and feedback at pin 2 would be referenced to ground in a DC system. They would no longer be floating artificially referenced to a redefined ground at half of the voltage. Take a moment and trace the circuit.

If you want to stop the circuit from working, go ahead and bypass C3 and connect the speaker directly to the connected emitters.

What? It stopped working? Don't say I didn't tell you so.

Lesson 58
The Electret Microphone

An electret microphone is a common component in audio systems. It is inexpensive and very sensitive. Microphones offer many design opportunities for this project.

Look carefully at the components inside the electret microphone (Figure L58-1).

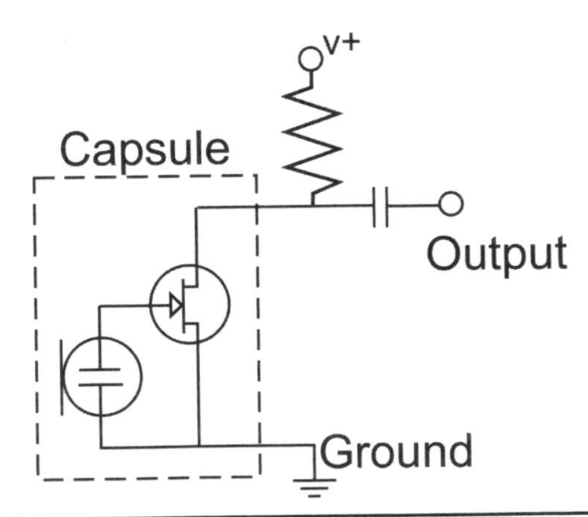

Figure L58-1

There are three types of microphones:

1. The coil microphone is similar to a speaker. It uses a magnet set inside a coil of wire. These can work independently of a powered circuit because they create their own voltage.
2. Piezo (crystal) microphones also create their own energy. These crystal structures respond to the compression and decompression of sound waves, releasing electrical energy. The crystals effectively translate the fluctuating energy into fluctuating voltage. This works in reverse, too. Feed voltage to the crystals and they vibrate. This effect is used to create quality, low-power speakers.
3. The electret microphone is an active capacitor. One of its plates is exposed to the air so that it vibrates as sound hits it. The plate's fluctuations disturb the steady DC voltage (pressure) by deforming the exposed plate ever so slightly. Those fluctuations in voltage match the sound. That becomes our signal. Because the electret microphone is an active capacitive component, it needs power to work.

The photo in Figure 58-2 shows one of the legs connected directly to the outer casing. That leg is ground.

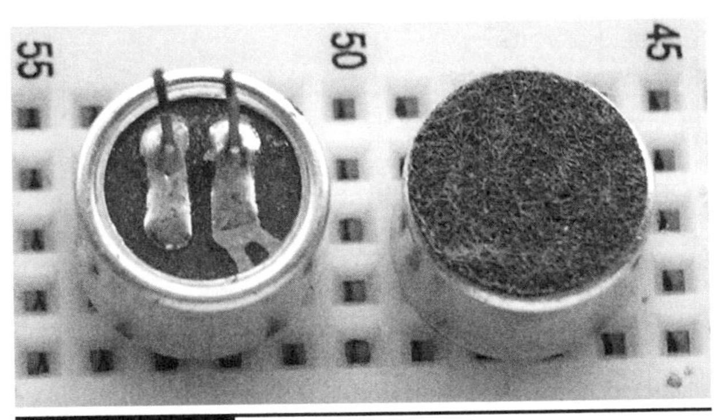

Figure L58-2

Build the Circuit

Set up Figure L58-3 as an independent circuit. Don't attach it to your Op Amp's input.

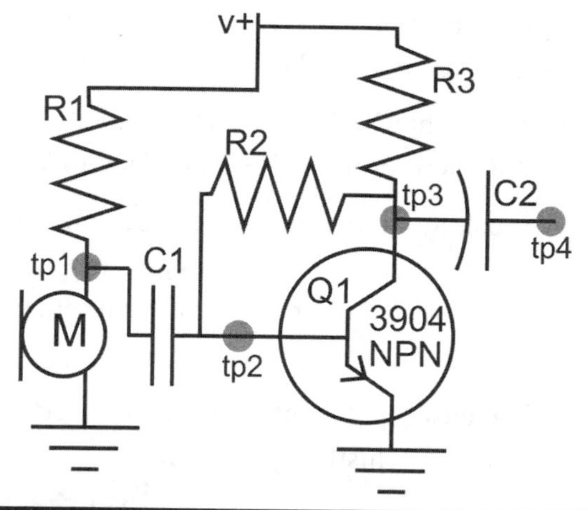

Figure L58-3

The parts needed are listed in the Parts Bin.

All of this looks new, but you're already familiar with it. Let's take a walk through.

PARTS BIN

- R1 & R3—10 kΩ
- R2—100 kΩ
- M—Electret microphone
- C1—.1 μF
- C2—4.7 μF
- Q1—2N3904 NPN

- R1. Because it is a single resistor, R1 only decreases the current, not the voltage.
- The microphone, like many capacitors, is balanced between DC voltage source and ground.
- C1 and C2 are audio couplers. They pass the AC signal but block the current.
- Q1 is your basic 2N3904 NPN transistor.
- R2 is the feedback loop, from output to input. It works exactly the same as in the Op Amp.
- The ratio of R2:R3 (100k:10k) dictates the gain created by this single transistor amplifier.

Checking It Out

Power it up and let's do some measuring. You will have to set up a speaker near the microphone while a recorded tone is playing. Check AC and DC voltages at all four test points and record these on Table L58-1.

TABLE L58-1 After You've Checked the Tone, Do It Again with Some Music

	DC Tone	DC Music	AC Tone	AC Music
TP1				
TP2				
TP3				
TP4				

You can use the Soundcard Scope's Channel 2 to see the tone's signal, but music output from the computer will occupy both channels. If you want to see the microphone's response to music, you'll have to use an outside sound source—something other than your computer.

To use the microphone as an input to the Op Amp, C2 acts as the output audio coupler for the microphone subsystem and the input audio coupler for the Op Amp circuit.

Lesson 59
Using the Speaker as a Microphone

An intercom has a microphone on one side and a speaker on the other. But then you would need two complete systems, each with a microphone input and a speaker output. You don't have a microphone? With some imagination and applied knowledge, we can do this with one system. You can use the speaker as a microphone. To do this, though, you have to understand a bit about how a speaker works. The speaker was designed as a speaker, not a microphone, as we will use it. It puts out a small signal—tiny.

When we use the speaker as a microphone, the following happens:

- Sound vibrates the cone.
- The cone moves the magnet.
- The magnetic field causes electron movement in the wire.
- This signal is used as the input to the Op Amp.

But still, the speaker is a poor microphone. It wasn't designed to be used as a microphone. But don't take my word for it.

Record this important information.

1. How much of a signal does the speaker produce on your DMM?
   a. Attach the DMM to the speaker. Set it to VAC. Red probe to V+ and black to V–.
   b. Speak into the speaker with your mouth about 2 inches (5 cm) from the cone. It's best to read from printed material rather than sit and repeat "Hellooooo." ______mV
   c. The best input signal you can produce is a "pucker" whistle. _______mV
2. How much of a signal does the speaker produce on your oscilloscope?
   a. Hook up your speaker to the test cord, and plug it directly into the computer's microphone input. Then open the Soundcard Scope. We are not using the oscilloscope probe. The signal is too small. If we used the voltage divider, it would not

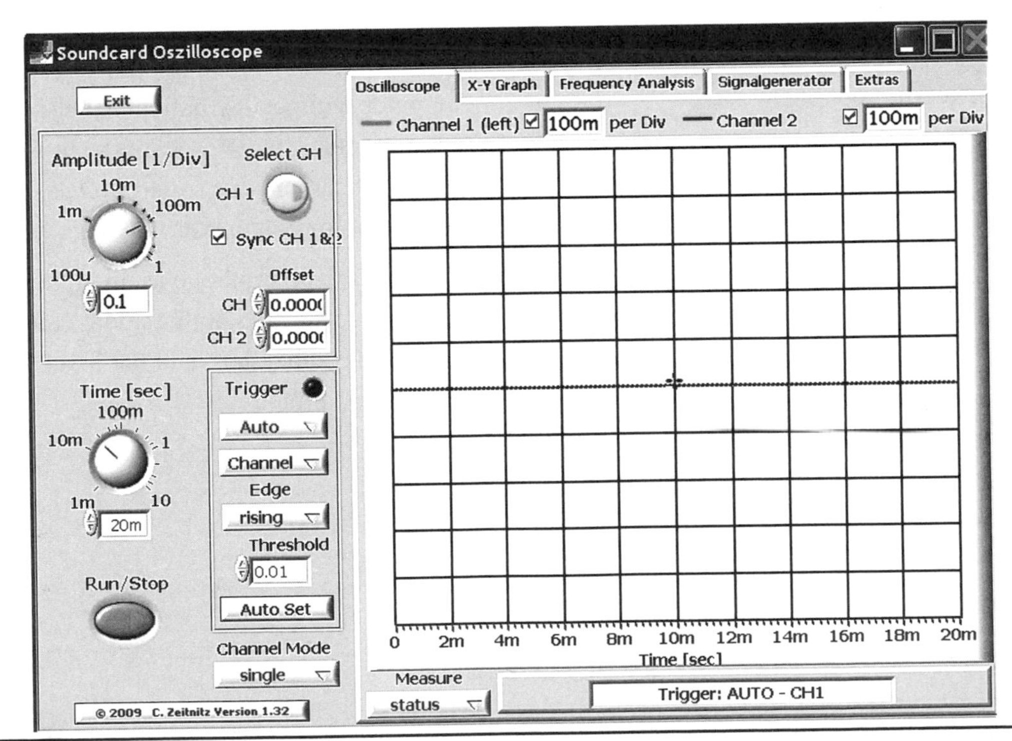

Figure L59-1

register. Use the oscilloscope setup shown in Figure L59-1.

b. Speak into the speaker. Make note of the maximum positive and negative voltages.

c. Now whistle into the speaker. You should see a beautiful sine wave. Did you know that your whistle is nearly a pure tone?

d. Draw a representation of your whistle signature onto the scope face provided in Figure L59-2.

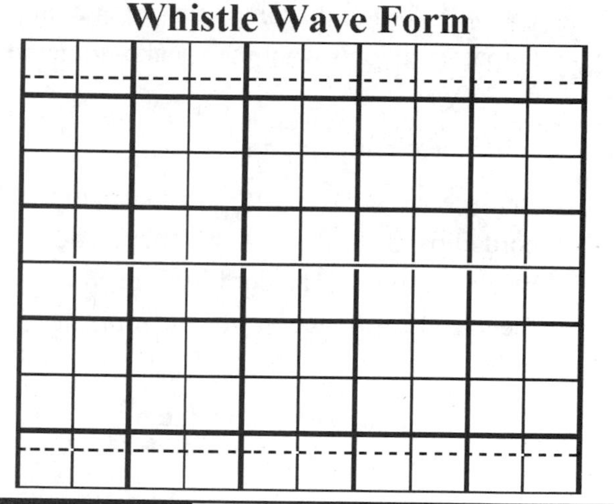

Figure L59-2

3. Applying the speaker in the circuit.

a. Now set this second speaker into your SBB as shown in Figure L59-3.

b. Turn the gain down all the way and use your pucker whistle.

c. As you increase the volume, the circuit might start screeching at you. Notch the gain backward until the screeching stops. The screeching is caused by feedback demonstrated in Figure L59-4. Sound from the output is reaching the input and multiplying itself.

d. You can solve this problem in two ways. The first is to give more distance between the output and the input. But right now, you can remove the output speaker. This is an easy solution for the moment. After all,

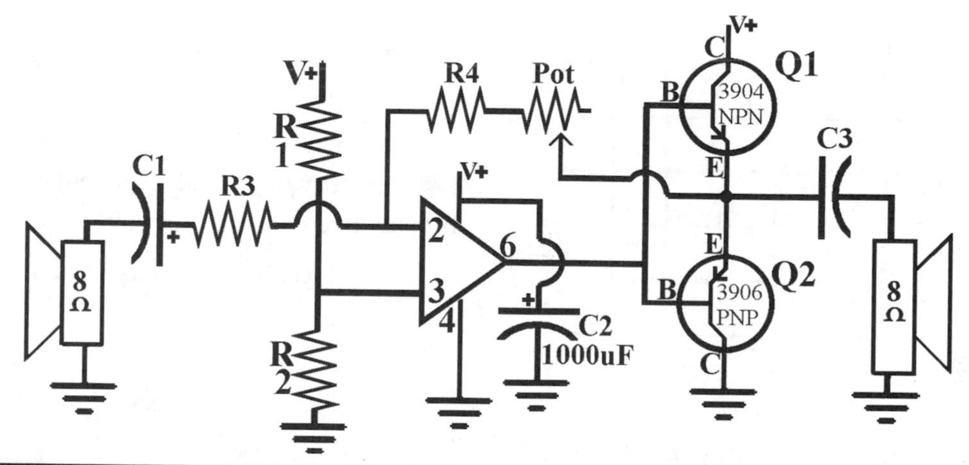

Figure L59-3

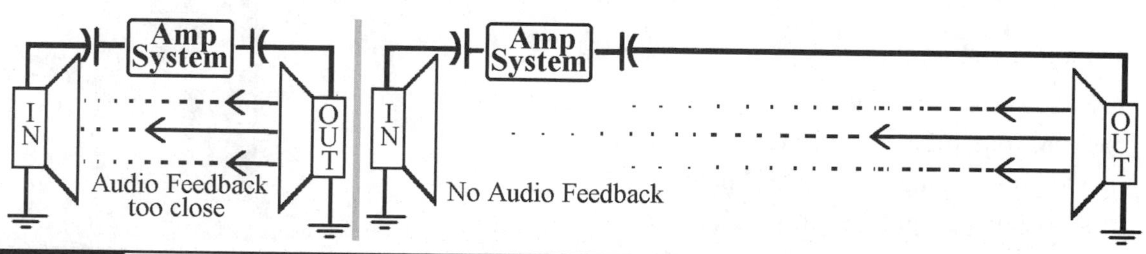

Figure L59-4

you will need to do some measuring at the output, and don't want the speaker to get in the way.

4. Scoping the circuit output.
   a. Replace the speaker with the connector cord clips, the red clip to C3 (neg), and black to ground. The cord is plugged into the sound card "line in," or microphone input.
   b. With the volume turned all the way down, whistle and look at the signal on the scope.
   c. It should be identical to the signal taken directly from the speaker earlier.
   d. Now turn the volume all the way up. Whistle again. Draw this on the scope face in Figure L59-5.

Amplified Whistle Wave Form

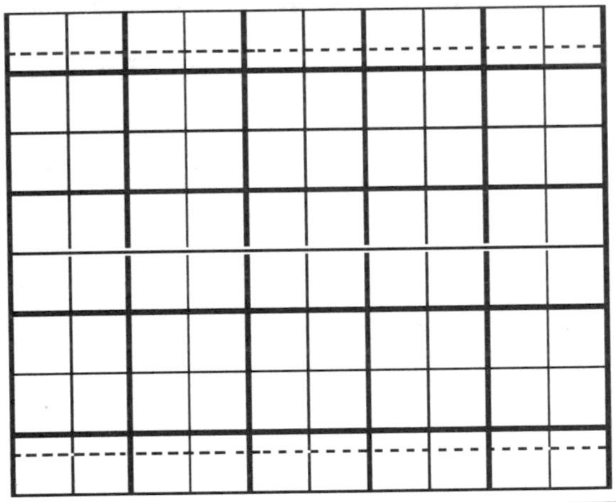

Figure L59-5

5. Check out the AC voltage of the maximum output of your whistling using the DMM. ______Vac Max.
6. Listen to the circuit output.
   a. Now before you put the output speaker back in, solder the ends of the dual-wire line to your speaker. That line should be at least five feet long. Place it back into the circuit.
   b. Place the input and output speakers as far from each other as possible.
   c. Whistle into the input speaker.

   Hmm. The output is very quiet, even at full volume.

   The last component will take care of that.

Lesson 60
Introducing Transformers and Putting It All Together

So you've seen that speakers aren't microphones. The small signals they produce need a different kind of preamp to get a clean signal from this subsystem to the next stage of the amplifier system. Right now, the output directly from the speaker is a paltry 1 millivolt. So the output from the amplifier is also very weak. Such a small signal needs to be cleanly amplified even before it is fed to the preamp. To restate, the signal needs to be preamplified before it gets to the preamplifier. Without having a two-stage preamplifier, there just isn't enough signal there to control the power amplifier.

Here I will introduce transformers as a method to preamplify the signal to the preamp. It is frequently done with microphones as well.

Here is a great place to use a transformer. See Figure L60-1.

Figure L60-1

What we have here is another basic electronic component. Transformers are used in homes in everything from wall adapters to microwaves. Larger versions are vital to the supply and distribution of electric power. You will use a miniature transformer like the one shown in Figure L60-2.

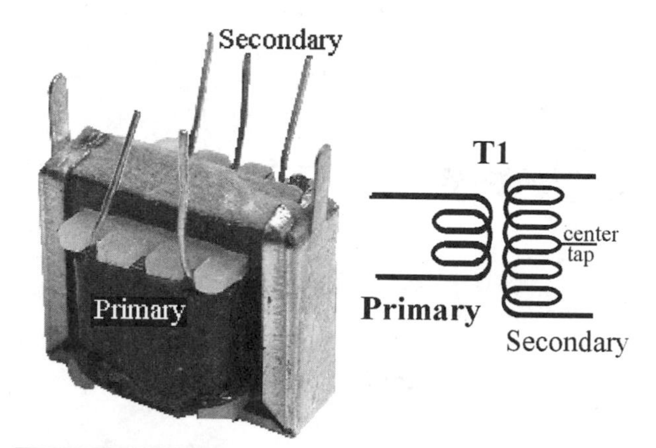

Figure L60-2

How a Transformer Works

Actually, there are two main questions here:

- How does a transformer work?
- What does it transform?

As you saw with the speaker, moving electrons create a magnetic field. What was not mentioned is that the reverse is also true. A moving magnetic field induces (encourages) electrons to move in a conducting wire. The moving electrons in one wire create a magnetic field that induces electrons in a nearby wire to move as shown in Figure L60-3.

Also notice that a second wire is too far away to be influenced by the magnetic field created in the first wire. Animated versions of these figures are available for viewing at www.mhprofessional.com/computingdownload.

Pretty fancy! But better yet, examine Figure L60-4. It depicts how this concept is applied so neatly in electricity and electronics.

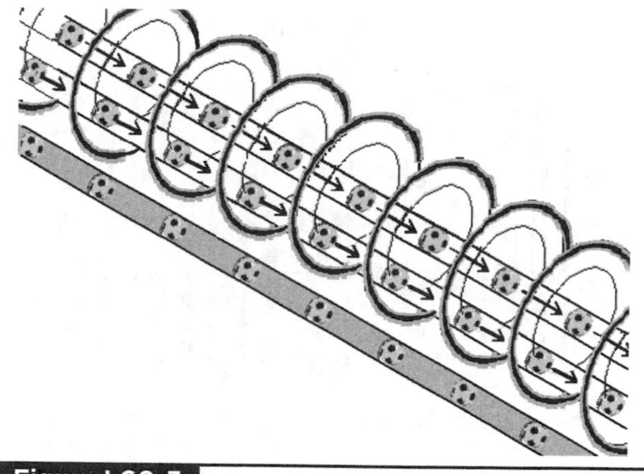

Figure L60-3

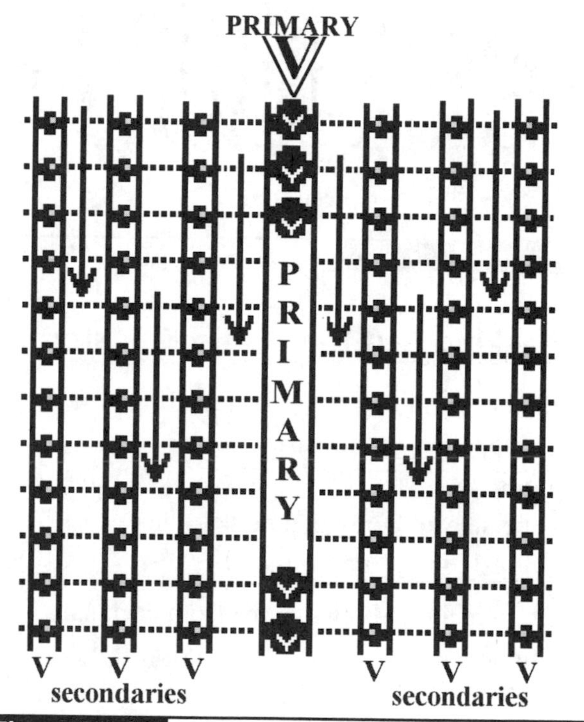

Figure 60-4

If we lay many wires next to the "primary" wire, it will induce (encourage) a voltage in each of the "secondary" wires. But wait. There's more. If each of those secondary wires is connected as one long wire, the voltage induced in each wire is added to the next. This is shown in Figure L60-5.

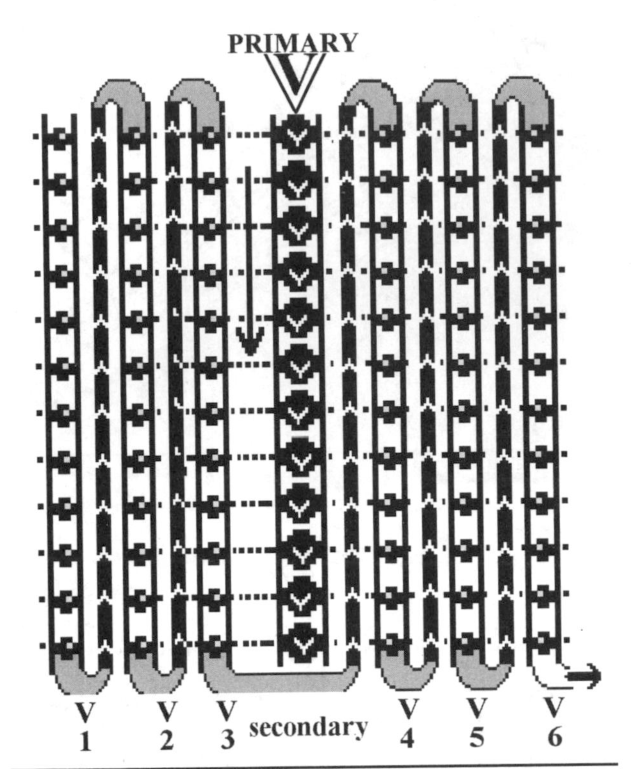

Figure L60-5

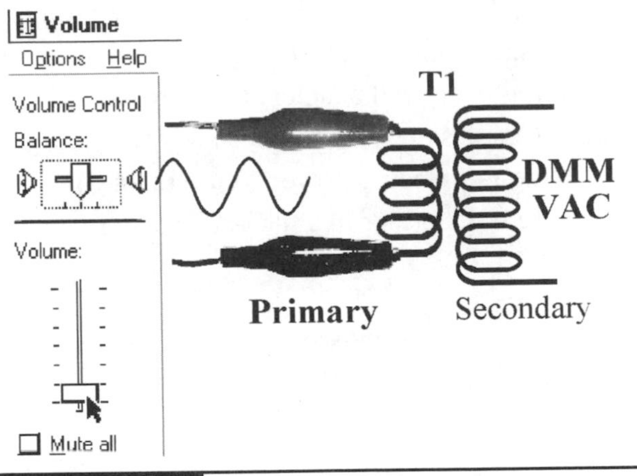

Figure L60-6

Checking Out the Audio Transformer

Pull out the miniature audio transformer. It should be wrapped in green. Various colors indicate various uses. Most likely there will be three wires on one side and only two on the other as shown previously in Figure L60-2.

If there are only two wires from each side, continue with the following exercise. It will become obvious which side is primary and which is secondary as you do the work.

The primary side has two wires. The secondary side has three wires. The middle wire on the secondary is referred to as the *center tap*. You can clip and tape it off. We won't use it at all.

Plug the test cord into the headphones output of your sound card. Use the software-based volume control to adjust the 1,000-Hz wave output to 5 mVac. Now attach the clips to the primary wires of the transformer as shown in Figure L60-6. Measure the output of the two outermost secondary wires (see Table L60-1).

TABLE 60-1 Output Measurement

Input to Primary	Output at Secondary	Ratio of Output VAC/ Input VAC
5 mVAC		
10 mVAC		
25 mVAC		
50 mVAC		
100 mVAC		

$$\frac{\text{Output Signal}}{\text{Input Signal}} \quad \text{Average Ratio} = \_\_\_\_\_\_\_\_$$

In transformers, the fixed gain is referred to as a *step-up* or *step-down* ratio. This ratio actually reflects the physical relationship of the number of primary to secondary windings. A step up of 8 is created when you have 20 windings in the primary and 160 windings in the secondary. A wall adapter steps the voltage down. For example, one that provides 9 volts from a 120-volt system literally has a ratio of 120 windings in the primary for every 9 in the secondary. That's close to a ratio of 13:1.

The ratio on the audio transformer should be much greater than 1:1. If it is less than 1:1, you have your input connected to the secondary side of the transformer.

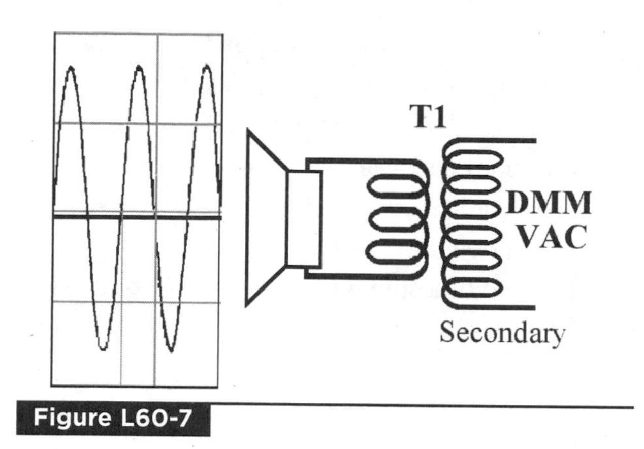

Figure L60-7

Now attach the speaker to the primary side of the audio transformer. Again, use the DMM to set the output at the secondary, measuring AC voltage as shown in Figure L60-7.

- Input at Primary = _____mVAC
- Output at Secondary = _____mVAC

Does this match the previous ratios? It should.

Insert the Audio Transformer into the Circuit

Insert the audio transformer into the circuit as shown in Figure L60-8.

Now test the entire system using the 1,000-Hz signal. Set the output of your sound card to 5 mVAC by first connecting it to the DMM. Use the software control to set the output signal strength precisely.

Measure the AC signal at different points indicated and record the readings in Table L60-2 provided.

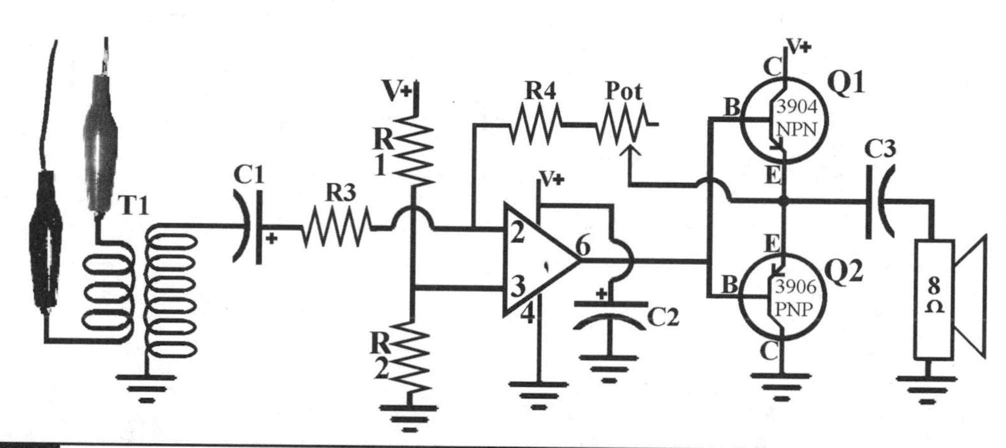

Figure L60-8

TABLE L60-2 AC Signal Measurements

1,000-Hz Sine Wave Tone	**5 mVac Gain = 1**	**5 mVac Gain = 11**	**10 mVac Gain = 1**	**10 mVac Gain = 11**
VAC TP input to reference				
VAC TP output to reference				
Favorite Music CD	**5 mVac Gain = 1**	**5 mVac Gain = 11**	**10 mVac Gain = 1**	**10 mVac Gain = 11**
VAC TP input to reference				
VAC TP output to reference				

Points to keep in mind.

1. This is not a high-fidelity system.
2. Any feed higher than 10 mVac will probably start sounding lousy because of clipping. The output signal is limited to 4.5 volts above the reference and 4.5 volts below the reference. You can't get more with this system. Try for bigger, and you lose most of the signal.

Parts list for reference:

- R1, R2 100 kΩ C1 = 4.7 μF
- R3, R4 1,000 Ω C2 = 1000 μF

C3 = 470 μF

The Intercom System

Remove the test cord and put the speaker to be used as a microphone into place as shown in Figure L60-9.

Now you can use it as an amplifier. Talk into the speaker set up as the microphone. Adjust the gain to get the best quality and volume. At this point, you will find it necessary to separate the two speakers by at least five feet of wire. If they still squeal, turn down the gain or put one speaker on the other side of a sound barrier like a box or door. Take a few more measurements (see Table L60-3).

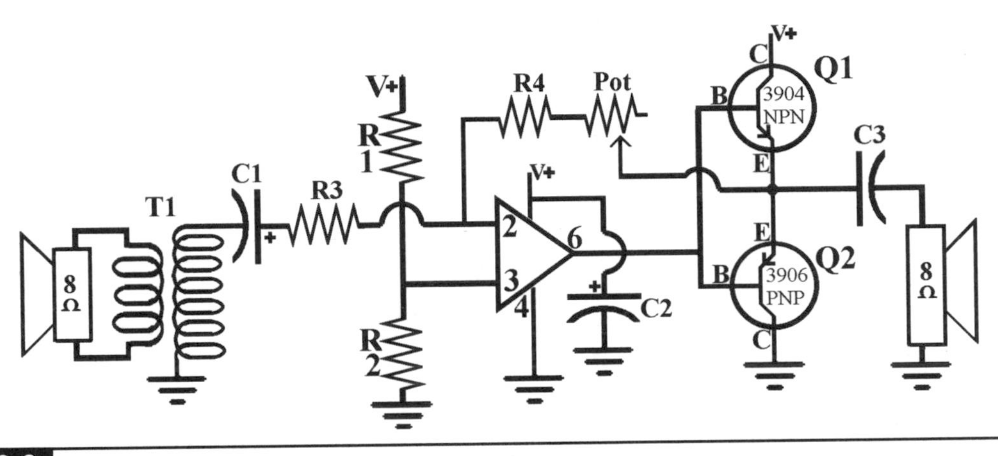

Figure L60-9

TABLE L60-3 Measurements

Pucker Whistle into Speaker	mVAC Gain = 1	mVAC Gain = 11	mVAC Gain = 1	mVAC Gain = 11
VAC TP input to reference				
VAC TP output to reference				

SECTION 18

Prototype and Design: Patience Has Its Rewards

1825: William Sturgeon invented the electromagnet.

1831: Michael Faraday's and Joseph Henry's electromagnetic phenomena research stimulated Sam Morse to devise a telegraph receiver.

1832: Morse conceived the idea of a single-circuit electromagnetic telegraph.

1835: Morse built the telegraph, only four years after Faraday explained magnetic induction.

1837: Morse solicited money and investments to build a test telegraph system. The great financiers of the age saw no future in sparks.

1838: The U.S. Congress turned Morse down. He tried to conjure up support in Europe with no success.

1843: Morse went on alone. He finally secured congressional support for a 41-mile (60-km) line between Baltimore, Maryland, and Washington, D.C.

1844, May 23: Morse sent the first telegraphic message, "Behold, What hath God wrought!"

Lesson 61
Systems and Subsystems

With an eye to upcoming design work, we make an in-depth tour of the amplifier's subsystems now and their possibilities. Currently, your breadboard should reflect the schematic in Figure L61-1. This setup is dealt with in Table L61-1.

The Systems We Take for Granted

There are two underlying systems that need to be recognized:

- The regulated and buffered power supply is vital. Figure L61-2 reminds us, and Table L61-2 discusses the power supply in detail.
- The audio couplers create an AC environment, necessary for the 741 Op Amp. These are non-negotiable.

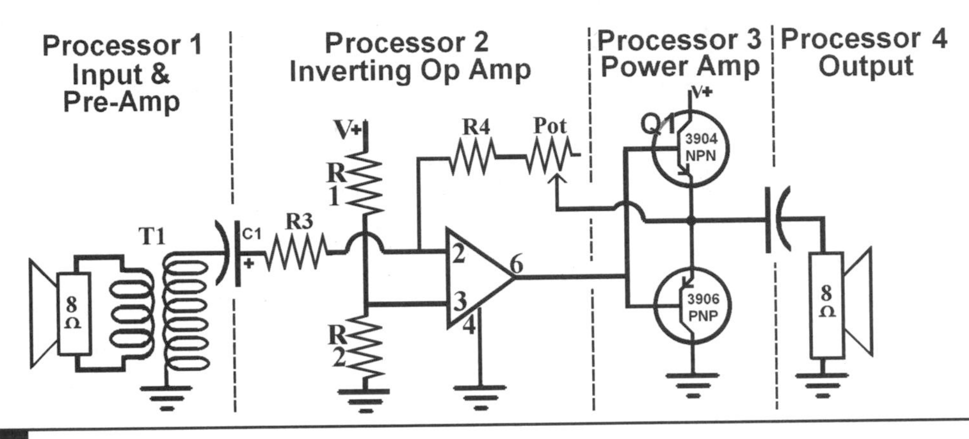

Figure L61-1

TABLE L61-1 The Basic System				
Processor #1a	**Processor #1b**	**Processor #2**	**Processor #3**	**Processor #4**
Input determines type of pre amp	Pre amp dependent on type of input	Op Amp: Analog inverting amplifier	Power amplifier adds current	Speaker
1 to 5 mVac produced by the speaker itself	The audio transformer's set gain increases voltage but not the power	Feedback loop controls gain: Range 1 to 11	Classic dual transistor "push/pull" setup	.5 watt 8Ω

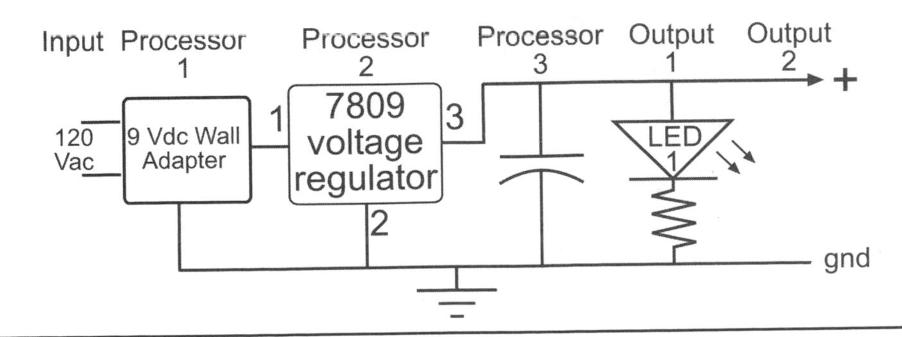

Figure L61-2

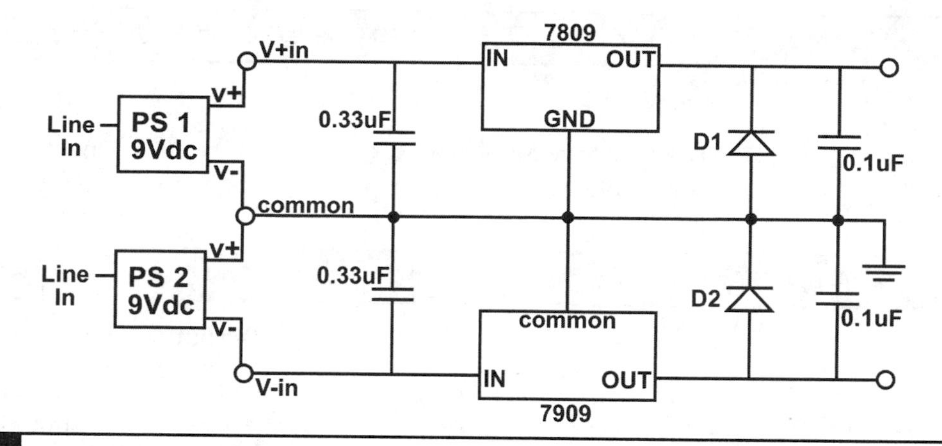

Figure L61-3

TABLE 61-2 Most Systems Take the Power Supply for Granted

Input	Processor 1 Wall Adapter	Processor 2 7809 (9v@1 amp)	Processor 3 Buffering Cap— 1000 μF minimum	Output 1 Regulated 9VDC	Output 2 LED
120 VAC Line Voltage	Wall adapter AC to DC (unregulated) with rated power output	Provides a steady 9 VDC. Cannot exceed rated output of wall adapter.	Provides extra power for sudden demand	Very stable 9VDC within wall adapter ratings	LED indicates that power is available

The power supply suggested in Figure L61-3 is best for increasing this circuit's power without having to modify either the circuit or its components.

Processor #1a: Initial Input— Speaker as Microphone

Table L61-3 details the speaker as a subsystem.

Processor #1b: Initial Input— Pre Amp (Audio Transformer)

The audio amplifier is used here because the speaker creates its own voltage. The ratio of the primary coil to secondary coil can be roughly determined by the resistance readings shown in Table L61-4.

TABLE L61-3 Any Single Component Can Be Considered as an Entire Subsystem

Input	Processor	Output
Sound waves as pressure variations move the paper diaphragm.	The diaphragm moves the magnet inside the wire coil, inducing electron movement.	Small voltages created in the coil match the sound input.

TABLE L61-4 The Audio Amplifier Has a Set Gain

Input	Processor	Output
Small signal fed into primary coil (.85Ω).	The secondary coil's wire is so fine and long that it has 60Ω. Very sensitive.	Voltage increases but current decreases. The power out is the same as the power in.

TABLE L61-5 Remember, the Heart of This System Is the Summing of the Two Signals

Input	Processor	Output
7 to 15mVac	The feedback loop allows adjustable gain up to 11 times.	AC signal connects directly to the matched transistors' bases.

TABLE L61-6 Recall that Power Is Stated as Watts: Watts = Voltage (x) Impedance [W = VI]

Input	Processor	Output
Varying AC signal controls the matched transistors simultaneously.	Classic "push/pull" setup. The varying voltage controls the valving action that draws current directly from the power supply.	Output AC signal maximum = .5 watt. The output is limited by the power available from the supply and the capacity of the transistors.

Processor #2: The Operational Amplifier

Our basic system is an analog-inverting Op Amp. Table L61-5 gives details about it.

Processor #3: The Power Amplifier

The power amp (Table L61-6) is the section that "adds" current to the existing voltage.

I won't be redundant and discuss the "speaker as output." It is exactly the opposite of the "speaker as input." Voltages fluctuations in the coil move the diaphragm back and forth over the magnet. The moving paper diaphragm creates sound waves as pressure variations in the air. This is essentially how any speaker works.

Variation on a Theme

Here are some suggestions just to get you thinking. These are expanded in Lesson 63.

Processor 1—Inputs

Any method that creates a fluctuation of at least 1 mVac can be used as an input to this system. Table L61-7 lists the four general categories of inputs available to us.

Processor 2—Op Amp

This is the best place to modify the signal. Table L61-8 suggests some possibilities.

TABLE L61-7 Potential Modifications for the Initial Input (Processor 1)

Sound	Light	Radio Frequency	Direct Line Input
Electret microphone (transistor pre amp)	Phototransistor (transistor pre amp)	Antennae or pick-up coil (audio transformer)	A signal created from another source like an MP3 player (might not need re-amp)
Coil microphone (audio transformer)	Light-dependent resistor (audio transformer)		
Crystal (piezo) microphone (audio transformer)			

TABLE L61-8 Even with All These Options, We Can Still Use the System as Just an Intercom

Add filters so the system works as	Incident indicator	Voltage comparator	Movement
High-pass Filter (blocks lower frequencies) Low-pass Filter (blocks higher frequencies) Mid-range Filter (removes the voice range, e.g., karaoke) Hi/Low Filter (filters all except the voice range)	The Op Amp outputs a digital Hi or Lo when an input lasts longer than a predetermined length of time.	Analog–Digital converter	A special input directly at the Op Amp can indicate both movement and direction.

Processor 3—The Power Amplifier

If you want more out of your system, you have to put more in. Size does make a difference. But providing a larger power output brings its own complications.

1. A bigger wall adapter will provide higher voltage and more amps: W = V (×) A.
2. To handle the larger power supply, you will need a bigger voltage regulator that can handle more voltage and more amps.
    a. A different package 7809 can handle more power. Then there are the 7812 and 7815. These obviously provide more voltage.
    b. The 7909? Something new! What does it do? Thought you'd never ask. It's a negative voltage regulator.

You can have 9 V+ on top and 9V– on the bottom: a spread of 18 volts. That's close to the limit for the 741 Op Amp.

If you increase the power supply, you need bigger, but still matched, transistors for the power amplifier section. The 2N3904 and 2N3906 are available in only the TO-92 package, and they are really limited in power. My suggestion would be the 2N2222 NPN switching transistor and its PNP complement, 2N2907A. In the TO-18 metal package, these are rated for operating steadily with a draw of 500 mA @ 40 volts (20 watts), with a maximum of 800 mA @ 60 volts (48 watts).

Processor 4—The Output

It's the end of this system, but it could just be the beginning of another.

1. Analog
    a. Speaker and sound
    b. Modulated power to a light source (IR LED or laser pointer)
2. Digital
    a. Hi or Low output as a trigger for another circuit (not a clock output)
    b. Animatronics (motor control)
3. Feed amplified signals and events to the computer.

Don't just sit there—start playing!

Lesson 62
Switching to the Two-Way Door Phone

Yes, you now have the first half of the intercom system. But with a flick of a switch, you can reverse the signals, making two halves of an intercom. Then you can finish up.

The Evolution of Switches

Up until now, you have been using momentary contact switches like the ones shown in Figure L62-1.

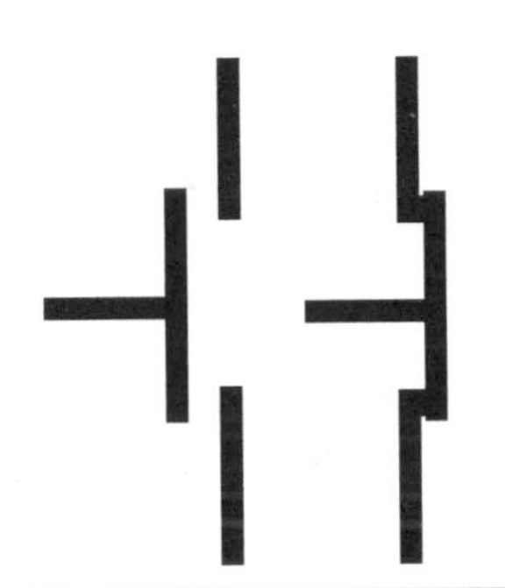

Figure L62-1

Normally open push buttons have contact only when they are pushed and held.

Normally closed push buttons have contact all the time, until they are pushed.

Examples of these can be found in items ranging from telephones to game controllers.

The single-pole single-throw (SPST) switch displayed in Figure L62-2 has definite on and off positions.

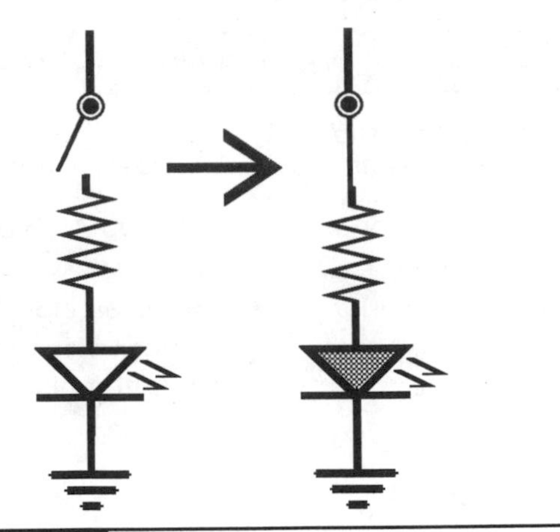

Figure L62-2

It is designed to be used with only one circuit pathway.

The single-pole double-throw (SPDT) switch splits one line into two paths. The example shown in Figure L62-3 shows how a single voltage can be used to power two different items. Sometimes there is an off position set into the middle.

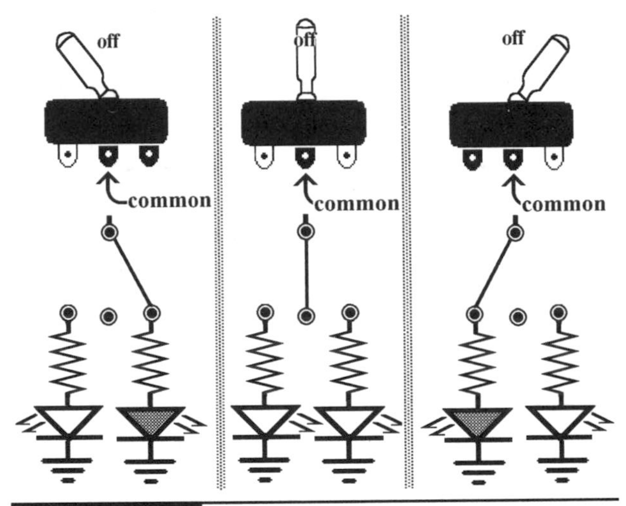

Figure L62-3

Notice that the center tab can have a connection to either side, depending on the position of the toggle.

The double-pull double-throw (DPDT) switch is like having two SPDT switches glued together side by side, sharing the same toggle. Figure L62-4 displays the commonly found package of the DPDT switch. Note the action of the switch in Figure L62-5. Except for the toggle, the two sides are completely independent of each other.

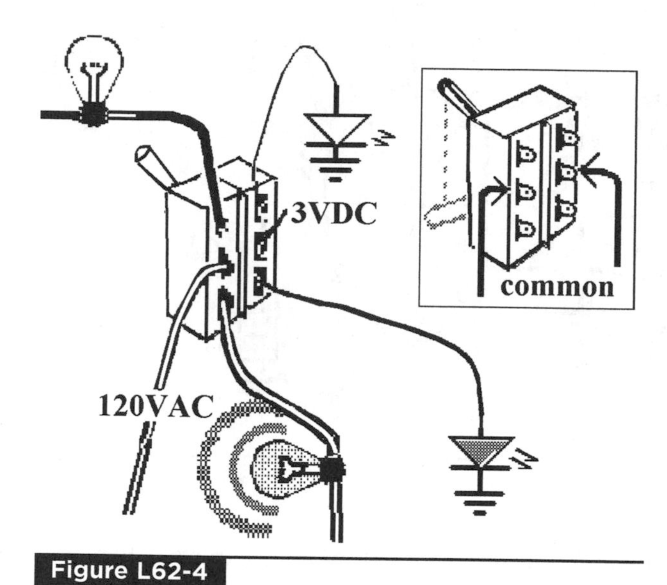

Figure L62-4

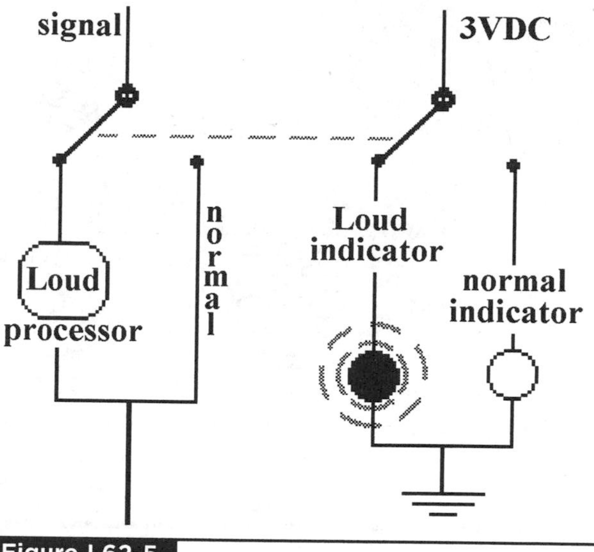

Figure L62-5

The DPDT is found frequently used on stereo systems. They are often paired with indicator lights to show the current function chosen.

But wait … aren't stereo systems just that? Stereo!

They have a left *and* a right signal. Each side is completely separate, too! Figure L62-6 shows how two DPDT switches can be made to work together in a big switch. It would be a double DPDT switch. But let's make life easier. Just call it a 4PDT switch. Figure L62-6 displays the four poles and double throw.

Now that's a switch! They don't come much bigger than that!

Making the Two-way Door Phone

So far, you have the complete system as shown in Figure L62-7.

It uses a speaker as a microphone and you can talk one way. By using the 4PDT switch, you can make it into a two-way system. This is done by rerouting the input and output signals. Figure L62-8 shows the numbering of the tabs for the 4PDT switch. Study the wiring diagram in Figure L62-9

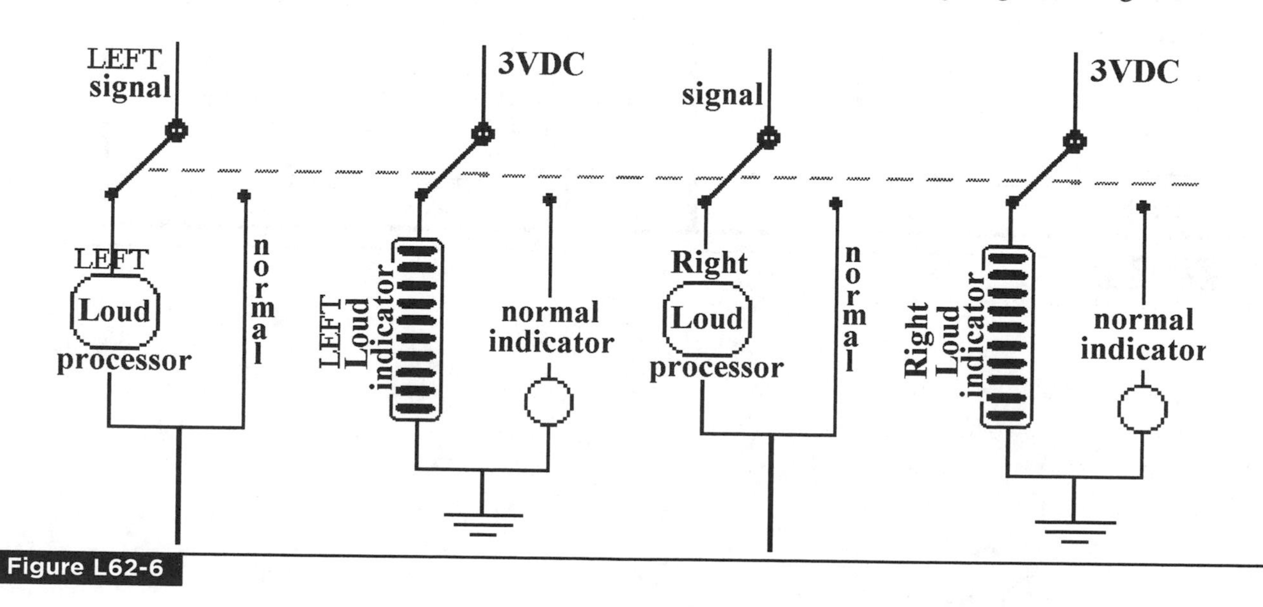

Figure L62-6

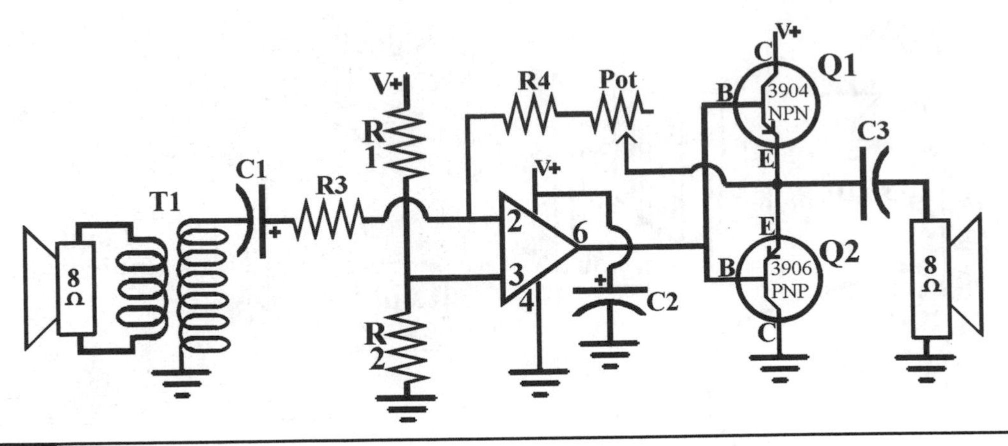

Figure L62-7

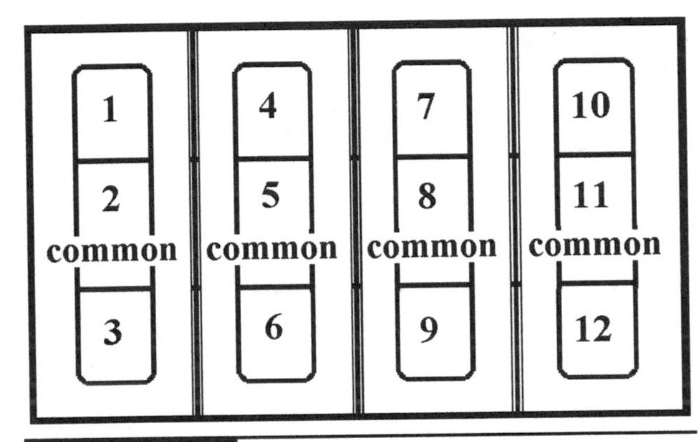

1	4	7	10
2 common	5 common	8 common	11 common
3	6	9	12

Figure L62-8

carefully. You don't want to make a mistake here. Each speaker will have four wires, two on each side.

Recognize that, depending on the position of the switch, only one on each side will be connected at any time. In one position, the speaker on the left acts as the microphone. In the other position, the speaker on the right acts as the microphone.

That is all there is to it!

There is only one switch, and you control it from the inside.

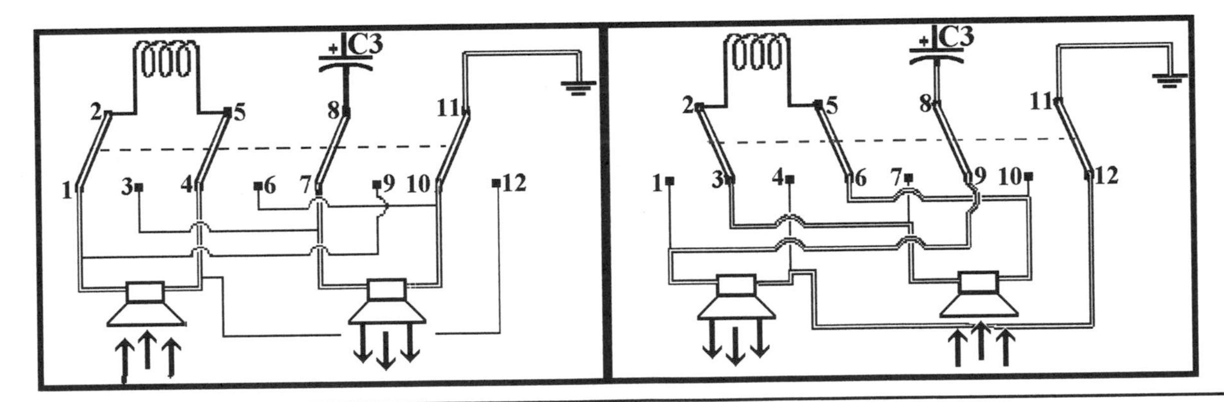

Figure L62-9

Lesson 63
Design and Applications: Exploring the Possibilities

Arthur C. Clarke formulated the following three "laws" of prediction:

1. When a distinguished but elderly scientist states that something is possible, he is almost certainly right. When he states that something is impossible, he is very probably wrong.
2. The only way of discovering the limits of the possible is to venture a little way past them into the impossible.
3. Any sufficiently advanced technology is indistinguishable from magic. Figure L63-1 is the old RCA logo "His Master's Voice." What we now consider simple used to amaze the world.

Here are five projects that provide examples of imaginative applications:

1. Amplitude modulation: Communication on a beam of light (Table L63-1)
   a. Laser listener: Remember that sound makes a window vibrate. Bounce a laser beam off the window and onto your phototransistor.

Figure L63-1

   b. Directly power the laser pointer with an amplified signal. A phototransistor and speaker will translate the beam back into sound.
   c. Communicating privately back and forth with two sets.
2. Animatronics: Talking and singing bivalve shell (Table L63-2)
3. Measuring density and vibrations: Water in a river, thickness of air pollution, heart monitor, or even a seismograph to measure traffic rumbling on a nearby road (Table L63-3).

TABLE L63-1 The High-Power Output Section Discusses Using the Power Amp as a Power Source

Input	Processor 1	Processor 2	Processor 3	Processor 4	Output
Sound source	Modulated laser beam pickup	Phototransistor or solar cell	Op Amp	Power amp	Speaker

TABLE L63-2 Making the World Your Oyster Is Easier than You Think

Input	Processor 1	Processor 2	Processor 3	Processor 4	Output
Sound source	Microphone and pre amp	Op Amp voltage comparator	Power amp	Power to a motor	Hinged movement

TABLE L63-3 This System Can Measure the Rate and Density of Any Flowing Fluid

Input	Processor 1	Processor 2	Processor 3	Line Out
Infrared LED or laser	Phototransistor or solar panel	Single transistor pre amp	Op Amp	Soundcard Scope and recorder

TABLE L63-4 This All Depends on the Input Antenna

Input	Processor 1	Processor 2	Processor 3	Output
Antennae loop (wire coil with an open center)	Audio transformer	Op Amp	Power amp	Speaker

For daytime and further distance, a laser is better input. To measure across distances, reflect the laser beam off a distant mirror. Make sure that all sections are solidly secured so they don't add to the shaking. This setup can be surprisingly sensitive. Also, the phototransistor can be protected from sunlight's IR by tucking it deep into a darkened tube.

4. Radio frequency pickup (Table L63-4): Use this as an electromagnetic energy detector.
5. Natural rhythms: Listen to the sunrise (Table L63-5).

Considerations for the Input

This circuit needs an input that creates at least a 1 mV fluctuation. Here are two suggestions that could help magnify that physical input if it is too small.

1. Specifically, when using a phototransistor during the daytime, you can protect it from the sunlight's IR by tucking it into a dark tube.
2. A parabolic surface works for both sound and light. Mini-umbrellas create a semi-rigid parabolic surface that works nicely for sound. Silvered Mylar film sold as emergency blankets at dollar stores reflect light. Figure L63-2 helps remind us that a parabola is not a semicircle. Large telescopes and small satellite dishes use the same concept for gathering and focusing.

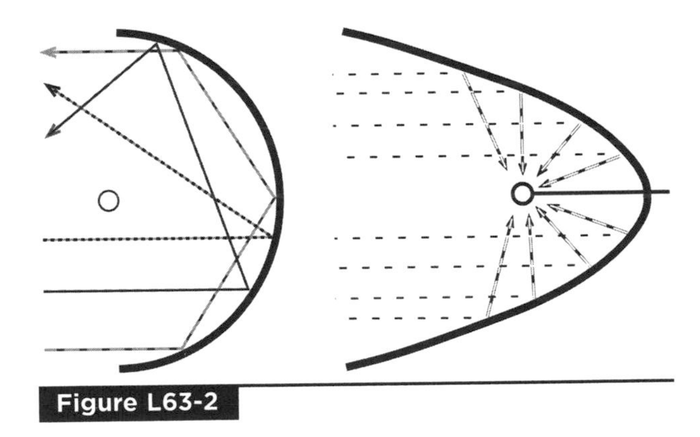

Figure L63-2

TABLE 63-5 Listen to the Sunlight Reflected Off Trees During a Windy Sunrise

Input	Processor 1	Processor 2	Processor 3	Processor 4	Output
Any reflected light	Lens to concentrate light	Phototransistor or solar cell	Op Amp	Power amp	Speaker

3. A lens inside a tube (Figure L63-3) works nicely for focusing a moderately strong light source. It is limited to light only and is not nearly as large as the parabolic surface.

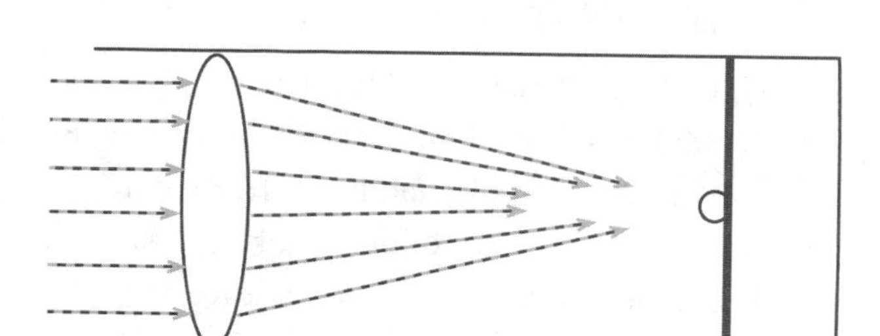
Figure L63-3

4. A cone made of rolled material (Figure L63-4) is the simplest of any focusing tools. Several factors make it less efficient than either of the previous options. The resulting signal is muddy, but it does work.

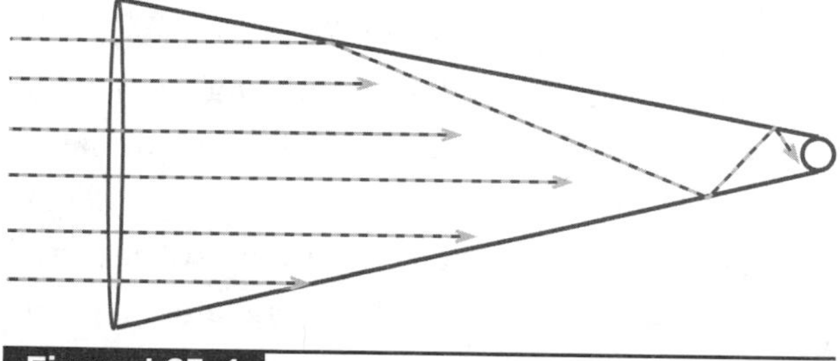
Figure L63-4

Consideration for the Output

The power amplifier's size is ultimately determined by your intended output. Even though a large power output will drive a smaller load, it could also burn it out. For example, a 2.0-watt output will eventually burn out a 0.5-watt 8 Ω speaker. After some heavy figuring using differential calculus, I suggest that a 2-watt speaker be used with a 2-watt power supply.

High-Power Output

1. Modulate power to directly power a laser pointer. Note the setup in Figure L63-5A. This works nicely over a few hundred meters.

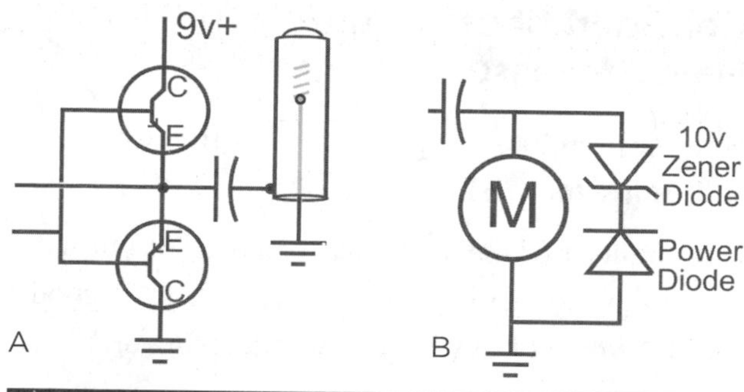

Figure L63-5

2. High-power digital signals can directly power motors. Figure L63-5B's setup can handle small motors. This setup provides short bursts in both directions. 5B's diodes prevent feedback, which would create disturbing power spikes that could damage the Op Amp. Remove the zener if you want movement in only one direction.

Low-Power Output

When I say low power, I mean "low power." To do this, the schematic in Figure L64-6 drops the power amplifier stage and places the second audio coupler directly after the Op Amp's output. A low-power output is necessary specifically for systems that have their own power supply.

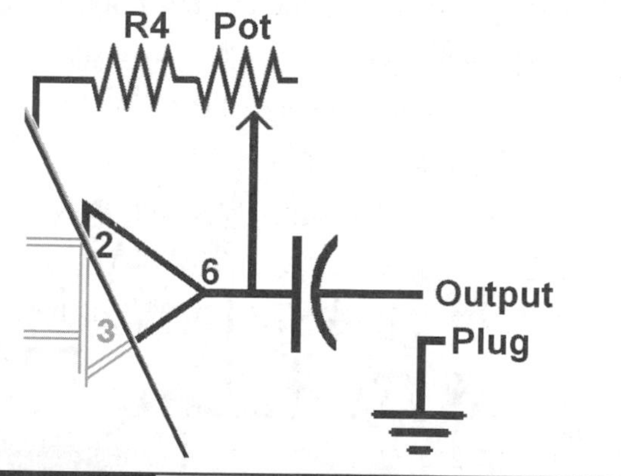

Figure L63-6

Considerations for the Main Processor

Brief descriptions are given here for the two different setups in Figure L63-7.

1. Figure L63-7A is the basic processor out with adjustable gain, maximum 1:11. This is good for voice or any basic mid-range signal.
   a. To listen to the sound of nature, or your friends at over 100 meters away, mount the electret microphone at the focal point of a simple parabolic reflector. This can be used as a basic audio amplifier. It is not high fidelity in any sense of the word. There are better Op Amps for this purpose. But it is excellent for voice communication.
   b. Blood flows. Treat this as an audio signal. What about a heart monitor? Use an IR LED and phototransistor that are both lensed sideways. Tape them onto the tip of your finger, on opposite sides, facing each other. Feed the output directly to the microphone input of the computer. Watch it on the Soundcard Scope. Record it using a high sampling rate and save it as a wave file (*.wav).
   c. A marvelous application is in your hands by simply replacing the electret microphone with a phototransistor. Make sure the lens is shiny enough to concentrate the diffused light. And I'm serious. At sunrise, point this device westward at trees in a soft breeze. There is no other sound like it as they come into the light. You can also listen to light reflecting off insect wings.
2. Figure L63-7B increases the gain to 1:1,000,000. The clipped output signals are as good as digital highs and lows. This circuit can be used for controlling small motors. Those motors, in turn, control a lever arm that lifts any small object on a hinge. This motion is counterbalanced by either gravity or a spring so the object returns to its original position when the circuit's output is low. There are lots of examples on the Internet.
3. Figure L63-8 is strictly a digital system. Its inputs can be physical, light, or sound. It will give a Hi output depending on the length and strength of the input signal. Also, this basic voltage comparator is a "voltage follower." This term refers to the fact that the output voltage state (Hi or Lo) matches that of the input. The input signal must last long enough and be strong enough to raise the voltage in the capacitor above half of V+ before the 1 MΩ resistor drains it. Vary the timing by changing the value of the capacitor. You can use the same single transistor amplifier from way back in Unit 2. Don't use the single transistor's pre-amp setup.

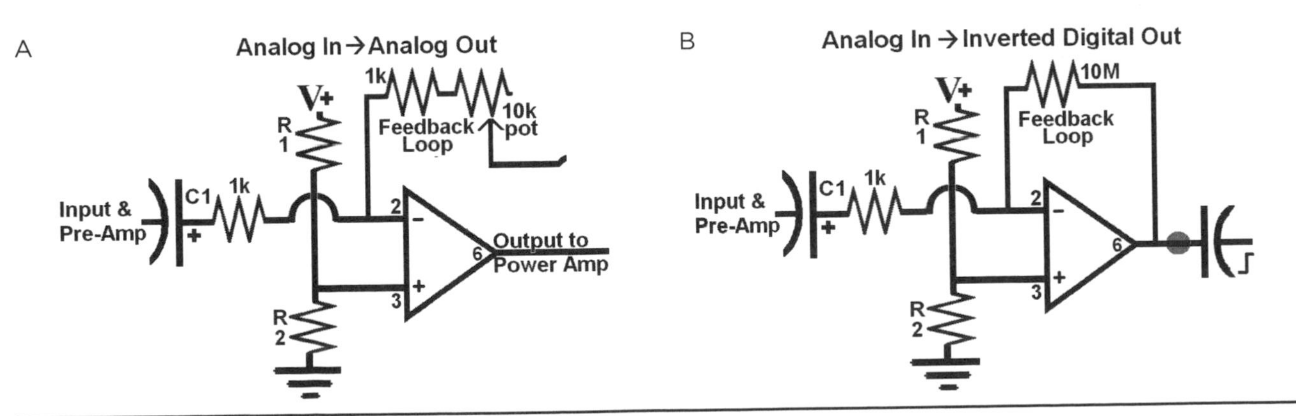

Figure L63-7

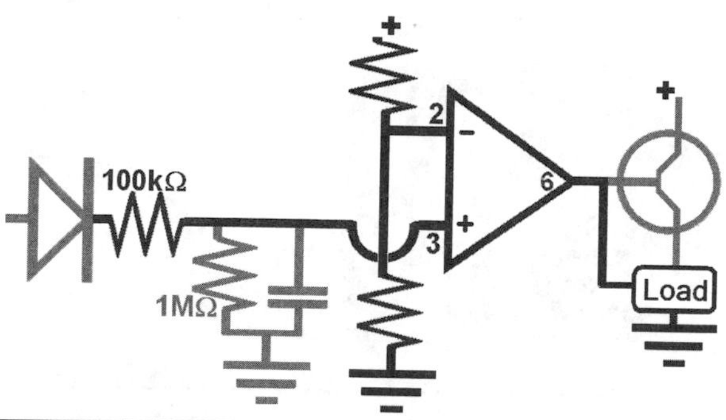

Figure L63-8

This high output is good as a trigger for starting any secondary system that depends on a Hi or Lo signal. Think wildlife and a digital camera. Or a basic security system.

4. Figure 63-9 presents an entirely different input to the circuit. Not sound. Not light. Electromagnetic radiation! This isn't the type of radiation most people think about when you say "radiation." It is a general term referring to the concept that electromagnetic energy moves in all directions—it *radiates* out from its source. In fact, we live in a world polluted by electromagnetic energy. All you need to sense it is a good antenna feeding into the audio transformer.

Figure L63-9

Use magnet wire that is 28 gauge or finer. Wind it at least 100 times around the outside of a 5-inch (13-cm) wide square of cardboard. Do not hold the antenna in your hand. You are a large capacitor and affect the small antenna. Place the antenna on the end of a plastic ruler. Do all of your initial testing with the audio transformer in place. Also, keep the volume very low when you start testing different things. My computer puts out about 4 mV noise. The microwave oven jumped from 0.0 volts to 15 mV as I turned it on. But the old TV leaked enough energy to boost the DMM's reading up beyond 150 mV. If I'd had earphones, I would have blasted myself through the roof.

Consider changing the shape of the antenna to meet your needs. The antenna shaped as a tube responds nicely to very small voltage fluctuations of any line passed through its open center. You can listen in on an active landline telephone conversation.

Lesson 64
Assembling the Project

See the Parts Bin for the project parts list.

PARTS BIN

- R1, R2—100 kΩ
- R3, R4—1 kΩ
- P1—10 kΩ
- C1—4.7 μF
- C2—1000 μF
- C3—470 μF
- T1—Audio transformer (1:50) green
- IC1—741 low-power Op Amp
- Q1—2N3904 NPN
- Q2—2N3906 PNP
- Speakers (2)—8 Ω
- Switch—4PDT

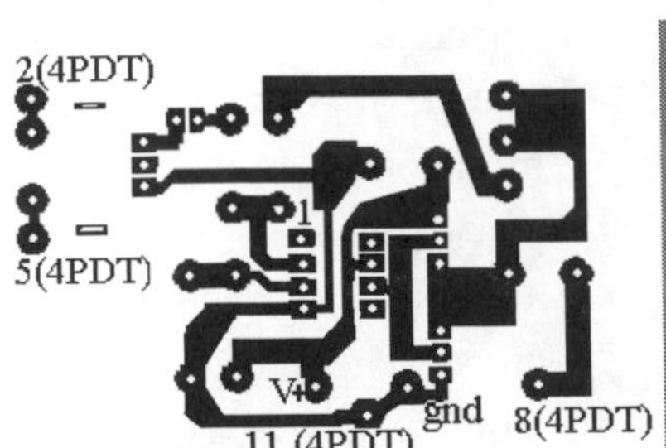

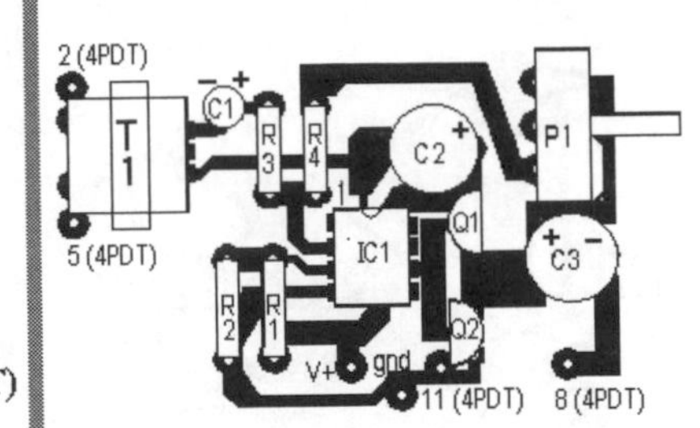

Figure L64-1

Figure L64-1 shows the PCB and parts placement for the door phone.

Assemble the project onto the printed circuit board.

The speaker wiring diagram is shown in Figure L64-2.

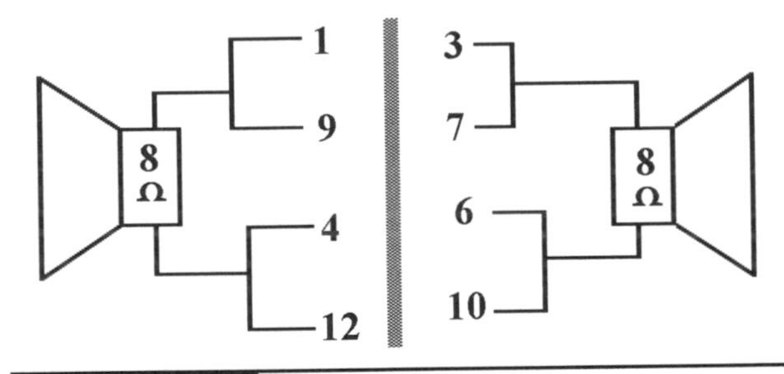

Figure L64-2

Remember to be careful as you do the assembly.

1. Solder the 8-pin chip seat for the 741. Do *not* solder the IC directly.
2. Note the placement of polar components. The only nonpolar components you have in this circuit are resistors.
3. The 3904 and 3906 transistors are not interchangeable in any circuit. Be sure they are properly placed. Overheating the transistors with a solder pen will destroy them. If the circuit doesn't work after soldering, check the transistors. Use the transistor check sheet provided in Table L51-2.
4. The potentiometer will not fit directly onto the PCB. Solder short wires to the legs of the potentiometer. Keep them short or they will act like antennas and be the source of a background hum that shouldn't exist.

What else is there to say? Seriously, if you've come this far, keep going. The big question is "Where to?" To begin with, get a subscription to *Nuts and Volts* or *Everyday Practical Electronics*. Or get some other Evil Genius books. Like I said earlier: Don't just sit there, start playing!

PART FIVE

Appendices

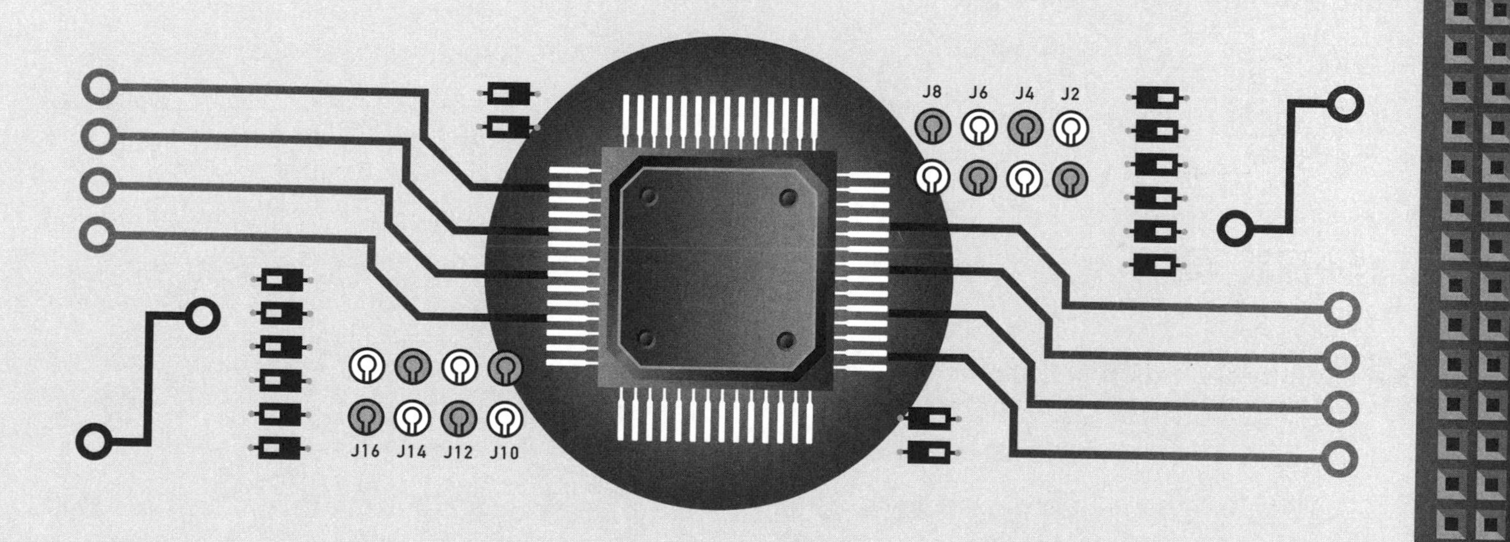

APPENDIX A

Common Component Packaging

YOU MAY HAVE NOTICED that sometimes the components you work with do not match the pictures of the components shown in the book. This section gives an introduction to common packaging used for electronic components. A single component such as the SCR can be packaged in a variety of formats. Each package provides an advantage of one form or another. It might be how it fits on a circuit board or how it handles power.

This is an overview meant to help you be flexible. Get used to reading the labels. Just because it looks different does not mean that it is different.

Transistors

Transistors have, by far, the most variety in their packaging, as shown in Figure A-1.

Many components are available in a variety of packages. Others, like the 2N3094 and 2N3906 transistors, are available only in the TO-92.

Potentiometers

Many potentiometers are being replaced by digital counterparts. However, they are still very common. Figure A-2 shows their potential packaging.

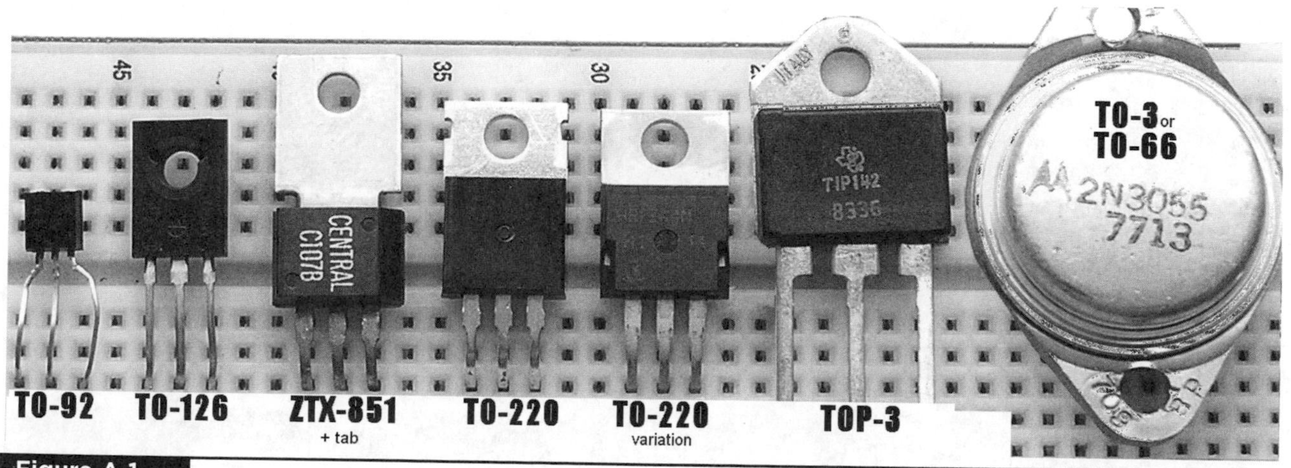

Figure A-1

Figure A-2

Transformers

Transformer packaging sizes, like any other component, are determined by the power requirements. The largest transformer, shown in Figure A-3, is a small power transformer.

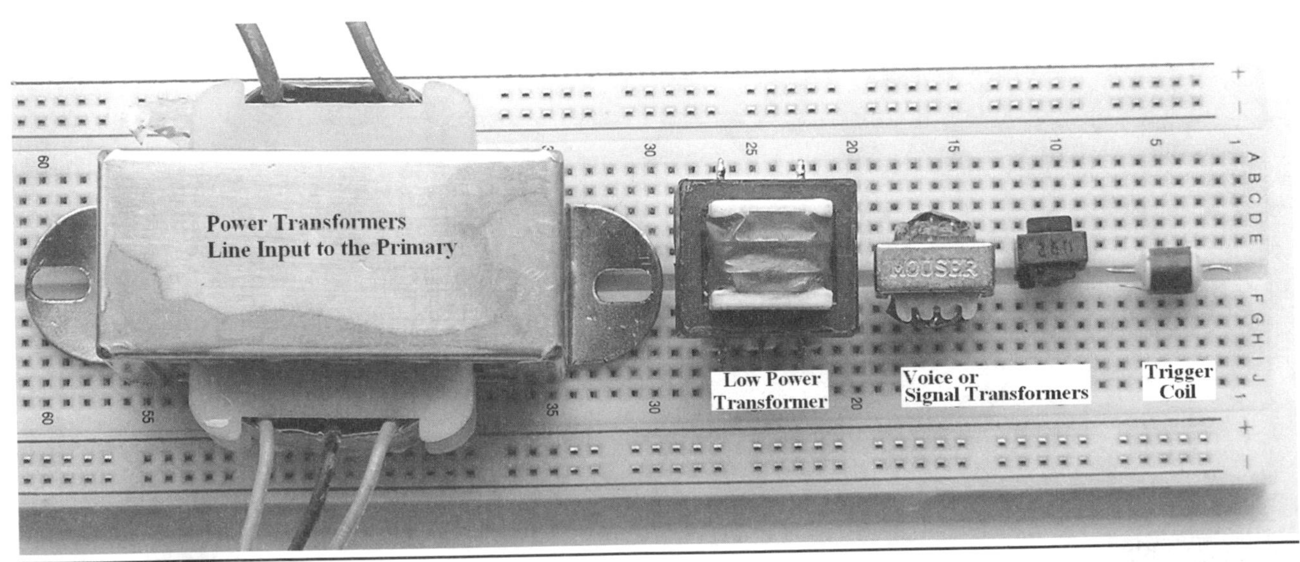

Figure A-3

APPENDIX B

Capacitors: Reading and Decoding

Capacitors tend to be confusing because there is no standard method of marking them. Even I get turned around when trying to decipher some creatively marked caps. If you are unsure, just use your DDM to verify what you think you see.

Reading Values

Each range of capacitors is packaged and referred to differently.

Remember! Coloring is nothing more than advertising for who made it.

- Electrolytic capacitors are easy. They have positive and negative connections. They must be placed properly. Polarity is clearly marked. Radial electrolytic capacitors have both leads from one end. Axial capacitors have a lead at each end. Figure B-1 shows that voltage ratings and capacitance are clearly marked.

TABLE B-1 The Farad Was Defined Over 200 Years Ago; Today, the Base Unit is the µF

Farad : F	Microfarad : µF	Nanofarad = nF	Picofarad = pF
1 F	= 1 million µF	= 1 billion nF	= 1 trillion pF
1 F	= 1,000,000 µF	= 1,000,000,000 nF	= 1,000,000,000,000 pF
	1 µF = 10^{-6} F	1 nF = 10^{-9} F	1 pF = 10^{-12} F
	1 µF	= 1,000 nF	= 1,000,000 pF
		1 nF	= 1,000 pF

Figure B-1

- Film capacitors are usually standard boxy shapes. Assume they are marked in nanofarads, but don't bank on nanofarads or a boxy shape all the time. Sometimes they are dipped in plastic and lose their boxy appearance. These are made of stacked layers of conductive film.
- Disk capacitors dominate the picofarad range. They are usually made of two circular plates, dipped into ceramic material and baked. Good examples of both appear in Figure B-2.

The markings of disk capacitors appear cryptic. The numbers read sort of like a resistor. Figure B-3 shows a range of values, and Table B-2 shows how to read them. Regarding the cap marked n22, the unit marker substitues as the decimal. So its value is .22nF.

Any capacitor rated above the intended voltage can be used in a circuit. For example, a capacitor rated at 600 volts can be used in a 9-volt circuit. There's no problem as long as maximum voltage does not exceed the capacitor's rated maximum.

Tolerance: Quality and Accuracy of Capacitors

The marking for accuracy of capacitors is noted on each package. Look at the pictures. On each, there is a prominent letter "K," "M," etc. Shown in Table B-3, these letters note tolerance.

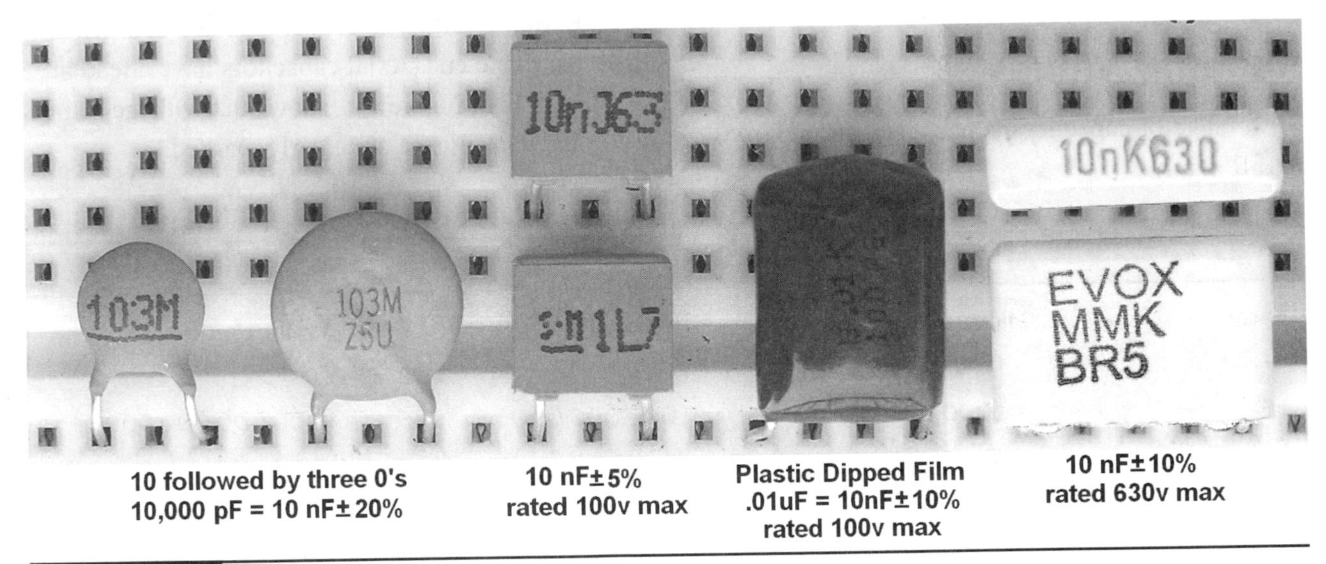

Figure B-2

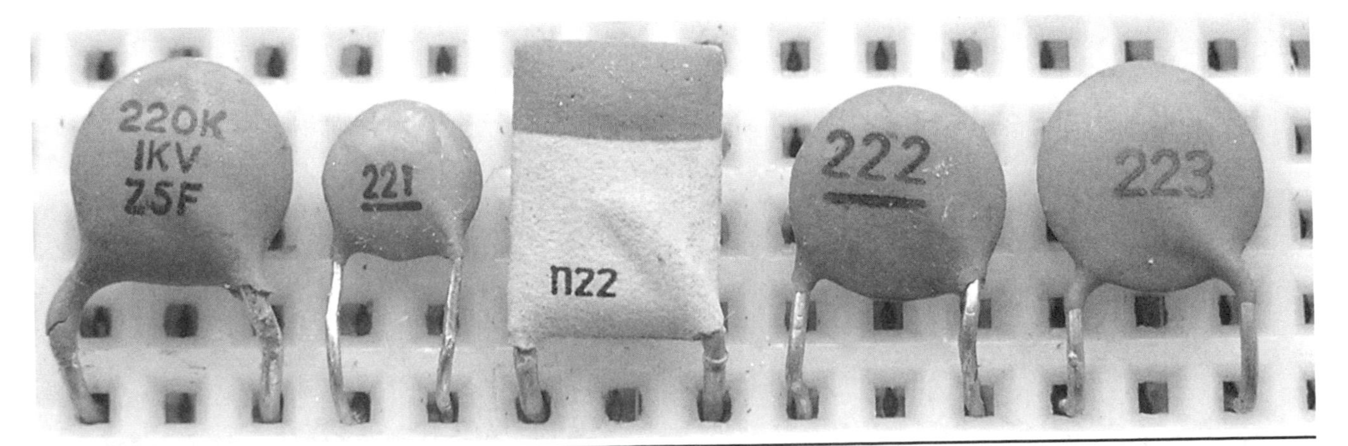

Figure B-3

TABLE B-2 The Third Number Denotes the Number of Zeros

Marking	Base Number	Number of Zeros	Results in pF	Results in nF
220	22	0	22 pF	.022 nF
221	22	1	220 pF	.22 nF
222	22	2	2200 pF	2.2 nF
223	22	3	22000 pF	22 nF

TABLE B-3 The Letter "K" Used Here Has Nothing to Do with 1,000

F	G	J	K	M	Z
±31%	±32%	±35%	±310%	±320%	-20% to +80%

APPENDIX C

Animations List

Related Animations

THESE ANIMATIONS CAN be found at www.mhprofessional.com/computingdownload

Part One

Figure L3-4

Figure L6-4

Figure L8-18

Figure l8-19

Figure L8-22

Figure L9-5

Figure L12-3

Figure L12-12

Figure L13-2

Figure L13-7 (a and b)

Part Two

Figure L15-1

Figure L16-1

Figure L16-2

Figure L16-3

Figure L16-4

Figure L16-5

Figure L16-7

Figure L16-9

Figure L19-4

Figure L20-1

Figure L21-2

Figure L21-5

Figure L22-7

Figure L23-5

Figure L23-6

Figure L23-7

Figure L24-1

Figure L24-2

Figure L24-3

Figure L25-2

Figure L27-1

Figure L27-2

Figure L27-3

Figure L27-4

Figure L28-1

Figure L28-2

Figure L28-3

Figure L28-4

Figure L28-6

Figure L28-7

Figure L32-2

Figure L32-3

Part Three

Figure L37-1

Figure L38-2

Figure L39-1

Figure L40-2

Figure L41-1

Figure L42-2

Figure L43-1

Figure L44-5

Figure L46-3

Figure L46-4

Figure L46-5

Part Four

Figure L49-1

Figure L49-2 (a, b, and c)

Figure L50-2

Figure L50-3

Figure L50-5 (a, b, and c)

Figure L52-2

Figure L53-1

Figure L53-2

Figure L53-3

Figure L53-4

Figure L53-5

Figure L53-7

Figure L53-8

Figure L53-9

Figure L54-1 (a, b, and c)

Figure L54-6 (a and b)

Figure L55-7

Figure L56-2

Figure L57-2

Figure L60-3

Figure L60-4

Figure L60-5

Figure L62-1

Figure L62-2

Figure L62-3

Figure L62-4

Figure L62-5

Figure L62-6

APPENDIX D

Glossary

AC
Alternating current (AC) is a current that pushes the electrons both forwards and backwards through a wire. AC is easily produced and transmitted for household and industrial use. Audio signals are AC because the sound wave creates a signal that pushes and pulls—creating a positive and negative voltage. AC is generally not used for digital electronic applications. *See DC (direct current).*

active
The state of a circuit, referring to whether the circuit is responding to an input. Alternatively, the other state is rest.

air capacitor
An air capacitor is an adjustable cap that works in the pF range, used for radio frequency tuning.

amp (ampere)
This is the basic unit of current (I) in electricity. It is a specific measure of electrons that flow past a point in one second.

amplifier
An amplifier uses a smaller signal to control the larger flow of voltage and current directly from the power source. *See pre-amp, power amp, Op Amp.*

amplitude
This is the strength of a signal. If the volume changes, the output amplitude changes. A 6-volt AC signal has six times the amplitude of a 1-volt AC signal.

analog
A signal that varies continuously over a given range, as compared to a digital signal where only discrete values of Hi (V+) and Lo (ground) are considered significant.

analog to digital (AD)
A device that changes a continuously variable signal, such as motion or electrical voltage, into digital values. Thus, it provides analog-to-digital conversion.

AND gate
Digital logic for the AND gate. When all inputs are Hi, the output is Hi.

anode
Commonly used term referring to the positive side of a polar component.

astable
An astable oscillator creates a continuous frequency as long as that portion of the processor is active. The frequency of the oscillator is set by the value of the components in the RC circuit.

battery
The term comes from a grouping of cells together in parallel or series. A 9-volt battery is made of a grouping of six 1.5-volt cells.

BCD (binary counting decimal)
A binary input is interpreted and then presented on a decimal seven-segment display output. The 4511 is a BCD interpreter.

bias
The general term means a tendency or prejudice towards a specific point of view. In electronics, a bias is created by using voltage dividers to create an artificial center point for an audio AC signal. If an audio system uses 12 volts, the bias is set to 6 volts. This method is easiest when only positive DC voltage is available.

binary
The counting system that depends on only two digits, used by computers and machinery because it is easier for these systems to sense an on/off setup, rather than partial voltages. Computers use a magnetic system of north/south polarity to store bits of information. *See bit, byte, digital.*

bit
A single piece of information that stores binary data. It is the individual piece (bit) that is turned on/off, or an area polarized north/south on magnetic media, or a dimpled/not dimpled spot on a CD/DVD disk.

bounce
Bounce is found in any mechanical switch connected to a digital system. Most digital circuitry is capable of responding in microseconds, but when a switch is thrown, the current change often fluctuates like a bouncing ball. If this fluctuation "bounces" above the digital response threshold of half of V+, it creates problems of counting an event twice instead of once. To take care of this problem, use a Schmitt trigger.

byte
A combination of 8 bits of information, capable of storing up to 256 pieces of data.

capacitor
A device composed of two conducting surfaces separated by an insulator called a dielectric. (1) It has the ability to store electrical energy and release it in an RC circuit—called sink and drain. (2) It can be used as a filter to smooth out a signal. (3) It can be used to block the flow of current, but still allow the changing signal to pass, as in an audio coupling—like a membrane across a drum. *See film capacitor, electrolytic capacitor, disk capacitor.*

cathode
Commonly used term referring to the negative side of a polar component.

cell
A single electrochemical generating unit. Usually zinc and copper are used in general-purpose and alkaline cells that produce 1.5 volts. Nickel and cadmium are used in rechargeable cells. They are ganged together in series or parallel to create a battery.

circuit
A course, pathway, tour, or journey. An electrical circuit is one that is a complete pathway and provides a path for the current.

clock signal
A clean digital signal that rises from zero to full voltage instantly in microseconds. A clock does not have any bounce. A circuit that produces a clock signal from a push button is often referred to as a Schmitt trigger.

CMOS
An acronym for complementary metal oxide semiconductor, pronounced see-moss, CMOS is a widely used type of IC.

conductor

A conductor carries electricity easily. It is made of any material that has unbound electrons that are able to freely flow when a voltage or pressure (electromotive force) is applied to it. All metals are conductors. Some nonmetals, such as silicon and carbon, also act as conductors. *See insulator.*

continuity

Continuity is a test done with the digital multimeter to check if there is a "continuous" connection between two points. Most multimeters are designed to give instant feedback in a YES/NO format: YES indicates there is a connection with very little resistance; NO indicates there is not a connection of very little resistance.

coulomb

Named for Charles de Coulomb (1736–1806), who devised a method of measuring electrical quantity. It is the quantity of electricity conveyed in one second by a current of one ampere.

coupling

A capacitor can be used for audio coupling, that is, to pass an AC signal but not pass any current. This is like sound passing through the head of a drum. No air passes through the drum head, but the signal still passes with ease and clarity. Audio coupling also applies capacitors as filters.

DC

DC (direct current) moves in only one direction. It may change voltage, but it does not reverse direction.

DCB

Decimal counting binary (DCB) happens when a processor has an input of a clock signal and changes its binary output by one step for each additional clock signal. The 4510 is an example of a DCB interpreter.

digital

(1) Relating to data that is represented in the form of discrete on's and off's, 1 or 0s; as opposed to analog (continuous voltage fluctuation). (2) An electronic circuit that uses digital logic in its input, processor, or output.

digital circuit

A circuit that has a processor and output that operate in only two states: on and off.

digital logic

The method of using analog and digital inputs to make digital decisions. The five most common gates are AND, OR, NAND, NOR, and NOT.

digital output

The output signal of a digital circuit represents a set value of conditions. The information is contained in a code consisting of a sequence of one or more discrete voltage or current levels, each representing a value of zero or one. The sequence 01000001 may then be interpreted as a byte with a value of 65 and then presented on your monitor as the letter A.

digital recording

Any recording in which the signal is first converted to digital format. A CD uses 16 bits of information for a single binary word, while a DVD uses 20 bits of information for a single binary word.

diode

A basic electronic device characterized by the ability to pass an electric current easily from positive to negative. They act as a one-way street. Three types of diodes are used in this course: (1) power diode designed for 1 watt at 400 volts, (2) the signal diode that passes miliamps, and (3) the zener diode.

DIP

Dual inline package. One of a series of acronyms referring to common packaging used in electronics. DIP packaging specifically has two rows of matching pins.

disk capacitor

Disk capacitors are designed to hold the smallest capacitance values in the picofarad (pF) range. They are usually light brown and disk shaped.

dumb

A dumb circuit does not have any way to determine if a specific event has occurred or when a particular number has been reached.

electrolytic capacitor

An electrolytic capacitor is a polarized capacitor, usually in the microfarad (μF) range. It has a positive and negative side. If it is put in backwards, it will not work effectively, and might fry itself (very messy).

electromotive force

(1) The amount of energy supplied by an electric current passing through a given source, as measured in volts. (2) The energy difference existing between the positive and ground that are provided by the same energy source. An electromotive force (electrons) passing through a wire will create a magnetic field. A magnetic field moving around a wire will create an electromotive force and push electrons.

electron

A negatively charged particle that is part of the structure of an atom. Free electrons flow in conductors to create current. An electromotive force—voltage—is needed to push electrons through a conductor. Electrons that are not bound to a specific atom or molecule are considered "free." A material with many unbound electrons is considered a good conductor, such as metals. A material that has all of the electrons tied up in chemical bonds will not conduct electricity. It is a good insulator.

energy

The ability to do work. Energy in electricity is measured in volts.

farad

The unit of capacitance named for Michael Faraday. The basic common unit is 1 μF, which equals 1 microfarad, that is, .000001 F.

film capacitor

Film capacitors are designed to hold the mid-range capacitance values in the nanofarad (nF) range. They are often shaped like a small box.

filter

A capacitor can act as a filter in two distinct ways. (1) It can be a buffer by acting as a reservoir of extra electrons. When a spike goes above the regular voltage, the sudden surge is softened because the capacitor fills up and dumps the extra back out slower than the actual surge. If there is a momentary drop in voltage, that is softened as well, because the capacitor discharges its content onto the line. (2) It can be a high pass or low pass filter by allowing only higher or lower frequencies to pass. Low-frequency sound passes through a .1 μF capacitor, while higher sounds are filtered out. High-frequency sounds pass through a .001 μF capacitor, while the lower sounds are filtered out.

force

In electricity, voltage (V+) is the unit used to measure the amount of electromotive force (push) in a circuit.

frequency

The number of cycles or events per unit time, commonly having units of sec-1 (hertz). In acoustics, this refers to the number of waves per unit time that are produced by a vibrating object.

gain
Gain is the basic measurement of the amplitude of signal output divided by the amplitude of signal input. It is set by resistor ratios to control feedback in an Op Amp.

gate
See digital, logic.

ghost
A ghost is troubleshooting a problem that cannot be defined, because an error is not regular. The most common causes of a "ghost" are (1) loose wires, (2) wires that cross and touch when they shouldn't, and (3) unused inputs to CMOS chips that have not been tied to V+ or ground.

heat sink
A heat sink is used to remove unused power from an amplifier or transistor. For example, if a 20-watt amplifier is being fed 20 watts of power but only using 5 watts because the music is quiet, 15 watts of power have to be dumped as heat. A heat sink allows the heat to move away from the amplifier chip. If the heat is not moved away, the amplifier chip will overheat and shut down, or overheat and melt.

hertz (Hz)
A unit of frequency equivalent to cycles per second. One kilohertz is 1,000 cycles per second.

hysteresis
Related to hysteria. In electronics, it refers to the output of a logic gate going hysterical as the voltage at one of the inputs passes slowly through the half V+ range. The gate tries to physically respond as both high and low at the same time, flip-flopping wildly. Sometime this results in a trailing blip of sound at the output.

IC
An integrated circuit (IC) is any miniature manufactured circuit made of many transistors and components working together in one package. The package is usually presented in DIP or SIP format. The technique for manufacturing ICs depends on using a photographic process. Tens of thousands of ICs are available on the market today.

input
The input is the action that starts the circuit. It is the first part of a system diagram.

insulator
An insulator cannot conduct electricity because it has all of the electrons bound into its chemical makeup. It does not have free electrons, so an electric force is not able to move through the material. *See conductor.*

invert
To turn upside down. For example, a NOT gate inverts a signal from Hi to Lo.

kilo
A measurement unit of 1,000 abbreviated as "K." For example 100 K ohms is easier to write than 100,000 ohms.

Kirchoff's law
Kirchoff's law is used to figure the resistance in a system that has many load levels set in parallel.

layout
The layout represents the trace of copper on the printed circuit board. Any layout is specific to one application. To change the specific circuit, you must effectively change the layout.

load
Any item that consumes electricity, such as a resistor, a speaker, household appliance, street lighting, or an industrial electrical motor.

logic
A logic circuit consisting of one or more diode gates connected to the base of a transistor, in which the AND or OR functions performed by the diode gates control the on-off status of the transistor.

mega
A measurement unit of 1,000,000 often abbreviated as "M." For example, 1 M ohms is easier to write than 1,000,000 ohms.

micro
A measurement unit of .000001 often abbreviated as "µ" or simply "u" referring to the Greek letter "mu." For example 47 µF is easier to write than .000047 F.

microprocessor
The central processing unit of a microcomputer, usually manufactured on a single semiconductor chip, containing anywhere from a few thousand to several million or more semiconductors on a single integrated circuit.

milli
A measurement unit of .001 often abbreviated as "m" and not to be confused with "mu" for micro. For example, 7 mW is easier to write than .007 watts. An average LED uses 20 mA.

monostable
A monostable oscillator is a "one-shot" timer—usually an RC circuit. When triggered by an outside event, it turns on a portion of a processor as the capacitor fills. As the resistor drains the voltage from the capacitor, the controlled processor turns off. The speed of the "one-shot" timer is set by the value of the components in the RC circuit.

NAND
Digital logic for the NAND date (not AND). When any input is Lo, the output is Hi.

nano
A measurement unit of .000000001 often abbreviated as "n." For example, 100 nF is easier to write than .0000001 F. Because film capacitors are produced widely in the nanofarad range, their measurements are often given in nF. .001 µF = 1 nF = 1000 pF.

NOR
Digital logic for the NOR gate (not OR). When all inputs are Lo the output is Hi.

normally closed (NC)
The two contacts are connected in an NC switch. Current passes easily because the two wires connect and there is no resistance. When activated, they become disconnected and the electrons cannot pass through the air.

normally open (N.O.)
The NO push button has infinite resistance between the two legs when the button is not pushed. When it is pushed, the two wires connect and there is no resistance.

NOT
Digital logic for the NOT gate. It inverts the input. A Hi input becomes a Lo output, and a Lo input becomes a Hi output.

NPN
An NPN transistor turns on more with more positive voltage to the base. The emitter is connected toward the ground. An NPN junction is formed by placing a thin layer of p-type semiconductor material between two layers of n-type semiconductor material. The p-type semiconductor becomes a conductor when a small voltage is applied to it.

ohm
A standard unit of electrical resistance, equal to the resistance of a circuit in which an electromotive force of 1 volt will maintain a current of 1 amp. Named for George S. Ohm.

Ohm's law

Simply stated, the amount of voltage pressure in a circuit is equal to the amount of current multiplied by the resistance of the load. V = I (×) R.

Op Amp

Shorthand for Operational Amplifier.

OR

Digital logic for the OR gate. When any input is Hi, the output is Hi.

oscillator

A circuit that creates a regular and predictable beat or frequency. *See astable, monostable.*

output

Output is the action resulting from the processing done by an electronic system. It is the last unit of a system diagram.

parallel

A parallel circuit happens when all the components or loads are connected to each other side by side. Each load is provided with the same amount of voltage, but the current is evenly shared between the loads. *See series.*

pico

A measurement unit of .000000000001 (10^{-9}) often abbreviated as "p." For example, 100 pF is easier to write than .0000000001 F. Because disk capacitors are produced widely in the picofarad range, their measurements are often given in pF. .001 μF = 1 nF = 1000 pF.

piezoelectric

This effect occurs in specific nonmetallic materials, such as quartz crystals, or in engineered ceramics and plastics. Piezos can be used as speakers and in sensors. (1) A voltage or electric field is produced in specific crystals when these crystals are compressed or decompressed rapidly. (2) When a voltage is applied to a piezo material, it will compress or decompress, relative to the size of the voltage.

PNP

A PNP transistor turns on more with less voltage to the base. The emitter is connected toward the voltage. A PNP junction is formed by placing a thin layer of n-type semiconductor material between two layers of p-type semiconductor material. The n-type semiconductor becomes an insulator when a small voltage is applied to it.

polarity

Some components need to be set in with attention to the direction of the current and voltage. These parts are "polar." If these parts are placed in backwards, the component will not function properly. They may even be damaged if voltage is directed through them the wrong way. Most electronic components have polarity. Resistors do not have polarity because the current and voltage can pass both ways.

power

Power is defined as the speed at which work is done. In electronics and electricity, power is measured in watts. 1 watt = 1 amp × 1 volt. Restated, that is the amount of electrons passing a specific point in one second, multiplied by the amount of force pushing the electrons.

power amp

A power amp is used to increase a signal's strength (amplitude) from the 1-watt range to whatever is needed for the speakers at hand. Common power amps provide anywhere from 20 watts to 300 watts. Power amps are rated by their range of response. The concern with power amps is the ability to provide enough power (12 volts × 25 amps = 300 watts) and being able to remove unused power through heat sinks.

pre-amp

A pre-amp is a low-power amplifier that increases a signal's strength (amplitude) from the milliwatt range to the 1-watt range. A pre-amp is often rated by its gain and clarity of signal. Most power amplifiers need to have a signal fed to them that is in the 1- to 2-watt range.

processor

(1) A device that interprets and executes instructions. (2) Referring to the center portion of a system diagram.

prototype

A prototype is a test product, built as closely to the real finished product as possible. A prototype is usually built in a fashion that allows easy and quick changes for testing purposes.

Q

Transistors are always noted with a "Q." Transistor is an awfully long word, and "T" is used for other things in electronics, like "transformer."

random

A result happening by chance. There should be no pattern in "random" events.

RC circuit

A resistor/capacitor (RC) circuit is used for capturing voltage and releasing it in a controlled manner. This allows the designer to control the time at which the digital input will sense the change above half voltage to below half voltage.

resistance

The fact of opposing or acting against. In electricity, resistance opposes the flow of current, generates heat, controls electron flow, and helps supply the correct voltage when set up as a voltage divider. Resistance depends on the material used, the length and cross-sectional area of the conductor, and the temperature. It is measured in ohms and follows Ohm's law. Resistance is measured in ohms, kilo-ohms, or megohms.

resistor

A component in an electric circuit that provides opposition to or limits the current flow according to Ohm's law. The carbon resistors used in this course are made by mixing carbon particles with a ceramic binder in measured amounts. These are baked into standard cylindrical shapes, marked with standard color-coding. They create resistance to current flow and act as a load in the circuit.

rest

At rest, power is provided to the system, but the system is not active. The 4011 chip needs microamps (.003 of a milliamp) to stay powered in its rest mode. At this rate, a 9-volt alkaline battery will power a 4011 chip for two to three years.

schematic

A plan, plot, and design of an electronic system, or portion of that system, using only symbols to represent the components and their connections to each other.

Schmitt trigger

A circuit that produces a clock signal from a push button is often referred to as a Schmitt trigger. It is designed to eliminate bounce. A common Schmitt trigger is available in the 4093 CMOS IC.

semiconductor

A material that has intermediate values of electrical resistivity, between the values for metals and insulators. It conducts electrons under certain situations, but resists their flow in other situations. The properties of such basic materials as silicon, carbon, and germanium are the heart of semiconductors. Semiconductors are the basic material of various electronic devices used in telecommunications, computer technology, control systems, and other applications.

series

A series circuit is one in which each component or load is joined end to end, successively with the next. The current flow is determined by adding the resistance of each component in the chain together. The current stays the same for each component, but the voltage drops in relation to the resistance of each load. *See parallel.*

SIP

Single inline package. One of a series of acronyms referring to common packaging used in electronics. SIP packaging specifically has a single row of pins.

smart

A smart circuit can use feedback to determine if a specific event has occurred. For example, a 4017 walking ring counter can be wired to sense when a specific number in a counting sequence has been reached.

solder

(1) An action that joins metal objects without melting them. The metal objects are fused together with an alloy that has been applied to the joint between them. (2) Any of several alloys used in this process. Electronics most often use a 60% tin plus 40% lead alloy.

state

(1) The state of a circuit is either "active" or "at rest." Active refers to the idea that the circuit has been turned on. At rest means that the circuit is waiting for an event to happen. A CMOS circuit at rest needs as little as a microamp of current to power the IC chip at rest. (2) Refers to the output of a logic gate. The output state is either Hi or Lo.

static electricity

Energy in the form of a stationary electric charge. This charge can be built up on your body when you walk across a nylon carpet. You reach for a doorknob, and a spark jumps the gap as an electrostatic discharge. Such a charge can be stored in capacitors and thunderclouds, or any nonconductive surface. It can be produced by friction or induction. This charge can build up huge voltage (but small current) and can fry many ICs.

system

(1) Referring to a large orderly arrangement of interconnected units such as a telephone system. (2) A system diagram that represents the inputs, processors, and outputs of an electronic unit, emphasizing their relationship in the circuit.

threshold

The amount of voltage needed for an input of a digital circuit to sense a change of state.

tie

To connect components together. For example, to tie pins 12 and 13 together means to connect them with a wire.

transistor

The basic electronic switch. These are analog switches—the voltage passed through the transistor is directly related to the amount of voltage applied to the base. It is an active semiconductor device, usually made from germanium or silicon, and possessing at least three terminals (typically, a base, emitter, and collector) and characterized by its ability to amplify current. They are used in a wide variety of equipment, such as amplifiers, oscillators, and switching circuits. *See NPN transistor, PNP transistor.*

Vcc

Vcc is another way of labeling ground.

VCO

VCO is shorthand for voltage-controlled oscillator. A complete VCO is housed in the 4046 CMOS IC.

Vdd

Vdd is another way of labling V+.

voltage

Voltage is the amount of "push" behind the current. It represents the potential energy or force in a circuit. The potential difference or electromotive force is measured in volts.

voltage divider

By using resistors as artificial loads, voltage can be set at a midpoint to any predetermined value desired. This technique is widely used in biasing audio systems and in sensors where an event causes one of the resistors to change value, therefore changing the midpoint voltage value.

watt

Power output is defined as how fast work is done. In electronics and electricity, power is measured in watts. 1 watt = 1 amp × 1 volt. Restated, that is the amount of electrons passing a specific point in one second, multiplied by the amount of electromotive force.

work

The measure of force times distance.

zener diode

A diode that "breaks down" at a specific voltage value, allowing current to move from cathode to anode. It can be used for voltage reference purposes or for voltage stabilization.

APPENDIX E

Make Your Own Printed Circuit Boards

THIS METHOD IS GREAT for prototypes and hobbyists.

I know that other people have developed this process. The only thing I can claim is that I independently developed this process and successfully used it in my classroom.

For your convenience, www.mhprofessional.com/computingdownload presents this section with large color photos providing clarity lacking in black and white.

This photo shows all of the materials needed to create your own printed circuit board (Figure E-1).

Figure E-1

Clean the copper clad material using scouring powder to remove the light plastic film that keeps the copper from tarnishing (Figure E-2). No fingerprints!

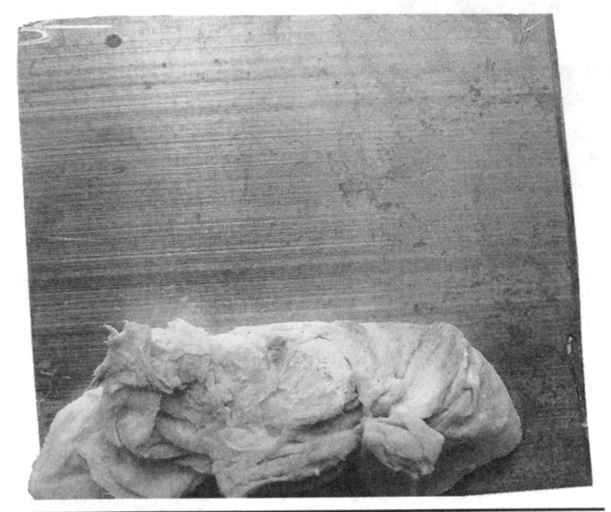

Figure E-2

I use photo-quality inkjet paper (Epson S041062). It is the best quality for the money ($17/100 sheets). The key is the clay base. This keeps the ink from absorbing into the paper fibers (Figure E-3).

Figure E-3

When looking at the paper, bend a corner of the sheet over on itself (don't crease it) to compare the brightness of each side. Don't think "white." That gets confusing. Compare "brightness." The brighter side is the one with the clay. Print onto the clay surface.

Remember, use a laser printer. The philosophy is to melt the plastic toner against the copper. I have had feedback that some brands of laser printers simply do not work, but I've had success with any laser printers I've used, including HPs, Lexmarks, and Canons.

This close-up shows the quality of traces that can be made using my inexpensive method (Figure E-4). Even beginners can create the .03-inch (.75-mm) width. With care and practice, you can get the .02-inch width just as regularly. The highlighted pad had paper debris covering the copper underneath, so it did not get etched properly.

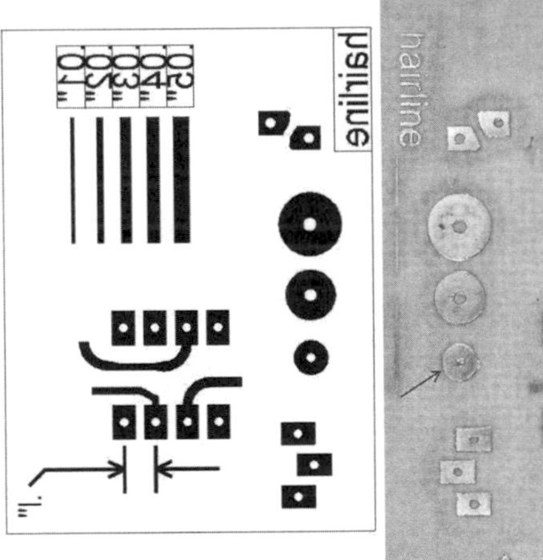

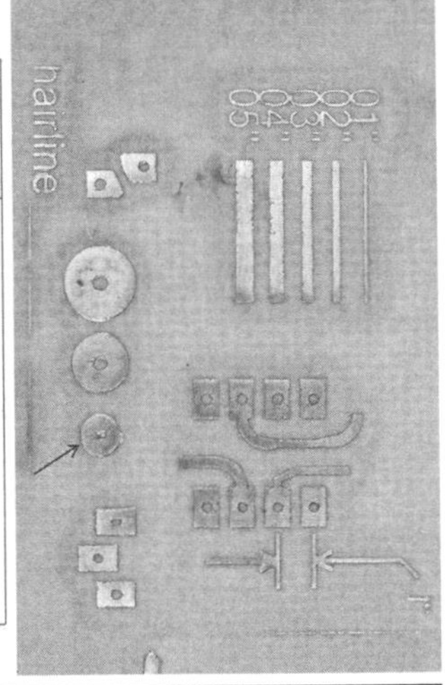

Figure E-4

With the image facing the copper side of the PCB material, fold the margins of the layout tightly around the edges of the PCB and secure with masking tape. Regular masking tape doesn't disintegrate when heated. The paper must be secured so that the image can't shift (Figure E-5).

Use a regular clothes iron (Figure E-6). Turn off the steam. Set to maximum temperature. Some irons are hotter than others. The paper should not discolor during this one-minute heating process.

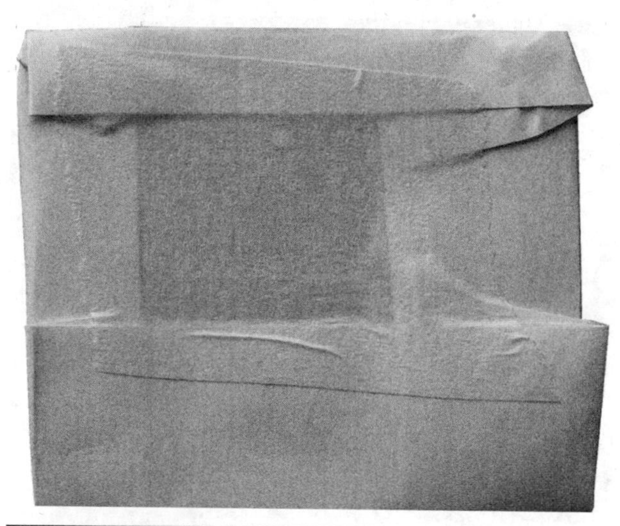

Figure E-5

Figure E-6

Use an ironing board or cutting board underneath. It should be a smooth, nonmetal, heat-resistant surface.

This photo attempts to show real strain in my hand as I apply pressure while moving in tight circles, about one per second for a minute.

Immediately apply pressure to the paper and copper clad board.

Without a vise at home, I place the hot PCB and paper between two small metal sheets. This then goes onto the floor, copper side up. I then place a book (to spread my weight evenly) on top of the setup. My foot is too big to show, but my

Figure E-7

daughter's Ken doll helps convey the idea (Figure E-7). Stand for a minute right on top of the board.

Use a vise for better results. Gently secure the board between metal plates and apply soft pressure. Don't squish. Just secure the board until it has cooled.

Place the cool wrapped board into water. Let it sit for a minute. This is important even though nothing appears to happen (Figure E-8).

Figure E-8

DON'T SKIP THIS STEP! This is the one step that everybody forgets and then complains afterwards that the processes didn't work.

Spread a few drops of dish soap around on the wet paper (Figure E-9). The soap helps the water soak through. The paper becomes translucent and expands where it's not attached to the PCB.

Figure E-9

Remove the tape from the back and peel the majority of the paper off. A good amount of paper does remain behind. Remember, you are separating the clay surface from the paper. This shows a good idea of what you can expect as you peel the paper off (Figure E-10).

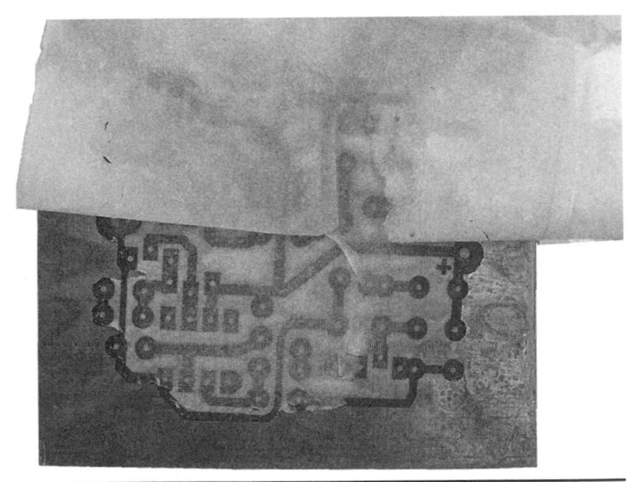

Figure E-10

Use a dripping wet paper towel to gently rub off the remaining paper (Figure E-11). The toweling will disintegrate as you do this. Don't rub too hard.

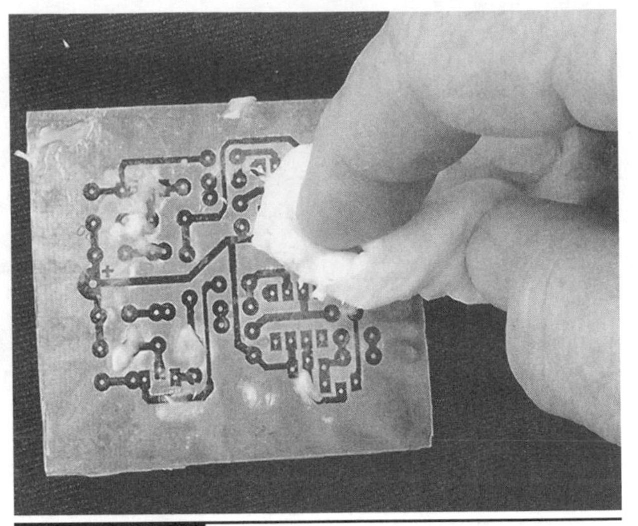

Figure E-11

You don't want to remove any of the toner now melted onto the copper.

This shows the dried results. Notice the paper remaining in some spots. Wet the area and use a sharpened point to clear it. Even fresh chemicals have difficulty etching through paper (Figure E-12).

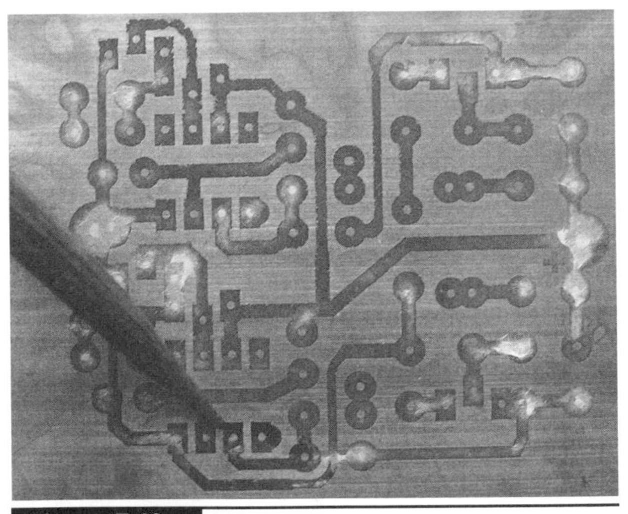

Figure E-12

These photos show progression from paper to finished PCB (Figure E-13).

- Before etching, carefully inspect the layout after the paper is removed and the board is cleaned.
- I used a "Staedtler Mars" permanent marker to reinforce areas that looked weak. Any color

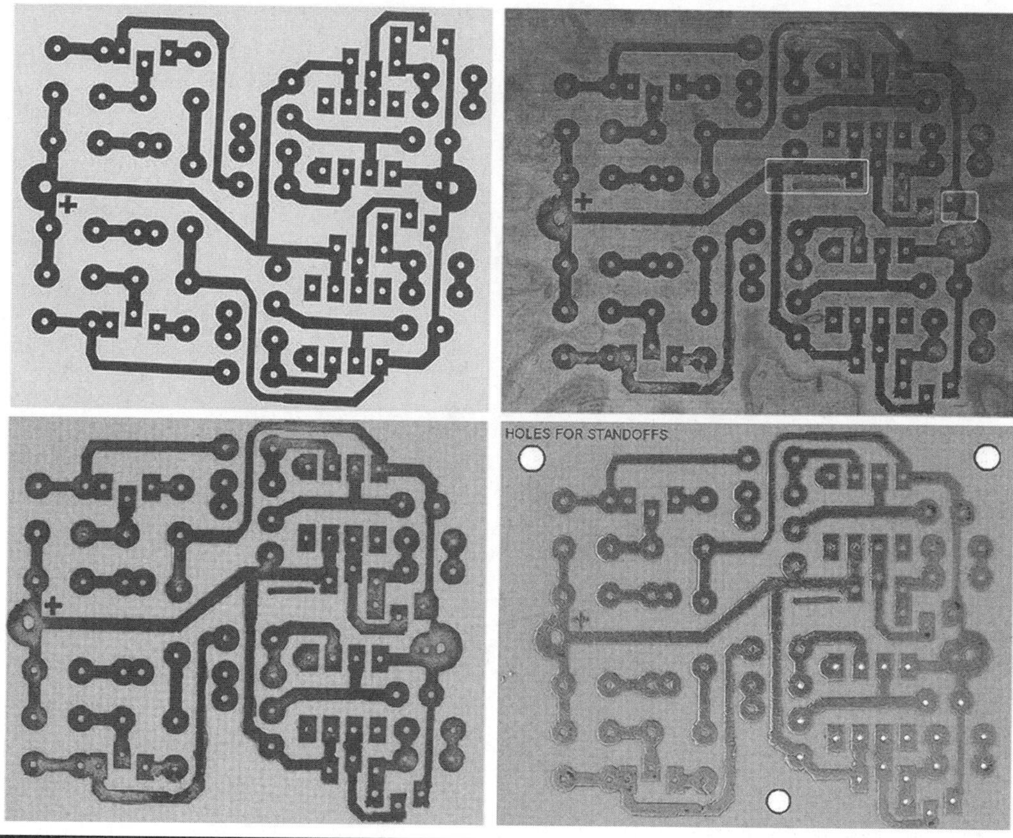

Figure E-13

works, but red highlights where it has been applied. I mention the brand name here only because the quality of the permanent ink does matter.

For etching, I prefer ferric chloride. It is inexpensive and doesn't lose potency with age. Read the safety information. Handled properly, it is safe enough. Use only plastic tools. Handled improperly, it is ***very nasty***.

- Use steel wool or finer grit sandpaper to remove the toner.
- This is a good board. If you have small breaks, those can be bridged with solder.
- Drill all holes with a #60 bit.

If you need to use support screws, be sure to drill the 1/8-inch-diameter holes before you populate the board.

Index

D

E

Q

R

S

W

Z